Geographies of Nationhood

Cartography, Science, and Society in the Russian Imperial Baltic

CATHERINE GIBSON

OXFORD
UNIVERSITY PRESS

OXFORD
UNIVERSITY PRESS

Great Clarendon Street, Oxford, OX2 6DP,
United Kingdom

Oxford University Press is a department of the University of Oxford.
It furthers the University's objective of excellence in research, scholarship,
and education by publishing worldwide. Oxford is a registered trade mark of
Oxford University Press in the UK and in certain other countries

First Edition published in 2022

Impression: 1

Published in the United States of America by Oxford University Press
198 Madison Avenue, New York, NY 10016, United States of America

British Library Cataloguing in Publication Data
Data available

Library of Congress Control Number: 2021952852

ISBN 978–0–19–284432–3

DOI: 10.1093/oso/9780192844323.001.0001

Printed and bound by
CPI Group (UK) Ltd, Croydon, CR0 4YY

EUI EUROPEAN UNIVERSITY INSTITUTE

This book has been published with a financial subsidy from the
European University Institute. This publication is based on an EUI PhD thesis defended
at the Department of History & Civilization in 2019.

Acknowledgements

The process of writing a book is very much like that of making a map. Neither work appears in isolation, and I would like to express my gratitude to the great number of people who contributed to the process of researching, writing, and publishing this book.

I would like to thank the organizations that have funded research for this book. I am very grateful for the financial support I received from the European University Institute in the form of a research grant from the United Kingdom to fund my doctoral studies and to the Department of History & Civilization for additional funding for archival missions, conference travel, Russian-language classes, and a book publication subsidy. My postdoctoral research at the School of Theology & Religious Studies at the University of Tartu was funded by the European Regional Development Fund and the Mobilitas Pluss Programme (Grant no. MOBJD517) and the Estonian Research Council (Grant no. PRG 1599). Many thanks to Irina Paert and the other members of the department for making me feel welcome and for their incredible support. I also benefitted from research stays and archival research trips supported by a Paderewski Grant for research in Poland from the European Centre in Natolin, a junior fellowship from the Herder Institute in Marburg, and an Erasmus+ mobility grant to spend a semester at the School of History at the University of St. Andrews. An Emerging Scholar Grant from the Association for the Advancement of Baltic Studies generously paid for additional archival research trips and digitized scans of archival materials when my plans to travel to Russia were cancelled due to the Covid-19 pandemic.

Thank you to the many librarians, archivists, and support staff across Estonia, Latvia, Russia, Belarus, Lithuania, Finland, Poland, Germany, and the UK who shared their knowledge and provided crucial assistance with locating and interpreting source materials. I am especially grateful to the National Archives of Estonia, National Library of Estonia, Estonian Literary Museum Archival Library, Latvian State Historical Archives, National Library of Latvia, Russian State Library, National Library of Russia, Russian Geographical Society, St. Petersburg Branch of the Archives of the Russian Academy of Science, Russian State Historical Archive, Slavonic Library at the National Library of Finland, National Historical Archives of Belarus, and David Rumsey Map Collection and Map Center at Stanford Libraries for granting me permission to reproduce archival documents and digitized materials from their collections. At the EUI library, I extremely grateful to Serge Noiret for the numerous book purchases and to Pernille Kjeldsen for sourcing rare books and articles for me through interlibrary loan.

Portions of Chapters 1, 3, and 4 first appeared as articles in *Nationalities Papers* and the *Journal of History Geography.* My thanks to the editors and publishers of these journals for permission to rework aspects of the material here. I owe special thanks to Matthew Kott, Mike Loader, and Siobhán Hearne as editors of volume *Defining Latvia: Recent Explorations in History, Culture, and Politics* (CEU Press, 2022), for granting me permission to reuse some parts of my chapter here.

I would like to take the opportunity to thank the numerous people who have supported me. I am deeply grateful to my supervisor Pieter Judson and second reader Pavel Kolář at EUI for providing continued encouragement, insightful guidance, and probing critique that helped me sharpen my argument and writing. Many thanks are due to my external advisor, Tomasz Kamusella, who first introduced me to the history of the lands between Russia and Germany and has shared his expertise, library, and friendship in a most generous way ever since. I am indebted to Steven Seegel for his advice, mentorship, and the many conversations which helped me to figure out what the story at the heart of this book should be.

The book's journey has been shaped by countless encounters with people who shared their kindness, hospitality, and assistance at various stages in the project: Vytautas Petronis and Darius Staliūnas in Vilnius, Inese Runce in Riga, Christian Lotz in Marburg, Hienadź Sahanovič in Minsk, and Father Serge at the Francis Skaryna Library and Museum in London. My work has gained much from discussions with Karsten Brüggemann, Mart Kuldkepp, Tomaš Nenartovič, Taavi Pae, Bernhard Struck, and Erki Tammiksaar, among others, who generously helped me track down references, shared advice on archives, advised on translations, or read and provided feedback on work in progress. The anonymous readers for OUP provided excellent critical readings and guidance for broadening the appeal of the book, which sharpened my sense of the subject's bigger themes and relevance. Thank you to the whole team at OUP, especially Cathryn Steele, for assistance in preparing the manuscript for publication. I wish to thank Desi Allevato for compiling the index. All shortcomings are my own.

My friends and colleagues at EUI exemplified collegiality at its best. I would like to especially thank Déborah Dubald, Martin Vailly, Louis Le Dourain, Ismay Milford, Kirsten Kamphuis, and Mikko Toivanen for your friendship and stimulating discussions across historical times and places. Thank you to my fellow members of the Peripheral Histories team—Susan Grunewald, Siobhán Hearne, Jo Laycock, and Alun Thomas—for your important work to promote scholarship across all corners of Eurasia. To the members of the online Russian and Soviet Cartography *kruzhok*—Nick Baron, Sofia Gavrilova, Martin Jeske, Alex Kent, Nadia Mamontova—my sincerest thanks for sharing your uplifting enthusiasm for map history during the isolating months of lockdown in 2020–21 when this manuscript was finished.

Lastly, I would like to thank my parents Paul and Fiona, my sisters Martha and Isla, and my grandparents Enid, Jim, Iris, and Renee for your support. You are a tonic and our sprawling conversations, on topics ranging from linguistics to geology, helped me to see connections in surprising places and keep sight of what fascinated me about maps in the first place. Most importantly, I thank my partner, Ostap, for standing steadfastly beside me through all the travel, international moves, precarity, and emotional turmoil of academic life. Your love, patience, and unwavering support matter more to me than you can possibly imagine.

Contents

List of Illustrations, Figures, and Maps

List of Plates

List of Abbreviations

Archives

DSHI	Herder Institute Document Collection (Dokumentesammlung des Herder-Instituts)
RA, EAA/ERA	National Archives of Estonia (Eesti Rahvusarhiiv)
ELMAL	Estonian Literary Museum Archival Library (Eesti Kirjandusmuuseumi Arhiivraamatukogu)
LVVA	Latvian State Historical Archives (Latvijas Valsts vēstures arhīvs)
NA RGO	Scientific Archive of the Russian Geographical Society (Nauchnyi arkhiv Russkogo geograficheskogo obshchestva)
NHAB	National Historical Archives of Belarus (Natsyianal'ny histarychny arkhiŭ Belarusi)
RGIA	Russian State Historical Archive (Rossiiskii gosudarstvennyi istoricheskii arkhiv)
SPbF ARAN	St. Petersburg Branch of the Archive of the Russian Academy of Sciences (Sankt-Peterburgskii filial arkhiva Rossiiskoi akademii nauk)

Structures of Archival Holdings

Latvian archival documents: *fonds* (f.), *apraksts* (apr.), *lieta* (l.), *lapa* (lp.), *verso* (v.).

Estonian archival documents: *fond*, *inventari number*, *säilitusüksus*, *lehekülg*. N.B. The contemporary referencing system used by the National Archives of Estonia omits these abbreviations (e.g. EAA.854.7.714.1461).

Russian archival documents: *fond* (f.), *opis'* (op.), *delo* (d.), *listok* (l.), *obratnyi* (ob.).

Belarusian archival documents: *fond* (f.), *vopis* (vop.), *sprava* (spr.), *arkush* (ark.), *advarotny* (adv.).

Notes on the Text

Dates

Dates pertaining to the Russian Empire prior to 31 January 1918 are given in the Julian (old-style) calendar, which ran thirteen days behind the Gregorian (new-style) calendar. Both dates are given when referencing letters between individuals in the Russian Empire and their correspondents abroad.

Translations, Transliterations, and Spelling

All translations are my own, unless otherwise indicated by the notes. Transliterations from Cyrillic Russian and Belarusian follow the Library of Congress romanization system, including the soft sign but without diacritics. Exceptions are made in cases where a standard English-language version exists, e.g. for tsars' names and the cities of St. Petersburg, Moscow, and Riga. To aid with tracing works in library catalogues, nineteenth-century publications in Russian, German, Estonian, and Latvian are rendered in modern orthography.

Names

In the period and geographical region examined in this book, places were typically known by multiple names in various languages. As these usages tell us a lot about territorial imagination, I preserve the toponyms used in the sources in question, for example German Reval, Russian Revel, or Estonian Tallinn. Contemporary Estonian or Latvian versions are provided in parentheses to help readers locate these places today. For consistency, I refer to the three Baltic provinces throughout by their German names—Estland, Livland, and Kurland. Likewise, multilingual individuals often went by different forms of their names during their lifetime. This was especially common among the Russia Empire's German-speakers and those with Baltic German ancestry. As names are important expressions of social, political, and national identifications, at the first mention I provide the most commonly used forms of names of key figures in the book. Thereafter, I refer to the person using the name that they published under or most frequently used to sign themselves off in the sources examined.

Introduction

In August 1872, St. Petersburg was abuzz with talk of statistics. More than 120 foreign experts and 250 specialists from the Russian Empire gathered in the imperial capital for the eighth meeting of the International Statistical Congress. As an experiment in scientific internationalism, the congress brought together experts from across the globe to discuss the implementation of uniform methods and themes for statistical inquiry, ranging from population censuses to trade, industry, medical, and criminal statistics. The delegates worked together to formulate proposals for common standards to facilitate comparability, such as to include 'spoken language' as a statistical measure in censuses, although in practice few states implemented the recommendations.[1]

As part of the congress, the organizing committee arranged two sessions on the uses of geographical and graphical methods for statistical research. The relationship between statistics and cartography had been a regular topic of discussion since the third International Statistical Congress in Vienna in 1857, when statistician Karl von Czoernig discussed his recently published ethnographic study of the Habsburg Empire. Yet, the avalanche of thematic and statistical maps published in subsequent years meant that the organizers felt that data visualization was a subject that warranted serious international attention in its own right.[2] The organizing committee hoped that these sessions would provide a forum to discuss 'the various systems of representing statistical facts to the eye by geometrical figures, lines, and diagrams, and by different colors, and by attaching such facts topographically to geographical maps by various colors, forms of type in printing, and other devices'.[3] The leader of the conference organizing committee in St. Petersburg, Petr Petrovich Semenov, head of the Central Statistical Committee of the Russian Empire and vice president of the Imperial Russian Geographical

[1] On the links between statistics, expertise, and internationalism, see Brian, 'Statistique administrative'; Randeraad, 'The International Statistical Congress'; Rodogno et al., eds. *Shaping the Transnational Sphere*.

[2] Mayr, *Gutachten über die Anwendung der graphischen und geographischen Methode*. On the international discussions about approaches to data visualization, see Göderle, *Zensus und Ethnizität*: 196–207; Palsky, 'The Debate on the Standardization of Statistical Maps'.

[3] *Report of the Official Delegates*: 41.

Geographies of Nationhood: Cartography, Science, and Society in the Russian Imperial Baltic. Catherine Gibson, Oxford University Press. © Catherine Gibson 2022. DOI: 10.1093/oso/9780192844323.003.0001

Society from 1873 to 1914, emphasized the value of maps as a research tool that gave meaning to the numerical data and illuminated bigger spatial patterns:

> Take the trouble to look at the statistical tables of the population in Russia: you will more or less find the population of the counties [*uezdy*], but you will not grasp at first the idea of a rational, and so to say, logical order in the greater or lesser density of population groups. Throw a glance at the map, this order will be revealed immediately....[4]

The delegates agreed on the communicative power of data visualizations, noting how numbers on their own could be boring and 'for this reason, all means of making reading and understanding easier have to be recommended'.[5] However, when it came to devising international standards for statistical maps, charts, and diagrams, the delegates eschewed uniformity, concluding instead that 'individual fantasy and genius and national tastes may be allowed to develop their own peculiarities'.[6]

The lively discussions at the 1872 International Statistical Congress highlight some of the key questions and tensions explored in this book surrounding the purpose, methods, and uses of ethnographic cartography. Were ethnographic maps a form of data visualization that furthered scientific knowledge or were they simply artistic projections of 'individual fantasy'? Could international or universal standards be developed or were ethnographic maps inevitably bound to the whims of 'national tastes' and the 'peculiarities' of individual mapmakers? Does a map merely aggregate data or is it qualitatively more than the summation of its parts, as suggested by Semenov, enabling us to see an overarching 'rational...and logical order' in the world? Do maps represent knowledge about the world or is knowledge a product of what is inscribed onto a map? At the heart of this book lies an exploration of the fraught encounters between science and technology, concepts of territory and space, and ways of thinking about ethnolinguistic communities. This book is the story of how the process of 'drawing' ethnolinguistic groups together became an essential component of thinking about nationhood and territory together.[7] Maps established practices among contemporary viewers about, in the words of Lorraine Daston and Peter Galison, 'what to look for and how to see it'.[8] Studying the history of ethnographic mapmaking thus gives us tangible historical traces of 'ways of seeing in the making'.[9]

[4] *Vos'maia Sessiia Mezhdunarodnogo Statisticheskogo Kongressa*: 30.

[5] *Vos'maia Sessiia Mezhdunarodnogo Statisticheskogo Kongressa*: 26.

[6] *Report of the Official Delegates*: 41.

[7] On the relationships between drawing and thinking, see Latour, 'Drawing Things Together'.

[8] Daston & Galison, *Objectivity*: 369.

[9] Daston & Galison, *Objectivity*: 369. On images as sources about ways of seeing in the past, see also Burke, *Eyewitnessing*.

Thinking Geographically about Nationhood

The history of cartography in the eighteenth and nineteenth centuries has been closely linked to imperial exploration and expansion, as well the role of science and technology as instruments of state power, control, governance, and surveillance.[10] As Charles Maier argues, the idea of mastering, controlling, bordering, and settling space was a defining feature of modern statehood.[11] Yet concepts of territoriality were not fixed and were continually being reworked across different times and places. For Russia's rulers, mastering space was of central importance due to the sheer size and physical location of the landmass under their control, which stretched halfway across the globe.[12] Muscovite mapmakers produced maps of topography, terrain, natural resources, and settlements as a way of charting expanding frontiers, extracting duties from populations, and locating economic resources.[13] Peter the Great's reign in the early eighteenth century introduced Western European astronomy, navigation, geodesy, and surveying practices to mapmaking.[14] In the second half of the eighteenth century, Catherine the Great inaugurated the first general land survey to register property boundaries in European Russia. The rule of Tsar Paul I at the turn of the nineteenth century marked the beginning of a gradual shift in responsibility for mapping from the Academy of Sciences to the military with the creation of the Map Depot in 1797. In 1812 Tsar Alexander I established the Military-Topographical Depot, which assumed responsibility for producing printed maps and exercising censorship over cartographic materials. During the second half of the nineteenth century, mapmaking gradually permeated almost every realm of the imperial administration as tool of scientific and rationalized state planning and policy.[15]

Cartography developed hand in hand with the efforts of imperial administrators and scholars to use science and technology to make the Russian Empire's territorial possessions and subjects legible. Statisticians, ethnographers, linguists, and physical anthropologists laboured to enumerate, describe, and classify the

[10] The literature on this subject is vast. Key works which have shaped this book include: Akerman, ed. *The Imperial Map*; Dunlop, *Cartophilia*; Edney, *Mapping an Empire*; Hansen, *Mapping the Germans*; Scott, *Seeing Like a State*, esp. pp. 11–52.

[11] Maier, *Once within Borders*.

[12] On the preoccupation with space in the Russian experience, see Burbank & Von Hagen, eds. *Russian Empire*; Smith, ed. *Beyond the Limits*.

[13] For an overview of early modern Russian cartography, see: Bagrow & Castner, *A History of Russian Cartography*; Kivelson, *Cartographies of Tsardom*.

[14] Golubinskii, 'New Technology'; Postnikov, *Russia in Maps*; Seegel, *Mapping Europe's Borderlands*: 23–64.

[15] For instance, on the uses of maps to determine landownership and inform agricultural and economic policy, see O'Neill, *Claiming Crimea*, esp. pp. 164–218; Laptev, *Russkaia Infografika*: 100–21. On the use of mapping by the local administration and police to implement urban segregation and zoning, see Hearne, *Policing Prostitution*: 154–68. On the use of mapping by the local administration and church authorities to monitor the location of schools and track religious conversions see Shor, 'Pravoslavnye sholy', esp. map on p. 514 and Gibson & Paert, 'Apostasy in the Baltic Provinces': 29–32.

empire's diverse populations into ethnolinguistic, religious, and racial groups.[16] Imperial administrators developed bureaucratic practices such as statistical surveys and censuses to count and categorize the empire's population.[17] Inadvertently, many of the scientific practices for describing and categorizing populations through censuses, mapping, and ethnographic museum exhibitions institutionalized during the imperial period were later taken up by nationalists as a foundation for nation-building.[18] The second half of the nineteenth century gave rise to a vogue for ethnographic mapmaking as a medium to bring together imperial desires to demarcate populations, legitimize territorial configurations, and as a way for imperial subjects to give voice to national aspirations.[19]

The term 'ethnographic map' refers to a broadly defined genre of thematic maps that began to be produced in mid-nineteenth-century Central and Eastern Europe to visualize the geographical territories inhabited by various ethnolinguistic groups. Mapmakers applied cartographic techniques initially developed for mapping physical geography, such as geology or soil types, to map the human social world.[20] The development of geographical methods and approaches for understanding ethnolinguistic groups had far-reaching implications for emerging ideas about nationhood.

First, compared to statistical tables or ethnographic descriptions, maps foregrounded space as a key factor in the debates on classifying and taxonomizing people. At a time when social and geographic mobility was limited for most of the Russian Empire's population, people spoke a continuum of dialects and the process of defining an ethnolinguistic community involved delineating the group in spatial terms.[21] Ideas about the internal composition and borders of a community of speakers of the same language lay at the heart of the process of defining ethnic groups. By presenting populations in spatial terms and inscribing them onto the landscape, the territories occupied by ethnolinguistic groups and the divisions between them were presented as occurring in nature and outside of history, even while their promoters often looked to history to justify their claims. The illusion of scientific objectivity afforded by mapping meant that the social construction of

[16] On the history of ethnography and anthropology in imperial Russia, see Anderson et al., *Life Histories of Etnos Theory*; Cvetkovski & Hofmeister, eds. *An Empire of Others*; Hirsch, *Empire of Nations*: 21–61; Knight, 'Seeking the Self'; Köhler, *Russische Ethnographie*; Mogilner, *Homo Imperii*; Rainbow, *Ideologies of Race*; Slezkine, *Arctic Mirrors*.

[17] Cadiot, *Le laboratoire impérial*; Darrow, 'Census as a Technology of Empire'; Holquist, 'To Count, to Extract'; Mespoulet, *Statistique et révolution*; Staliūnas, 'National Census'.

[18] On the 'unintended consequences' of bureaucratic practices on emerging national consciousness in the Habsburg Empire, see Stergar & Scheer, 'Ethnic Boxes'. On the role of censuses, maps, and museums as instruments of nation-building, see Anderson, *Imagined Communities*: 163–85.

[19] Labbé, 'La carte ethnographique'; Seegel, *Mapping Europe's Borderlands*: 138–44; Göderle, *Zensus und Ethnizität*: 196–207.

[20] Cook, 'From False Starts to Firm Beginnings'.

[21] It is important to acknowledge that certain segments of imperial society were more mobile, such as resettlers, labour migrants, soldiers, and nomadic populations. See Siegelbaum & Moch, *Broad is My Native Land*.

ethnicity was overlooked in the scramble to assert historical claims to distinguish autochthonous or indigenous inhabitants from colonizers or migrants, and the implications these labels had for notions of belonging to a homeland and claiming territorial rights.

Second, mapmaking is essentially an exercise in aggregating and synthesizing vast amounts of data to produce a generalizing overview. As Martin Rudwick notes, in the case of geological maps the coloured areas 'represent a distributional *extrapolation* from the evidence of major exposures across intervening areas of the map. In other words, the colours denote an implicit belief that the relevant-rock type would be found in the intervening areas, under the concealing cover of soil and vegetation.'[22] The process of creating an ethnographic map followed a similar logic. Such maps consolidated information from ethnographic descriptions, data tables, or material objects of folk culture, and translated them into spatial terms. Maps reduced complex forms of identification to different areas of human 'types' depicted as flat plates of tightly tessellating colour. In doing so, mapping depicted space as a key factor in the process of ordering the human world and presented this geographical way of differentiating people as natural and rooted in territory. It was precisely this drive towards ordering, generalization, and essentialism inherent to cartography that made ethnographic maps into a powerful medium for communicating ideas about nationhood. Yet these same factors also made maps especially prone to political manipulation for various ends. Although mapmakers throughout history have often presented their maps as rational and objective works of science—enchanting us with tantalizing claims to show the world "as it really is"—the process of drawing borders between populations and imprinting them on particular territories created deeply politicized images of peoples and lands.[23] In the course of translating data into the medium of cartography, mapmakers divided communities, erased certain segments of the population, and promoted the aggrandizing claims of other groups.

Finally, the process of making an ethnographic map forced mapmakers to think about the intersections between various ways of classifying people. In the Russian imperial context, Aibulat Psianchin distinguishes between ethnic maps (*etnicheskie karty*), which depicted the distribution of ethnic groups in the past and present, and ethnographic maps (*etnograficheskie karty*) portraying aspects of an ethnic group's material, linguistic, and folk culture. Historically, there was much overlap between these two genres and nineteenth-century ethnographic maps often incorporated elements of both approaches.[24] Moreover, ethnographic

[22] Rudwick, 'The Emergence of a Visual Language': 161.

[23] On the relationship between cartography, politics, and power, see Harley, *The New Nature of Maps*: 53–81; Pickles, *A History of Spaces*: 107–23.

[24] Psianchin, *Istoriia etnicheskoi kartografii*: 23–9.

maps often appeared under other guises. Mapmakers sometimes chose to view ethnicity through the lens of language, religion, socioeconomic status, or history, reflected in the title of their maps as linguistic, dialectal, religious, or historical maps of 'tribes' (*plemia*). As data on 'nationality' per se was rarely collected in the Russian Empire, mapmakers were forced to search for proxies and use available data on religion, language, and other socioeconomic factors as indicators of nationhood.[25] Efforts to map ethnolinguistic groups were thus frequently undertaken in conjunction with other types of mapping, such as social cartography, topographical mapping, and border-drawing.

The Baltic Provinces and their Inhabitants

Specifically, this book examines how and why mapmakers invested substantial amounts of time, money, and effort into mapping the inhabitants of the Russian Empire's Baltic provinces. The Baltic provinces consisted of three administrative provinces located in the Russian Empire's north-western littoral—Estland/Estliand, Livland/Lifliand, and Kurland/Kurliand—as they were known in their respective German and Russian language names. These territories largely correspond to present-day Estonia and the western two-thirds of Latvia. The region was incorporated into the Russian Empire over the course of the eighteenth century. At the end of the Great Northern War, the Swedish Empire formally ceded the former duchies of Estland and Livland to Peter the Great's Russia in 1721. The vassal state of the Duchy of Courland and Semigallia was added at the end of the eighteenth century as part of the lands annexed by the Russian Empire at the third partition of the Polish-Lithuanian Commonwealth in 1795.

The term 'Baltic' has a complex history and it is important to explain at the outset how the concept is used in this book. Today, we associate the word with the 'Baltic states', meaning the three independent states of Estonia, Latvia, and Lithuania located on the north-eastern shore of the Baltic Sea. Historically, however, the concept of 'Baltic' went through several earlier incarnations. In Baltic and Slavic languages, the etymological root *balt* is related to 'white' or 'marshland' (*boloto*) and was originally associated with the sea—*Mare Balticum*. It was only with the incorporation of the three provinces into the Russian Empire in the eighteenth century that the term 'Baltic' was transposed ashore and was used, in addition to its maritime meaning, to designate the territories on the eastern coast of the Baltic Sea. Over the course of the eighteenth century, 'Baltic' became associated with these territories newly incorporated into the Russian Empire and characterized by a common historical heritage in the medieval Livonian territories of

[25] As will be discussed in subsequent chapters, the Baltic provinces were a notable exception where nationality statistics were gathered periodically. See Plakans & Wetherell, 'The 1881 Census'.

Terra Mariana, by the region's German-speaking noble ruling class, and by the influence of Lutheranism as the majority religion.[26] It is important to emphasize that, contrary to contemporary usage, the present-day territory of Lithuania was not considered 'Baltic' during this period. Instead, today's Lithuania was grouped with the lands of present-day Poland and Belarus as part of the Russian Empire's North-Western Territory and thus lies outside of the main area of this book's focus. The geopolitical concept of 'Baltic' that is familiar to us today—i.e. including Lithuania—only entered widespread usage in the late 1910s, with the notable difference that Finland (formerly the Russian Empire's Grand Duchy of Finland) was widely perceived as a fourth Baltic state at this time.

The idea of the three Baltic provinces of Estland, Livland, and Kurland as a defined region was consolidated in the nineteenth century with the administration of the three provinces under a single Governor-General from 1801. In the first half of the nineteenth century, newspapers and learned societies framed their geographical scope as covering all three provinces.[27] Nevertheless, as the conservative Baltic German publicist Georg Berkholz observed as late as 1882, identification with the wider Baltic region was far from ubiquitous, writing: 'Even though we answer, for example, when travelling abroad, when asked about our nationality, we do not like to say, "I am a Balt", but rather in a more particular way: an Est- or Liv- or Kurländer.'[28] The identification of the Baltic's elites first and foremost with their province rather than with a broader pan-Baltic regional identity was reinforced by the separate noble corporations (*Ritterschaften*), to which the ranks of the nobility for each of the three provinces as well as the island of Oesel (Saaremaa) belonged.

In this book, I refer to the region as 'Baltic', a term which entered English in the territorial sense in the early 1840s.[29] However, the English translation can also be misleading when we talk about the nineteenth century as a whole since it merges two ideologically divergent conceptions of the region. In brief, in the first half of the nineteenth century the three provinces were referred to in official Russian parlance as *ostzeiskie gubernii*, a Russified version of the German-language collective term for the three provinces—*Ostsee Provinzen*, which was derived from the German-language name for the Baltic Sea, meaning 'Eastern Sea'.[30] In the 1860s and 1870s, in the context of debates about the so-called 'Baltic Question' in

[26] On the conceptual history of the term 'Baltic', see Piirimäe, 'The Baltic'.

[27] Examples include: *Ostsee-Provinzen-Blatt* (est. 1823), *Kur-, Liv-, und Esthlandisches Provinzialblatt* (1828–38), *Das Inland* (est. 1836), *Gesellschaft für Geschichte und Alterumskunde der Ostseeprovinzen Russlands* (est. Riga 1834).

[28] Berkholz, 'Geschichte des Wortes "baltisch"': 529. See also Hackmann, 'Wo liegt das Baltikum?'.

[29] Eastlake, *A Residence on the Shores of the Baltic*; Jansen, 'Baltus': 41.

[30] Berkholz disliked this Russian version of the German name, which he called 'the quite crude transferred form—*ostzeiskii*'. Berkholz, 'Geschichte des Wortes "baltisch"': 520.

the Russian Empire, the Russian concepts of *pribaltiiskie gubernii* (Baltic provinces) and *pribaltiiskii krai* (Baltic territory) were gradually introduced.[31]

To complicate matters further, 'Baltic' was not only used to denote a region of the Russian Empire but also in an ethnolinguistic sense in several distinct ways. Firstly, the term 'Baltic languages' (*baltischen Sprachen*) was coined in 1845 by the Prussian linguist Georg Heinrich Ferdinand Nesselmann as a common name for Latvian, Lithuanian, and Old Prussian, which he identified as belonging to a distinct Indo-European 'language family'.[32] Estonian-speakers, by contrast, were not included as their language was (and still is) classified by linguists as Finno-Ugric. Thus, the area inhabited by 'Baltic peoples' (i.e. Latvian- and Lithuanian-speakers) did not overlap with the territorial concept of 'Baltic', pertaining to the three Baltic provinces. Secondly, in reaction to the growing tide of Russian nationalism in the 1880s, German-speakers in the three Baltic provinces began to collectively refer to themselves as Balts (*Balten*) or German-Balts (*Deutschbalten*), with the stress falling the latter part of the word to distinguish themselves from Germans living in the newly founded German state.[33] The more common English-language term 'Baltic German' (*Baltendeutsch*) is rejected by most German Balts and their descendants because of its twentieth-century associations with *Ostforschung* (research of the east) and Nazi policies.[34] In Russian-language sources from the 1860s and 1870s, we also encounter the term 'Balts' (*ostzeitsy*), a pejorative term used to refer to the Baltic German landowners who continued to organize socio-economic life in the Baltic provinces around the manorial economy system.[35] The complex semantics of the term 'Baltic' might be confusing at first for readers unfamiliar with the region. However, a sensitivity to these layers of meaning is crucial for understanding the imagined geographies and political ideas inherent in naming places and peoples in the chapters that follow.

Estland's provincial capital of Reval (today's Tallinn) lay just 350km west of St. Petersburg. However, geographical proximity to the empire's capital did not mean that this region was perceived by imperial administrators as part of the imperial centre. The Baltic provinces were treated by the tsarist government as an imperial borderland, with their own set of legal, cultural, social, and economic particularities.[36] Based on guarantees made by Peter the Great in

[31] Karjahärm, 'Terminology Pertaining to Ethnic Relations': 28. *Pribaltiiskii* did not immediately supplant *ostzeiskii* in Russian-language sources, but gradually permeated discussions of the region.

[32] Nesselmann, *Die Sprache der alten Preussen*.

[33] Von Hirschhausen, *Die Grenzen der Gemeinsamkeit*: 358.

[34] Wezel, 'Introduction: German Community—German Nationality?': 2.

[35] Karjahärm, 'Ostzeitsy i Baltiitsy'. Soviet historiography continued to use the term *ostzeitsy* to refer to the German-speaking nobility, burghers, and clergy. See Dukhanov, *Ostzeitsy: iav' i vymysel*; Dukhanov, *Ostzeitsy: polityka ostzeiskogo dvorianstva*.

[36] On the Baltic provinces in nineteenth-century Russian imagination, see Brüggemann, *Licht und Luft*.

the early eighteenth century, the region's German-speaking nobles and landowners retained autonomy over religion, courts, trade, commerce, and land ownership and management. The provinces' administrative and legislative affairs were managed by noble assemblies (*Landtag*), where proceedings were conducted in German and presided over by the Baltic Governor-General, who acted as an intermediary between the nobility and the tsarist government. While German remained the language of the elite and educated classes, as well as of administration and education until the 1880s, the peasantry mostly spoke Estonian (in Estland and northern Livland) or Latvian (in southern Livland and Kurland). Urban centres were inhabited by significant numbers of Russian-speaking soldiers, merchants, Orthodox clergy, and imperial officials. Yiddish could be frequently heard in major cities as well as in smaller towns in Kurland, which was located within the Pale of Settlement to which Jewish inhabitants were legally restricted in the Russian Empire. Lutheranism was the majority religion across the region, although other confessional groups were present in significant numbers, including Orthodox, Jews, Catholics, and Old Believers.

Another important factor that distinguished the Baltic provinces from the rest of the Russian Empire was that from 1816 to 1819 Tsar Alexander I abolished serfdom in this region, almost fifty years prior to the rest of imperial Russia in 1861. However, the Baltic German landowners remained reluctant to sell land to peasants and tenant farmers were forced to pay labour dues to the estate owners on whose land they worked.[37] In the middle of the nineteenth century, peasant laws were passed to introduce money rents and encourage landowners to sell farmland. Nevertheless, social cleavages between the Baltic German landowners and Estonian- and Latvian-speaking peasantry, the limited and uneven development of agriculture and the manorial economy, and a series of famines and peasant revolts created tense relations between peasants and landlords.[38] The particular socioeconomic structure of Baltic society faced mounting criticism from reform-minded intellectuals in the empire by the mid-nineteenth century. By the 1880s, the imperial government under Tsar Alexander III took a harsher stance and introduced policies to curb regional autonomy and more closely integrate the Baltic provinces with the empire, including the introduction of Russian as the language of administration and education.[39]

Despite their relative autonomy over local governance for much of the nineteenth century, the Baltic provinces played a key role in the administrative, social, and cultural life of the wider Russian Empire. The region's Baltic German elites served in many key positions within the imperial bureaucracy, military, diplomatic service,

[37] Lust, 'The Impact of the Baltic Emancipation Reforms'.
[38] Lust, 'Feeding the Landless'.
[39] Thaden, *Russification*: 54–75.

and academia.[40] The University of Dorpat (today's Tartu), founded in 1632 by the Swedish king Gustavus Adolphus and reopened in 1802 after a period of closure, was one of the most important intellectual centres in the whole empire. Students from all over the Baltic provinces and further afield studied at the university, which produced many of the empire's leading scholars.[41] By the end of the nineteenth century, the early abolition of serfdom and widespread network of Lutheran schools providing elementary education in the Baltic provinces meant that the population had the highest literacy rates in the whole of the Russian Empire.[42] The Baltic provinces were one of the most urbanized and industrialized regions of the Russian Empire and on the eve of the First World War, Riga was the third largest city in the empire after St. Petersburg and Moscow with a population of just over half a million.[43] The Baltic provinces were also of key strategic significance for the empire. Water transportation along the Western Dvina (Daugava) river and, as well as the development of the railways from the 1870s, connected the imperial hinterland to the ice-free ports of Riga, Reval/Revel' (Tallinn), Windau/Vindava (Ventspils), and Libau/Libava (Liepāja).

The particular socioeconomic, cultural, and political circumstances of the Baltic provinces, as Karsten Brüggemann argues, meant that the region functioned as an important 'mirror' and 'test site' for tsarist administrators and Russian intellectuals. Imperial rule over the Baltic territories forced them to confront a wide range of different ideas about the nature of imperial governance over the empire's vast domains and to reflect on the meaning of Russian national identity in the context this multi-ethnic and multi-confessional empire.[44] On the one hand, the beautiful natural landscape, favourable climate, and relative proximity to the imperial capital meant that the Baltic provinces were an imperial borderland that appeared quite familiar. Many of St. Petersburg's elite frequently travelled to the region and the seaside resorts in Hapsal/Hapsalu and Pernau/Pärnu flourished as long-standing popular summer vacation spots for the Romanov family.[45] On the other hand, in the eyes of many of the empire's intellectual elites in the nineteenth century, the Baltic provinces also posed fundamental challenges to the stability of imperial rule and the identity of Russia as an empire. The region was perceived by many Russian intellectuals as standing apart from the inland, Orthodox, and Slavic Russia proper to the east. Over the course of the nineteenth century, the status of the Baltic provinces as an imperial borderland and the loyalty of the region's German-speaking Lutheran landowners would be increasingly called into question.

[40] Armstrong, 'Acculturation'; Gornischeff, *Venemaa suhted Preisimaaga*; Henriksson, *The Tsar's Loyal Germans*; Rabow-Edling, *Married to the Empire*.

[41] Beecher, *Ivory Tower of Babel*: 61–76.

[42] Raun, 'Literacy in the Russian Empire'

[43] Raun, 'The Revolution of 1905': 455.

[44] Brüggemann, *Licht und Luft*: 17.

[45] On the Romanov's visits to the Baltic provinces, see Isakov & Shor, *Vlastiteli rossiiskoi imperii.*

Situating the Baltic in Russian Imperial and European History

There is a rich historiography on ethnographic cartography in Central and Eastern Europe, which is dominated by case studies examining the use of maps to construct various national territories.[46] These works provide valuable insights into processes of nation-building through the activities of local elites to construct ideas about a national space, frequently drawing on Thongchai Winichakul's concept of the 'geo-body'.[47] Studies abound on the diverse uses of maps to persuade or compel locals population to think in national terms, as well as how these maps were used as the basis for demands for autonomy or sometimes even a future, though not yet realized, state. This nation-centric approach is echoed in approaches to writing the history of cartography in the region in general, which—as in the Estonian and Latvian case—is presented as a national history on the basis of cartographic collections in national libraries and archives.[48] However, the dominant focus on national approaches to thinking about territorial communities and histories of mapmaking has meant that alternative ways of imagining human collectives and of organizing space which existed in the past are often overlooked.

An important intervention into the dominant nation-state paradigm in the history of cartography in Central and Eastern Europe was made by Steven Seegel in his transnational account of the contested efforts to map the borderlands between the Russian Empire, Austria-Hungary, and Prussia (later, imperial Germany). Seegel examines the processes of mapping this region both as a tool of imperial governance and as an instrument of nation-building by borderland elites. His work brings a cartographic perspective to the vibrant body of scholarship on the territories of present-day Lithuania, Belarus, Poland, and Ukraine located at the crosscurrents of various imperial and nation-building projects to show how this borderland region emerged in the eighteenth and nineteenth centuries as a 'site of overlapping territorial fantasies'.[49]

[46] Hansen, *Mapping the Germans*; Kalemaj, *Contested Borders*; Kordiš, 'Peter Kozler'; Kotenko, 'Ukraine's First Ethnographic Map'; Petronis, *Constructing Lithuania*; Wilkinson, *Maps and Politics*.

[47] Winichakul, *Siam Mapped*. On the application of Winichakul's concept of a national 'geo-body' to the history of cartography specifically, see Pickles, *A History of Spaces*: 115–23.

[48] On Estonia, see Varep, *Jooni Eesti kartograafia ajaloost*; Maridsle et al., 'The Development of Estonian National Cartography'. On Latvia, see Štrauhmanis, *Latvijas kartogrāfijas vēsture no XIII gadsimta*; Štrauhmanis, *Latvijas kartogrāfijas vēsture: bibliogrāfiskais rādītājs*; Caune, *Latvijas zemju robežas*. For two recent contributions to the history of cartography in Estonia and Latvia, see Pae, *Eesti Rahvusatlas*; Barbans, *100 gadi 100 kartēs*.

[49] Seegel, *Mapping Europe's Borderlands*: 133. See for instance: Snyder, *The Reconstruction of Nations*; Kate Brown, *Biography of No Place*; Ackermann, *Palimpsest Grodno*; Safronovas, *The Creation of National Spaces*; Staliūnas, ed. *Spatial Concepts of Lithuania*; Weeks, *Vilnius between Nations*; Nenartovič, *Kaiserlich-russische*.

By contrast, the Baltic provinces have received surprisingly little attention in this broader literature on 'new imperial history' and 'spatial history' probing the relationship between empire and nationhood in nineteenth-century Central and Eastern Europe.[50] Within the broader field of Russian imperial history, the Baltic province's degree of autonomy, particular set of socioeconomic, cultural, and political circumstances, and the fact that many source materials are in German, has led to a tendency to examine the region's nineteenth-century history in isolation from the wider Russian imperial space, similar to the Grand Duchy of Finland. Meanwhile, from a broader perspective of European history, the Baltic provinces are often only dealt with tangentially in the context of wider Russian and German historiographies, while not being treated as fully part of either. Latvia's historical connections with the other former territories of Polish-Lithuanian Commonwealth, specifically the former Duchy of Courland and Semigallia and the Voivodeship of Polish Livonia, are also rarely mentioned.[51]

The distancing of the Russian Empire and imperialism as factors in nineteenth-century Baltic history has been shaped by historiographical trends of the twentieth century, which prioritized the development of national histories for the newly independent states of Estonia and Latvia. The years of Soviet occupation only increased the motivation of scholars in the post-Soviet period of reinstated independence to work on topics that distinguished Estonian and Latvian history from that of its eastern neighbour. In recent years, scholars such as Karsten Brüggemann and Bradley Woodworth have produced important work that counters the prevailing nationalist and isolationist tendencies in nineteenth-century Baltic historiography by situating developments in the region within a wider Russian imperial perspective as part of the patchwork nature of imperial rule over its vast territories.[52]

This book builds on the work of scholars such as Seegel—in cartography—and Karsten Brüggemann and Bradley Woodworth—on imperial approaches to Baltic history—to argue that the history of the Baltic provinces needs to be understood within a broader framework of Eastern European imperial borderlands. What follows is thus deliberately a history of the Baltic provinces in a loose administrative sense. I ask what we can learn about the history of the Baltic if we incorporate the Russian Empire and a wider transnational perspective into what we see, look for, and listen to.[53] The areas inhabited by various ethnolinguistic groups often cut across administrative borders and regions. Mapmakers produced maps to

[50] Gerasimov et al., 'In Search of a New Imperial History'; Baron, 'New Spatial Histories'.

[51] For example, the territories of present-day Latvia are not included in: Snyder, *The Reconstruction of Nations*. For a reflection on this historiographical absence, see Zajas, *Absent Culture*.

[52] Brüggemann & Woodworth, eds. *Russland an der Ostsee*; Brüggemann, *Licht und Luft*.

[53] For a discussion of how the 'transnational turn' might be applied to Estonian history, see Brüggemann, *On Saints, Migrants and Communists*.

subvert established imagined geographies and encourage their readers to see connections between regions. Tracing the production and circulation of cartographic visual and material culture draws in a cast of actors from beyond the border of the Baltic provinces. I place particular emphasis on tracing the entanglements and cross-border interactions between the Baltic provinces and the neighbouring provinces of St. Petersburg, Pskov, and Vitebsk, as well as the North-Western Territory to the south. For instance, although Vitebsk province lay to the south-east of the Baltic provinces, the western districts of the province were inhabited by a substantial number of Latgalian-speakers and, on this basis, this region was later incorporated into the independent state of Latvia after 1920.

By examining the development of ethnographic mapmaking outside of institutions located in the imperial capital in St. Petersburg, this book highlights the importance of border regions in the intellectual geography of the Russian Empire. It examines the role of the empire's German-speakers as an imperial population who were both the executors and objects of ethnographic mapping, enabling us to challenge straightforward asymmetrical distributions of power between colonizers and colonized, and between foreign imperial rulers and subjects.[54] In addition, the particular socioeconomic and political context of the Baltic provinces as an imperial borderland with exceptionally high literacy levels leaves us with a well-documented flourishing of cartographic print culture and map literacy that expands outside of the imperial administration, military, and learned scholarly establishments. Mapmakers in the Baltic engaged in dynamic ways with ethnographic mapmaking and cartographic print culture to address a wide range of locally relevant issues. They turned to the drawing board to reflect on the complex and reciprocal nature of centre and periphery relations, and entangled forms of imperial, regional, and national identity at the edges of empire.

Approaches to Map Production: Social Experiences, Sites, Entrepreneurship, and Reception

This book tells the stories of people who made and read ethnographic maps. Following an approach which Matthew Edney has termed 'processual' histories of mapmaking, it is as much about mapmakers and consumers as it is about the contents of the maps themselves.[55] When it comes to identifying the key protagonists in mapmaking, research on the history of cartography in the Russian Empire has generally approached this question from a top-down perspective. Existing literature centres on the work of a small circle of 'geographical superstars' who

[54] On concepts of 'foreign rule' in Baltic history, see Brüggemann, 'Foreign Rule and Collaboration'.
[55] Edney, *Cartography*: 10.

were either closely linked to the military or were members of the Russian Geographical Society and Academy of Sciences.[56] The anniversary publications of the Russian Geographical Society were especially influential in entrenching the narrative of cartography as the work of 'great men of science' and canonizing 'founding fathers' of the discipline.[57] Turing our gaze to the early twentieth century, we tend to associate ethnographic maps with the outputs of renowned cartographers such as Poland's Eugeniusz Romer, Hungary's Pál Teleki, or Yugoslavia's Jovan Cvijić, who used ethnographic cartography to project visions of a future Central and Eastern Europe following the First World War.[58] The legacy of this cohort of 'map men', as Steven Seegel terms them, in popular telling elides with the history of cartography as a profession inextricably linked in modern European history with imperialism, colonialism, and patriarchy, and as a discipline forged by the class and gender politics of the age.

These famous cartographers who have drawn our attention, undoubtedly influential though they were, represent only a small proportion of those involved in map production. By focusing on university- or military-trained cartographers, this top-down approach not only excludes whole groups of extraordinary individuals who form part of the history of mapmaking but it also perpetuates an incomplete image of what we think an ethnographic mapmaker should be and look like, namely: male; educated at university or in the military in geography, surveying, mathematics, statistics, or linguistics; well known; mythologized. This book reflects on the work of canonical figures in Russian imperial ethnographic cartography, such as Peter von Köppen and Aleksandr Rittikh, and asks how the light they cast has outshone those who were working on the social and geographical margins of imperial society and who pushed the boundaries of cartographic inquiry and mapmaking practices.

To do this, I decided to think about my source materials using the concept of 'mapmakers' rather than 'cartographers' to expand the analytical horizons of my research to include those who had little or no professional training in cartography or surveying, yet were involved in ethnographic map production in some capacity. Drawing inspiration from Joan W. Scott's call for histories of 'radically different social experiences', I understand 'mapmaker' in a broad sense to include those with 'social experiences' of ethnographic mapmaking.[59] Most of the individuals whose stories are told in this book did not self-identify as cartographers. Rather, the proliferation of mapmaking across multiple spheres of imperial life meant that ethnographic mapmaking was carried out in conjunction with the

[56] Livingstone, *The Geographical Tradition*: 7.

[57] Tokarev, *Istoriia russkoi etnografii*; Semenov, *Istoriia poluvekovoi deiatel'nosti IRGO*; Berg, *Vsesoiuznoe Geograficheskoe Obschestvo*; Agafonov et al., *Russkoe Geograficheskoe Obschestvo*.

[58] Seegel, *Map Men*. On the role of geographers and cartographers in the reshaping of Central and Eastern Europe, see also: Górny, *Kreślarze ojczyzn*; Duančić, *Geography and Nationalist Vision*.

[59] Scott, 'Gender': 1055.

work of imperial bureaucrats, physicians, churchmen, teachers, or museum and exhibition curators.[60] Individuals experimented with ethnographic mapmaking through their adjacent interests in statistics, ethnography, or linguistics, collaborated with family members on cartography projects in the home, or sought to profit from the printing and sale of maps in the context of imperial Russia's burgeoning commercial printing and publishing industry. Refocusing our gaze enables us to rethink what mapmaking is and who was able to participate in map production. Working outside of the norms of the learned establishment, these people turned the tools and conventions of the field against itself to challenge the narratives of both imperialism and nationalism.

Map cartouches typically bear the name of a single, almost always male, figure who was responsible for overseeing the project. Sometimes the margins of maps contain hints in fine print indicating the collaborative nature of cartographic authorship and manufacturing, perhaps mentioning the name of a compiler, funder, illuminator, engraver, printer, or publisher. However, in addition to the names printed on the map, many unattributed, anonymous, and 'invisible hands' were involved in the various stages of mapmaking, including surveyors, statisticians, local correspondents, family members, print workers, colourists, and book binders, to mention just a few.[61] These individuals are rarely acknowledged on the finished publications and their ideas, skills, and labour remain embedded in maps as part of underwritten 'silences'.[62] Yet, they often played a key role in determining a map's final appearance. Approaching the history of ethnographic mapmaking as a history of processes and social experiences of map production provides an important corrective to the myths of sole cartographic authorship. While the sources usually do not allow us to directly hear the voices of these 'invisible hands' gathered at the edges of conventional narratives of map history, correspondence between mapmakers, printers, engravers, and illuminators offer glimpses into the perspectives of those involved in map production and the diverging interpretations of actors involved. Moreover, while lower-class people may not have left behind a large body of written records, we can still learn a lot about their everyday lives by paying attention to descriptions of their actions.[63] Such stories provide us with an alternative way of thinking about the significance and meaning of maps for a broad social spectrum of imperial society and question established hierarchies between those doing the mapping and those being mapped.

In searching for the stories of people who engaged in ethnographic mapmaking, my understanding of the places and sites of map production in the Russian Empire expanded considerably. In the course of researching and writing this

[60] For a similar discussion of the spread of 'popular cartography' in the French-German borderlands, see Dunlop, *Cartophilia*: 4.

[61] Güttler, 'Unsichtbare Hände'.

[62] On reading silences in maps, see Harley, 'Silences and Secrecy'.

[63] Gerasimov, *Plebian Modernity*: 9.

book, I gradually moved from my initial entry points into the topic in the archives of the Russian Geographical Society and Academy of Sciences in St. Petersburg to examine diverse sites of mapmaking.[64] Following the hints and paper trails left in correspondence, invoices, receipts, and acknowledgements led me to examine the mapmaking activities taking place within the provincial bureaucracy, a small-town parsonage in rural Kurland, private printing houses in the Riga Old Town, and by populations in the Estonian-Latvian border region after the First World War. This approach is shaped by scholarship in the history of science which has emphasized the importance of place and locality in the production of scientific knowledge, itself part of the wider 'spatial turn'.[65] I ask how ethnographic mapmaking was understood in these difference places and how place impinged on the content of the maps produced there.

Centring social experiences of mapmaking and sites of map production also led me to think about the mechanisms for financing ethnographic map production. Printing an ethnographic map in nineteenth-century imperial Russia was an expensive and laborious undertaking. Not everyone had equal access to mapmaking opportunities as class, financial, and gender discrimination and inequalities created barriers to participation. The economic and consumer-driven forces affecting mapmaking and publishing have often been overlooked in the history of cartography, yet they are an important part of the historical context for understanding maps.[66] In the mid-nineteenth century, the Academy of Sciences and Russian Geographical Society played a key role in funding the production of ethnographic maps, but gradually provincial bureaucracies and private commercial publishing houses of varying sizes joined the marketplace. Developments in printing technology, paper manufacturing, and rising literacy rates led to a growing market demand for cheap paper products by the turn of the twentieth century and the increasing commodification of cartographic print culture.[67] Paying attention to the flows of money allows us to trace how maps embodied the varying ideas, assumptions, and intentions of funders, printers, publishers, and booksellers.

[64] For a similar argument about the need to expand research sites for studying physical anthropology in the Russian Empire, see Mogilner, *Homo Imperii*: 2.

[65] Chambers & Gillespie. 'Locality in the History of Science'; Livingstone, *Putting Science in Its Place*; Finnegan, 'The Spatial Turn'; Livingstone & Withers, *Geographies of Nineteenth-Century Science*; Mayhew & Withers, *Geographies of Knowledge*.

[66] An important exception in this respect is Mary Pedley's work on the commercialization of the mapmaking industry in eighteenth-century Britain and France: Pedley, *The Commerce of Cartography*: 12.

[67] On the commercialization of printing and the development of modern capitalism and consumerism in the Russian Empire, see Marker, *Publishing, Printing*; Ruud, *Russian Entrepreneur*; Steinberg, *Moral Communities*; Rowley, *Open Letters*; McReynolds, *The News Under Russia's Old Regime*; Fehrenbach, *Bildfabriken*. On the development of the paper industry, see Mashkina, 'The Pulp and Paper Industry'.

Finally, in conceiving of the history of 'social experiences' of ethnographic maps, I examine issues of cartographic audiences and reception. Understanding the production of ethnographic maps in their commercial contexts opens lines of enquiry into the popular consumption of cartographic material in late imperial Russia. This book charts the changing audiences and consumers of ethnographic maps in conjunction with the expanding commercial sphere of cartographic publishing, advertising, and print consumption in the Russian Empire.[68] Literacy not only increased across the second half of the nineteenth century, but cartographic literacy did as well, leading to what Valerie Kivelson calls a growth in 'map-mindedness'.[69] In doing so, I build on Martin Brückner study of the rise of mass 'geographical literacy' in early America and Theodore Porter's work on emergence of 'statistical thinking' and 'trust in numbers' as features of modern public life to consider how these broader societal phenomenon also applied to maps as forms of ethnographic data visualization.[70]

Considering audience reactions is also crucial from a methodological perspective when it comes to assessing the meaning, significance, and impact of a particular map for contemporaries. It is commonplace to refer to certain maps as being especially impactful or powerful, but examining reader responses allows us to pinpoint how and why particular maps resonated with audiences, while others did not. Readers could lavish praise or challenge the meanings communicated on a map with their harsh critiques.[71] Maps triggered responses from audiences at a 'non-rational level' by stirring emotions and provoking outpourings of wonder, pride, or anger.[72] Examining readers' sensory responses, and how mapmakers sometimes deliberately designed their maps to bring out a particular affective response from readers, is crucial for understanding developments in ethnographic mapmaking in Russian imperial society.[73] The emergence of ethnographic mapmaking in the mid-nineteenth century was closely linked to visual and material cultures of scientific image-making, as well as early developments in data visualization.[74] Debates about measurement, data collection, and scientific methods for visualizing ethnic groups developed alongside discussions about map aesthetics, art and design, and early ideas about the psychology of impactful

[68] On public audiences for science in nineteenth-century society, see Fyfe & Lightman, *Science in the Marketplace.*

[69] Kivelson, *Cartographies of Tsardom*: 27.

[70] Brückner, *Geographic Revolution*; Porter, *The Rise of Statistical Thinking*; Porter, *Trust in Numbers.*

[71] On the influence of map reviewing and cartographers' reputations, see Withers, 'On Trial—Social Relations of Map Production'; Withers, 'Map Making, Defamation and Credibility'.

[72] Romaniello & Starks, eds. *Russian History Through the Senses*: 2. On the role of emotions in cartographic design and reception, see Iturrioz & Wachowicz, 'An Artistic Perspective for Affective Cartography'.

[73] For a discussion of cartography and the senses beyond the visual, see Orlova, 'Karty dlia slepykh'. On designing 'affective' forms of data visualization, see D'Ignazio & Klein, *Data Feminism*: 84–8.

[74] On visual culture in the history of science, see Daston and Lunbeck, eds. *Histories of Scientific Observation*; Daston & Galison, *Objectivity.*

forms of visual communication.[75] Examining maps from the perspectives of their audiences also turns up many instances where audiences brought their own assumptions, prejudices, and agendas to their reading of ethnographic maps. Different audiences held varying culturally conditioned ideas about what constituted a 'good' and 'valuable' map. They could read meanings into maps which contrasted starkly with how the map was originally conceived by its maker. Maps were often reused, reworked, or transposed into a new context and given different meanings.

Maps and their Sources

My aim is not to provide a comprehensive overview of all ethnographic maps of the Baltic provinces. Instead, the argument is formed through a discussion of five case studies that enable us to unpack different aspects of the history of the Baltic provinces through the visual and material culture of maps. I examine maps as things that can inform our understanding of crucial social, political, economic, and cultural developments in Russian imperial society at different points in time.[76] Viewing ethnographic cartography as part of this broader chronological picture enables me to examine how ethnographic mapping was used by various actors and emerged as a cartographic genre that was particularly pliable to different political contexts and motives. Maps are highly intertextual objects, and taking this broad chronological scope allows me to trace how mapmakers referenced, built upon, reworked, challenged, and confronted the maps produced by earlier generations. Rather than thinking of maps in terms of their relative orientation towards science or politics, or the question of their degree of objectivity or subjectivity, it pushes us think instead about ethnographic maps in terms of their allegiance to, or deviance from, a set of historically contingent collective values, norms, and practices.

In telling a history of the Baltic provinces through ethnographic maps, I pay special attention to three types of cartographic source materials that have often been overlooked by scholars of Russian imperial cartography. Firstly, I draw inspiration from scholarship on Early Modern, indigenous, and educational cartography to expand my scope of research to include hand-drawn manuscript maps.[77] Examining hand-drawn manuscript maps produced in an age often associated heavily with print culture takes us outside of official institutional settings, and into alternative

[75] Cosgrove, 'Maps, Mapping, Modernity'; Murawska-Muthesius, *Imaging and Mapping*: 16–62.

[76] On writing the history of the Russian Empire through an object, see Sunderland, *The Baron's Cloak*.

[77] Kivelson, 'Early Mapping'; Akerman, ed. *Decolonising the Map*. On indigenous cartography in Soviet Russia, see *Zemlia obetovannaia*. On hand-drawn and embroidered maps made by school pupils, see Schulten, 'Emma Willard'; Schulten, 'Map Drawing'; Tyner, *Stitching the World*.

places of map production, such as homes and schools. Doing so reveals how ethnographic mapmaking was a much more widespread activity in the nineteenth and early twentieth centuries compared to if we only examine milieus of mapmaking linked to government institutions, academia, and learned societies who had resources to print and distribute printed maps. Secondly, I compare drafts, unfinished works, corrections, marginalia, notes on the reverse side, and different editions of maps to trace decisions made during the mapmaking process and to uncover the backstories behind the completed maps. Maps often underwent dramatic transformations during the production process due to the incorporation of new data, political or funding constraints, or a change of publication venue. As a result, the final map could often turn out remarkedly different from the mapmaker's original intention. The drafting process was a crucial, if often obscured, part of the mapmaking process. These various rounds of corrections provide us with important evidence of mapmakers as mediators in the 'art of transmission' of ethnographic knowledge from one medium to another, from data table or descriptive text to map.[78] Thirdly, I pay attention to contemporary descriptions of maps that have not survived in archives or libraries, as well as mapmaking projects that were planned, and sometimes even initiated, but never completed for various reasons. The history of ethnographic mapmaking is full of unsuccessful or unrealized endeavours. These failed or truncated projects can tell us much about the daily grind, practical difficulties, inevitable mishaps, and personal life events that made mapmaking a complicated and messy process.[79] Understanding why some maps never came to light is an important counterbalance to the idea of linear technological progress that we are led to believe if we only examine collections of surviving, published maps.

In addition to cartographic sources, I draw on a wide range of documents related to the provenance, production, reception, and uses of ethnographic maps. An important source are map commentaries, which were frequently published to accompany maps in the nineteenth century yet have often received less attention from historians than the maps themselves. However, as Alexander Schunka argues, these commentaries 'lay at the decisive point of intersection between the gathering of information (either directly at the location or from other maps or literary texts) and the next step in which this information congealed into a cartographical form of knowledge as presented in the completed map'.[80] These commentaries reveal mapmakers' motivations, political intentions, and the economic factors

[78] On notes and drafts as key sources for the history of ideas, see Blair, 'Note Taking'; Pugliano, 'Artisanal Writing'.

[79] For a discussion of the concept of 'failure' in history of science, see Gooday, 'Re-writing the "Book of Blots"'. The emphasis on the 'hard work' of mapmaking resonates with Pieter Judson's argument that nation-building and efforts to legitimize spatial imaginations were protracted and laborious. Judson, 'Constructing Nationalities': 4.

[80] Schunka, 'Cartographic Commentaries': 281.

impacting map production and dissemination. I read these commentaries in conjunction with mapmakers' personal papers, memoires, correspondence, and published writings. I consulted archival materials about mapmaking within the imperial administration and by members of learned societies, including the minutes of meetings, protocols, and instructions. Censorship reports, newspaper advertisements, reviews, and petitions are used to interrogate issues of audience and reception.

Scope and Structure

From today's perspective, the geographical scope of this book is transnational, spanning parts of present-day Estonia, Latvia, Russia, and Belarus. To a lesser extent, the lands and former inhabitants of present-day Finland, Lithuania, Poland, and Germany also form part of the story. However, from a nineteenth-century perspective, this book primarily concerns peoples and territories that were part of the same empire. It is only as a result of the turbulent geopolitics of the twentieth century that the source materials are now scattered across many different states. To understand the world of the nineteenth-century Baltic littoral, historians are now required to cross many borders to traverse a region which was once much more politically and administratively interconnected than it is today. Consequently, this book draws on research conducted in archives, libraries, and specialist map collections across Estonia, Latvia, Russia, Belarus, Finland, Lithuania, Germany, Poland, and Britain, as well as multiple digital collections. One of my major aims is to recreate the world of the nineteenth-century Baltic littoral by piecing together parts of the region's history that have hitherto been examined in isolation in separate national historiographies or have slipped through the cracks between these national narratives.

This book traces the use of cartographic methods to visualize ethnic groups in the Russian Empire's Baltic provinces from the early experiments in ethnographic mapmaking in the 1840s to the use of mapmaking in the post-war boundary negotiations in 1919–20. Each chapter in this book examines a period of heightened political tensions in which concepts of nationhood and territory were called into question and mapmakers turned to ethnographic cartography to advocate for a particular way of thinking about the relationship between people and space: in the aftermath of the European revolutions of 1848, the Polish-Lithuanian Uprising in the Russian Empire (1863–4), the founding of the German Empire at the western edge of the Russian Empire (1871), against the backdrop of the emerging Estonian and Latvian nationalist movements, and the aftermath of the First World War and the revolutions of 1917. Whereas the majority of literature on Russian imperial cartography stops in 1914, this book extends to cover the period of the First World War, revolutions of 1917, and the subsequent regional conflicts as part of a broader 'continuum of crisis' facing the Russian Empire.[81] The chapters are organized so that the argument broadly arcs from a discussion of the close relationship between ethnographic cartography and various scales of imperial governance in Chapters 1 and 2 towards an examination of the proliferation of ethnographic mapmaking as part of more widespread social, cultural, economic, and political activities by a diverse array of civil society actors across Chapters 3, 4, and 5.

A major aim of this book is to show that while mapmakers often deliberately positioned their maps within a larger cartographic genre, this was not a static genre. Understandings of what an ethnographic map was and should be underwent substantial modulations over the eighty-year period under examination. The label 'ethnographic map' was applied to different cartographic objects by mapmakers living and working in different periods, locations, and institutional contexts. What follows is an interrogation of how mapmakers understood ethnographic mapping across different times and spaces and what drove their impulses and desires to apply a geographical lens to questions of nationhood.

[81] Holquist, *Making War.*

1

Networks of Cartographic Influence, Patronage, and Reception

In 1851 the Imperial Russian Geographical Society published an ethnographic map of the western part of the Russian Empire. The *Ethnographic Map of European Russia* consisted of four sheets which could be joined together to form an arresting wall map (Plate 2). Mid-nineteenth-century wealthy and literate audiences in imperial Russia would have been familiar with visualizations of the empire's diverse inhabitants, either from popular albums of anthropological drawings depicting how people lived and dressed or from the beautiful ethnographic drawings that were such a proclivity of Muscovite maps.[1] However, the *Ethnographic Map of European Russia* opened up a whole new spatial way of thinking about the relationship between ethnicity and territory. Readers could trace their eyes around the patchwork of colours depicting thirty-eight ethnographic groups stretching from Warsaw to the Ural Mountains and from the far north of the Grand Duchy of Finland to the shores of the Black Sea. The map pulled readers in by conveying a powerful narrative about the magnificence of the Russian Empire ruling over such varied and geographically dispersed peoples. Petr Petrovich Semenov, the eminent geographer, statistician, and long-serving head of the Imperial Russian Geographical Society between 1873 and 1914, pronounced the map as one of the crowning moments of the Society's early years.[2]

The *Ethnographic Map of European Russia* was the Russian Empire's take on the burgeoning interest in ethnographic mapping that had begun to grip Europe. Within the space of a decade, a wave of thematic maps depicting the distribution of ethnolinguistic groups appeared in print across the territories of the German Confederation and Habsburg Empire.[3] Influenced by the experiments with data visualization and thematic mapping by Prussian geographer and explorer Alexander von Humboldt, scientists, geographers, and linguists explored various ways of applying cartographic

[1] On ethnography and visuality, see Frierson, *Peasant Icons*; Vishlenkova, *Vizual'noe narodovedenie imperii*; Knight, 'Russian Ethnography and the Visual Arts'. On ethnographic figures on maps, see Kivelson, *Cartographies of Tsardom*: 171–93; Sach, 'Symbols, Conventions, and Practices'. The pictorial tradition of depicting ethnographic groups on maps continued throughout the imperial period, especially on educational maps such as: Tomasik, *Nagliadnaia karta Evropeiskoi Rossii*.

[2] Semenov, *Istoriia poluvekovoi deiatel'nosti IRGO*: 111.

[3] On the early development of ethnographic mapping in Central and Eastern Europe, see Seegel, *Mapping Europe's Borderlands*: 138–48; Hansen, *Mapping the Germans*: 55–63; Göderle, *Zensus und Ethnizität*: 196–207; Labbé, *La nationalité, une histoire de chiffres*: 19–40.

Geographies of Nationhood: Cartography, Science, and Society in the Russian Imperial Baltic. Catherine Gibson, Oxford University Press. © Catherine Gibson 2022. DOI: 10.1093/oso/9780192844323.003.0002

techniques developed to map physical geography to map human phenomenon.[4] Notable early examples of the ethnographic map genre include the ethnographic plates in Constant Desjardins's *Atlas of Europe* (1837) and Heinrich Berghaus's *Physical Atlas* (1847), as well as sheet maps such as Karl Bernhardi's map of the German language (1843), Joseph Vincenz Haeufler's linguistic map of the Austrian Monarchy (1846), and Heinrich Kiepert's nationality map of Germany (1848). These maps stimulated vibrant discussions about nationalities questions among Central Europe's intellectual elites, gaining further momentum following the revolutionary wave that spread across Europe in 1848–9. The fervour surrounding ethnographic mapping in turn prompted various state-sponsored mapmaking initiatives, such as Karl von Czoernig's *Ethnographic Map of the Austrian Monarchy* (1855–7). Czoernig's map provided Habsburg administrators with empirical evidence to understand the multiethnic logic of the imperial space, while also making a strong statement about the importance of the Habsburg Empire as a unifying power over a territory that could not be straightforwardly divided along national lines.[5] Ethnographic mapping flourished under this spatial turn in defining and categorizing human groups. Ethnographers and statisticians turned to geographical approaches and methods to help answer questions about where different groups lived and where the borders between them lay.

Despite the strong Germanic intellectual influence on these early ethnographic maps, the vogue for mapping ethnic groups was not confined to Central Europe. To the east, mapmakers in imperial Russia were simultaneously experimenting with using cartography to think spatially about the empire's inhabitants. Yet, the impetus behind ethnographic mapmaking in Russia was also shaped by circumstances particular to the Russian Empire and cannot be straightforwardly classified as either proto-nationalist or as a top-down initiative closely aligned with the imperial government. Instead, this chapter examines how approaches to ethnographic mapmaking in imperial Russia in the 1840s built upon earlier ethnographic traditions and were shaped by the institutional structures that supported and funded map production.

Our guide to unravelling the development of ethnographic mapmaking in the Russian Empire can be found in the name inscribed onto the cartouche of the *Ethnographic Map of European Russia*—Peter von Köppen, also known by his Russian name as Petr Ivanovich Keppen. Telling the story of ethnographic mapmaking through Köppen allows us to examine what motivated individuals in the Russian Empire to begin making ethnographic maps, how prospective mapmakers secured political and financial support for their projects, and who they envisaged as the audiences for their maps. At its heart, this chapter seeks to understand

[4] On Humboldt's immense influence on the development of scientific data visualization, see Godlewska, *Geography Unbound*: 121–6; Schulten, *Mapping the Nation*: 80–6.

[5] On Czoernig's map, see Stachel, 'Die Harmonisierung national-politischer Gegensätze'.

what ethnographic mapmaking meant for Köppen and his contemporaries in mid-nineteenth century imperial Russia.

Peter von Köppen's Horizons of Experience

To understand how Peter von Köppen came to make the *Ethnographic Map of European Russia* in 1851 we first need to paint a picture of the kind of person he was. Köppen was born in 1793 in Kharkov, present-day Kharkiv, then part of the Russian Empire's Slobodsko-Ukraine province.[6] His father, Johann Friedrich von Köppen, was one of several thousand Prussians who accepted an invitation from Empress Catherine II to work in the Russian Empire. In 1786 Johann Friedrich moved from Brandenburg to Kharkov to take up the position as head of the medical district. Köppen's mother Caroline Friedrich Schulz was born in Hamburg but had moved in 1773 as a small child to Kharkov, where her father worked as director of an educational institute for the nobility. As one of the empire's 'Russian-Germans', Köppen's upbringing in a German-speaking Lutheran family in the Ukrainian provinces shaped him as a particular kind of 'imperial person' who comfortably existed in the worlds of German culture and Russian imperial society without contradiction.[7]

Köppen's education and early career were typical of many young men of a certain social standing from the Russian Empire's provinces. Köppen attended gymnasium in Kharkov and as a youth he joined the provincial land surveyors from 1805–9. Although we have little information about this early experience of topographical mapping, it is likely that it laid an important foundation for Köppen's own cartographic work later in life. Köppen's father planned to send him to be educated at a German-language university in the Holy Roman Empire or Prussia, but Johann Friedrich's untimely death in 1808 and the eastward movement of Napoleon's army meant that these plans were never realized. Instead, Köppen enrolled at the newly established University of Kharkov in 1809, where he pursued a broad education in law, statistics, history, physics, chemistry, botany, and zoology. Following graduation in 1814, Köppen moved to St. Petersburg to pursue a career in the civil service.

Over the course of his lifetime, Köppen worked in a diverse array of jobs: in the postal department and Department of Education, as editor of the main German-language newspaper in St. Petersburg (*St. Petersburgische Zeitung*), as a silk farming sub-inspector for the Ministry of Interior, and as an expert on agriculture in Crimea

[6] Information about Köppen's life is taken from: Keppen, 'Biografiia P. I. Keppena'; Sukhova, 'Petr Ivanovich Keppen'.

[7] On the concept of the 'imperial person' as way of thinking about the complex biographies and loyalties of the empire's 'Russian-Germans', see Sunderland, *The Baron's Cloak*: 7.

for the Ministry of State Properties. These administrative roles brought Köppen to different corners of the Russian Empire. While working for the postal department in 1819, the Ministry of Internal Affairs sent him to audit over 320 postal stations in the western part of the empire, stretching from Narva near the Baltic Sea shore to Sevastopol' on the Black Sea coast. Köppen worked for periods in Crimea in the late 1820s, conducted research expeditions to the Lower Volga region in 1829 and to Tula, Orel, and Voronezh provinces in 1850. In the course of his imperial duties, Köppen traversed many of the regions of 'European Russia' which would later appear on his ethnographic map.

While Köppen maintained one foot in the tsarist civil service as a ministerial expert and one of the so-called 'enlightened bureaucrats' active in the decades prior to the Great Reforms of the 1860s, his other was firmly planted in the small circle of St. Petersburg learned society.[8] Upon moving to St. Petersburg, Köppen was introduced to many key intellectuals of the period, including the well-known linguist and philologist Friedrich von Adelung, whose daughter, Alexandra Fedorovna, Köppen married in 1830. Köppen was a member of numerous learned societies and in 1843 Köppen was elected as a full member of the Imperial St. Petersburg Academy of Sciences as an expert on statistics. He was one of the founding members of the Russian Geographical Society and between 1845–51 he directed the Society's Statistical Section. The breadth of Köppen's scholarly interests is reflected in his publications, which span such diverse topics as population statistics, a commentary on the ninth revision (imperial censuses of the taxable population), ethnography, winemaking and the wine trade, archaeology, and geography.[9] In 1852, at the age of 59 and following several bouts of illness, Köppen retired to his estate at Karabag/Karabakh (present-day Bondarenkovo) near Alushta on the southern coast of Crimea.

Köppen began making ethnographic maps in his 50s after a long career as an imperial civil servant. Over the course of his lifetime he amassed a vast collection of statistical information on the inhabitants of imperial Russia gathered from different sources. His experiences as an imperial servitor and cosmopolitan scholar led him to mapmaking out of a curiosity to see what patterns would emerge if he brought together the data he had collected in visual form. Ethnographic mapmaking presented itself to Köppen as an attractive way to make sense of the intricate complexity of imperial life that he observed around him.

[8] On Köppen's role in the St. Petersburg circle of 'progressive bureaucrats', see Lincoln, *In the Vanguard of the Reform*: 80–101; Loskutova, 'Early Research on Insect Pests': 249.

[9] Of especial importance for Köppen's work on ethnographic mapmaking are his publications on the Lithuanian languages and population statistics: Keppen, *O proiskhozhdenii, iazyke i literature*; Keppen, *O sushchnosti statistiki*; Keppen, *O chisle zhitelei v Rossii*; Keppen, *Predvaritel'nyie svedeniia o chisle zhitelei v Rossii*; Keppen, *Deviataia reviziia*.

Forging a Social Network of European Mapmakers

In June 1822, a chance meeting occurred in Neusatz (Novi Sad) between Köppen and another of Central Europe's future ethnographic mapmakers, Pavel Josef Šafařík, as he was known in Czech, or as Pavol Jozef Šafárik in Slovak. Their encounter would have a profound impact on the transimperial nature of discussions about ethnographic mapmaking in subsequent decades. The meeting between Köppen and Šafařík took place during Köppen's travels through the Habsburg Empire and German Confederation in the early 1820s as a hired travel companion for a wealthy young lieutenant. By April 1823 Köppen and the lieutenant had parted ways in Vienna when Köppen grew frustrated with his companion's persistent money troubles. Nevertheless, the trip still proved very fruitful for Köppen. During his travels he was able to develop contacts with many prominent scholars from the region, such as pan-Slavist ideologist Ján Kollár, philologist Jacob Grimm, writer and scientist Johann Wolfgang von Goethe, and geographer Carl Ritter.[10]

While passing through Neusatz in the summer of 1822, Köppen hoped to make the acquaintance of the renowned expert on Slavic ethnography and linguistics, Pavel Josef Šafařík. At the time, Šafařík was working as headmaster of a local Serbian Orthodox gymnasium. In fact, the two men almost did not cross paths. In one of history's near misses, the first time Köppen passed through Neusatz Šafařík was out of town. However, when Köppen doubled back a couple of days later the two were able to meet. Köppen's and Šafařík's 'corresponding interests', to borrow Anne Secord's term, ignited an intellectual spark in one another.[11] The two men quickly discovered that they faced common challenges when it came to studying multilingual imperial populations. Köppen and Šafařík subsequently exchanged letters over several decades about their respective research on ethnography, linguistics, and later, ethnographic mapmaking.[12]

While several scholars have pointed out the intellectual links between Köppen and Šafařík, examining their correspondence allows us to trace the circulation of ideas and knowledge about ethnographic mapmaking and visual thinking across the Habsburg and Russian empires at the level of interpersonal relationships.[13] Foregrounding the social dimensions of ethnographic mapmaking is important

[10] On Köppen's contacts with foreign scholars, see Potepalov, 'Puteshestvie P. I. Keppena'; Keppen, 'Biografiia P. I. Keppena': 110–11.

[11] Secord, 'Corresponding Interests'.

[12] For a discussion of 'epistolary geography' as an approach for studying relationships between twentieth-century mapmakers, see Seegel, *Map Men*: 6–8.

[13] On the Vienna-St. Petersburg 'axis' of ethnographic cartography, see Seegel, *Mapping Europe's Borderlands*: 133–57. On Köppen and Šafařík specifically, see Petronis, *Constructing Lithuania*: 176–84. My analysis of Köppen's and Šafařík's letters draws methodological inspiration from the examination of social relations between British and German mapmakers during this period in Withers, 'On Trial—Social Relations of Map Production'.

for highlighting how scholars in the Russian Empire did not simply transpose a Central European model into a Russian setting. Köppen's decision to make ethnographic maps after 1845 was the culmination of ideas formulated as part of a transimperial network of scholars, which Köppen had played an integral role in developing.

Köppen's and Šafařík's letters first broached the subject of ethnographic mapmaking in 1836. By this time Šafařík was living in Prague and Köppen had returned to St. Petersburg after spending two years in Odessa. Note that this was the year prior to the publication of Constant Desjardins's *Ethnographic Map of Europe* in 1837, often regarded as one of the earliest ethnographic maps.[14] However, approaching the history of cartography as a strict linear chronology of firsts, breakthroughs, or 'milestones' does not convey a sense of how in practice mapmaking was often the product of a slow percolation of ideas over several years or even decades.[15] As prospective mapmakers, Köppen and Šafařík spent a lot of time and effort thinking through and discussing their ideas prior to actually making their maps. As we shall see, it was one thing to have an idea for a map but quite another to obtain the financial and material resources to bring it to light.

In 1836 Köppen was still a decade out from making his first ethnographic map, but his ideas about mapmaking were already developing through his work on several other cartography projects. In March 1836, Köppen led the project to compile a map of southern Crimea for the Governor-General of Novorossiia and Viceroy of Bessarabia province, Prince Mikhail Semenovich Vorontsov. Köppen was approached for the job as a civil servant with knowledge of Crimea rather than for his particular cartographic expertise. Nevertheless, while the map itself was a topographic map designed to be used by the military and travellers, we find hints that Köppen was already beginning to think about how to map multilingual territories. For instance, his fieldwork in Crimea to prepare the map involved identifying the location of so-called 'German colonies' of Mennonites and 'Russian villages'. In the commentary accompanying the map, Köppen detailed the methodological challenges and solutions he adopted to record oral versions of place names and transcribe Tatar language from the Arabic script into Cyrillic.[16] Another indication of Köppen's burgeoning interest in cartography came the following year when the Military Topographic Depot solicited Köppen's expertise by inviting him to review the *Detailed Map of the Western Part of Russia* (1821–39), a topographic map completed under the leadership of the chief of the Military Topographic Depot, Friedrich Theodor Schubert (known in Russian as Fedor Fedorovich Shubert). Köppen's lengthy review for the Academy of Sciences

[14] Krokar, 'Constant Desjardins'.
[15] Friendly, 'Milestones in the History of Data Visualization'.
[16] Keppen, *Ukazatel' k karte iuzhnogo Kryma*: 6, 9.

demonstrated his thorough knowledge of eighteenth- and early nineteenth-century maps of Russia.[17]

Over the second half of the 1830s, Köppen sent Šafařík regular updates about his work on the map of Crimea. Meanwhile, Šafařík discussed his own plan to produce a linguistic map of the Slavic languages. Šafařík complained that he was struggling with mapping the Slavic areas of the Russian Empire. Imperial borders posed a problem for scholars interested in studying Slavic languages. The geopolitical division of Slavic-speakers among different empires meant that any attempt to study this language group was, by nature, transimperial in scope. However, scholars often struggled to access information about Slavs living in other empires. As Šafařík complained to Köppen in 1836, the Russian Empire's Slavs 'are here so unknown as the Chinese as we do not and cannot have any journals or catalogues'.[18] Šafařík's work was hampered by the fact that he had no access to the index map of the *Detailed Map of the Russian Empire* (1801–16), the first publicly available sheet map of the Russian Empire and Šafařík's main cartographic reference.[19] Moreover, as experience showed, the circulation of information across imperial borders was also prone to logistical hassles. On one occasion a parcel sent by Köppen to Šafařík containing books was stolen by the young man entrusted with the transportation.

As a solution to the lack of information available to Šafařík about the geography of the Russian Empire's Slavs, Šafařík encouraged Köppen to publish his research. Šafařík wrote to Köppen that: 'A treatise on Russian peasant dialects, with a more accurate characterization of their peculiarities and a careful delimitation of their area or space (this can be best done on an ethnographic overview map), would certainly be greatly desired not only by me, but by all literature-lovers.' Šafařík urged Köppen to publish his research in German in a widely distributed scholarly journal, accompanied by a small language map, which Šafařík could then use as a reference for drawing the boundaries of the Russian Empire's Slavs on his own map.[20] This exchange reveals how Köppen and Šafařík understood maps not only as a method of data visualization but also as tools of science communication. A map could be transported much more easily than heavy tomes of statistical and ethnographic descriptions. Maps made ethnographic knowledge more portable and mobile, facilitating information sharing across imperial and linguistic borders.[21]

Köppen did not immediately act on Šafařík's suggestion to make a linguistic map as he was busy at the time with other civil service responsibilities. These included

[17] Köppen, 'Über die beim militärisch-topographischen Depot erscheinen neue Karte des europäischen Russlands' (unpublished manuscript). SPbF ARAN f. 30, op. 1, d. 108.

[18] Letter from Šafařík to Köppen, 8/20 July 1836, in Francev, ed. *Korespondence Pavla Josefa Šafaříka*: 343. Further references to the correspondence between Köppen and Šafařík come from this edited collection.

[19] Letter from Šafařík to Köppen, 8/20 May 1836: 351.

[20] Letter from Šafařík to Köppen, 8/20 July 1836: 353–4.

[21] Letter from Šafařík to Köppen, 2/14 May 1847: 405.

a project initiated by the Ministry of State Properties in 1843 to produce an *Economic-Statistical Atlas of European Russia*.[22] As the first climate and soil map of the Russian Empire, the atlas was perceived as a key resource for the Ministry's work on the development of applied sciences, such as forestry and entomology.[23] Köppen was appointed to oversee the creation of the atlas's climate map. However, he found that he was overburdened with work and passed the responsibility onto Adolph-Theodor Kupffer, a leading expert on geomagnetism and meteorology from Livland, who Köppen had collaborated with since the 1830s to gather meteorological observations. Nevertheless, Köppen appears to have remained connected to the atlas project and presented Kupffer's draft map to the scientific committee of the Ministry of State Properties in May 1847.[24]

During this time, Šafařík continued to pursue his project to create a map of the Slavic language area, which he published in 1842. Šafařík's map of the *Slavic-Settled Territories* depicted the distribution of Slavic and non-Slavic languages across Europe. The map was made with contributions from the cartographer and engraver Václav Merkla and was issued as a companion to Šafařík's book on Slavic ethnography.[25] An important source for information about Latvians and Lithuanians was Köppen's own research on this subject, as well as the work of several other notable linguists from the Russian Empire.[26] In the context of the early 1840s, Šafařík's map resonated with emerging ideas about Pan-Slavism since it constructed an image of a greater Slavic ethnic territory spanning imperial borders. Šafařík took Daniel Gottlieb Reyman's *Topographical Detailed Map of Central Europe* (1806–8) as his base map, yet translated the German place names into their Slavic (mostly Czech) versions. In doing so, Šafařík used mapping to make a strong statement about the strength of the Slavic presence in Central Europe. Šafařík's map garnered significance as an important landmark in cartographic thinking about the use of mapping as a discourse of political opposition to the emerging notion of a Pan-German *Mitteleuropa* and in support of an alternative Pan-Slavic vision.[27]

Scholars in the Russian Empire recognized the importance of Šafařík's work and the Slavicist Osip Maksimovich Bodianskii swiftly published a Russian translation in 1843. In his translator's foreword, Bodianskii attached special importance to Šafařík's map. He marvelled at how the visualization of ethnolinguistic data meant that the findings 'could be grasped, so to speak, with a single glance'. Bodianskii regarded

[22] *Khoziaistvenno-statisticheskii atlas Evropeiskoi Rossii*. A detailed account of the production of this atlas is given in: Loskutova, 'Svedeniia o klimate'.

[23] On Köppen's work as an advisor for the Ministry of State Properties during this period, see Loskutova & Fedotova, 'The Rise of Applied Entomology': 143–6; Loskutova, 'Early Research on Insect Pests': 236–7.

[24] Loskutova, 'Svedeniia o klimate': 130.

[25] Šafařík, *Slovanský Narodopis*.

[26] Köppen, *Über den Ursprung*. On Šafařík's contacts with other scholars in the Russian Empire, see Petronis, *Constructing Lithuania*: 178.

[27] Murawska-Muthesius, 'Mapping Eastern Europe': 20.

the map as an innovative form of visual communication, as 'to acquaint oneself with it does not require any puzzling or strenuous effort, but merely careful visual consideration alone'.[28] Bodianskii apologized to his Russian readers that the map text was not translated as the process of engraving and printing a new map would have delayed the publication of the Russian translation by two years. However, unlike with prose, Bodianskii was confident that most of his readership would already have a trained eye for reading the visual language of cartography. He assumed they would be familiar with Latin letters and the conventions of map reading, so interpreting the map would not pose a problem. Like musical or mathematical notation, cartography was perceived as a medium that facilitated communication between speakers of different languages.

Bodianskii saw great potential in Šafařík's map. He acknowledged that some inaccuracies were inevitable in a work of this enormity and concluded his foreword with a call to publish a Russian-language map building on Šafařík's work as soon as possible.[29] Others in the Russian Empire, however, held Šafařík's work in lower regard. The Slavicist Izmail Ivanovich Sreznevskii argued that Šafařík's map contained so many mistakes that it was neither useful nor important for scholars. Sreznevskii noted that many of the same errors were later reproduced by German geographer and cartographer Heinrich Berghaus in the ethnographic plate of his *Physical Atlas* published in 1847. Sreznevskii also criticized the small scale of Šafařík's map, which meant that it was very generalized and the information was too cramped to be of any real scientific use.[30] Nevertheless, the intellectual discussions about Šafařík's map in the Russian Empire laid the foundations for Köppen to secure funding for his own cartographic projects.

Upon the publication several years later of Köppen's own masterpiece, the *Ethnographic Map of European Russia* (1851), Köppen wrote to Šafařík with an informal dedication: 'May it remind you of our many years of friendly relations from time to time and tell you that time and space could not shake my respect for you.'[31] Köppen's and Šafařík's correspondence places social relations at the heart of ethnographic map production, presenting an important counterpoint to the highly individualistic impression of cartographic authorship conveyed by the names printed on map cartouches. Operating within a close-knit community, mapmakers forged deep bonds with one another, be it as close friendships or intense rivalries. In the case of Köppen and Šafařík, their dialogue was sustained over the decades due to their aligned interests and vivid engagement with each other's work. They sent regular progress updates to one another and commiserated over common challenges they faced. The wide-ranging expertise and sheer volume of data required to make an ethnographic map, especially one spanning imperial

[28] Shafarik, *Slavianskoe narodopisanie*: v. [29] Shafarik, *Slavianskoe narodopisanie*: vi.
[30] Sreznevskii, 'Etnograficheskaia karta Evropeiskoi Rossii'.
[31] Letter from Köppen to Šafařík, 4/16 May 1852: 411.

borders, was beyond the scope of a single person. The spirit of ethnographic mapmaking was intensively collaborative and intertextual from the beginning.

The social relations which bound together this network of mapmakers across empires reveals how the Russia Empire's scholars were not simply imitating or playing catch up with their Central European counterparts. Instead, the transmission of ideas about ethnographic cartography more closely corresponds to the concept of a 'cultural gradient', as proposed by Catherine Evtuhov and Stephen Kotkin, and of ideas 'radiating out from a multiplicity of centres and moving in different directions'.[32] So-called 'Russian-Germans', such as Köppen, functioned as key conduits for the exchange of ideas about ethnographic mapping and brought the Russian Empire into the fold of this German-language intellectual space. As will be examined in the following sections, they also played a pivotal role in the adaptation and application of these ideas about ethnographic mapmaking to address circumstances specific to the Russian Empire.

Follow the Money: Financing Köppen's Ethnographic Atlas and Maps

Making an ethnographic map was an expensive undertaking. Paper, lithographic printing plates, and ink had to be purchased. The wages of engravers, print workers, illuminators, and bookbinders needed paying. Köppen may have looked beyond the borders of the Russian Empire to discuss ideas about ethnographic cartography, however, he turned his gaze closer to home to search for financial support for map production using his contacts and knowledge of the imperial administration. The pecuniary trails of Köppen's ethnographic mapmaking activities—encompassing large quantities of funding requests, meeting minutes, account books, invoices, and receipts—underscore how prospective mapmakers in the mid-nineteenth century navigated between various funding mechanisms available to them.[33] The reams of paperwork documenting efforts to raise money remind us that the history of ethnographic mapmaking also forms part of a broader story about procuring funding for scientific research and publication.

In the second half of the 1840s, several organizations in St. Petersburg began financing the production of ethnographic maps. Topographic mapmaking in the Russian Empire had been centralized at the end of the eighteenth century and in 1812 the Military Topographic Depot was established as part of the General Staff

[32] Evtuhov & Kotkin, eds. *The Cultural Gradient*: 4.

[33] For an economic approach to the history of mapmaking in eighteenth-century Britain and France, see Pedley, *The Commerce of Cartography*. On the role of money in the history of science, see Andersen et al., 'The Money Trail'.

to exercise control over the censorship and distribution of maps.[34] Ethnographic cartography, however, was not subject to the same degree of specialist military regulation and censorship. Rather than being a top-down and coordinated endeavour of a single body, ethnographic mapmaking was spread across multiple institutions. Moreover, the small scale of most ethnographic maps and the fact that they were often published as plates within journals or as fold-out appendices to books meant they did not fall under the remit of the Military Topographic Depot. Instead, they were examined by the general censor and subjected to the same process of preliminary censorship that applied to all printed publications in the Russian Empire prior to the 1865 censorship reform.[35]

Riding on the wave of interest in Šafařík's work among intellectuals in the Russian Empire, in April 1846 Köppen submitted a proposal to the Committee of the Historical-Philological Section of the Imperial St. Petersburg Academy of Sciences to produce three copies of an ethnographic atlas of European Russia. The Academy had been founded in 1724 by Peter the Great as a centre of science and education in the empire.[36] The Academy was the logical choice for Köppen to approach. Prior to the gradual shift in responsibility for cartography from civilians to the military in the second half of the eighteenth century, the Academy's Geographical Department functioned as a key organ of mapmaking, including overseeing the production of the first full atlas of the Russian Empire in 1745.[37]

To convey a sense of his vision for the ethnographic atlas to his fellow academicians, Köppen produced three sample ethnographic maps of Novgorod, Tver, and Iaroslav provinces. These prototypes were made on sheets of Schubert's *Detailed Map of the Western Part of the Russian Empire* (1826–39), the same map that Köppen had reviewed for the Military Topographic Depot back in 1837. While the maps were expensive, they were not difficult to procure as in the 1830s the Military Topographic Depot had established a dedicated shop, the Geograficheskii Magazin, on the central boulevard of Nevskii Prospekt in St. Petersburg to generate additional income from the sale of maps.[38] Köppen underlined the names of villages in different colours on the base maps depending on the majority ethnic group living there. He then joined up the place names to create shaded areas showing the distribution of a particular group.[39] General Karl Karlovich Ventsel', an inspector of state land surveying who was promoted to military and civilian governor of Irkutsk province in 1851, helped Köppen access data and

[34] Jeske, 'Topographische Karten im Zarenreich'. I thank Martin Jeske for kindly sharing his manuscript in preparation.

[35] On the abolishment of preliminary censorship in 1865, see Ruud, *Fighting Words*: 137–49.

[36] On founding mission of the Academy of Sciences, see Gordin, 'The Importation of Being Earnest'.

[37] Seegel, *Mapping Europe's Borderlands*: 29–36.

[38] Jeske, 'Topographische Karten im Zarenreich', n.p.

[39] SPbF ARAN f. 30, op. 1, d. 465, l. 3ob–4.

prepare the maps. Ventsel' would later become one of Köppen's regular cartographic collaborators, helping him with subsequent ethnographic maps as well.[40] To expand this approach to make a full ethnographic atlas covering all of European Russia, Köppen proposed to combine sheets from two different topographic map series: fifty-one sheets of Schubert's 1:420,000 map series and an additional thirty-nine sheets of the older and slightly smaller scale 1:840,000 *Detailed Map of the Russian Empire* (1801–16) to supplement the eastern regions not covered on Schubert's map.[41]

Köppen approached the process of creating the ethnographic atlas as a composite material object. He did not start with a blank sheet of paper, but rather repurposed two existing series of printed topographic maps and superimposed a hand-coloured ethnographic thematic layer. The atlas production process was closely tied to Köppen's understanding of the ethnographic atlas as a scholarly reference work. The atlas would exist only in manuscript form and be accessible to a small circle of academicians for consultation in the libraries of the Academy of Sciences and Russian Geographical Society.[42] However, in parallel to the atlas, Köppen wanted to make a small-scale printed ethnographic map on four or six sheets to be sold to a wider audience. Whereas the atlas would be printed in Russian, Köppen wanted to print the sheet map in both Russian and German editions. Köppen perceived the atlas and sheet map as different forms of output from the same ethnographic mapping project. He had a clear idea about how he wanted to create ethnographic maps in different formats to reach these groups of readers: a large-scale academic reference atlas for his peers in the Academy of Sciences and public-facing smaller-scale commercial map publication for domestic consumption and for distribution to a wider international community of scholars who were literate in German.[43]

Köppen's atlas proposal was favourably received by the Committee of the Academy's Historical-Philological Section. The Academy's Secretary, Pavel Nikolaevich Fuss, wrote to Köppen with the good news that the Academy had agreed to finance Köppen's atlas, estimating that the cost 'would hardly rise above 123 silver roubles'.[44] Köppen was instructed to begin work immediately and was issued an advance of 100 roubles from the Academy's library budget to purchase the necessary printed maps from the Military Topographic Depot. However, when it came to the printed ethnographic map, the Academy was more hesitant. Concerned about the potential high costs, the committee asked Köppen to provide a detailed budget for the project. As the Academy feared, Köppen's calculations

[40] Keppen, *Ob etnograficheskoi karte*: 25.

[41] On the process of making these two topographic map series, see Jeske, *Ein Imperium wird vermessen*: 285–8, 294–338.

[42] SPbF ARAN f. 30, op. 1, d. 465, l. 4ob & 7ob.

[43] On the Soviet production of various standards and formats of maps for different audiences, see Postnikov, 'Maps for Ordinary Consumers'.

[44] SPbF ARAN f. 30, op. 1, d. 465, l. 7ob.

projected that an original four-sheet map would cost ten times the price of the atlas. The price for engraving the lithographic plates alone came to 400 roubles, including the cost of purchasing the copper. Based on a print run of 220 copies, Köppen estimated an additional 140 roubles would be required for paper, 30 for printing, 440 for hand colouring the maps, and 130 for drawing.[45] However, Köppen also announced to the Academy that he had already secured a pledge of 1,140 roubles from the newly founded Russian Geographical Society to produce a Russian-language four-sheet ethnographic sheet map with an accompanying commentary.[46] This was a staggering amount of money, amounting to more than 10 per cent of the Society's annual operating budget. Köppen attempted to use the pledge of support from the Geographical Society to leverage additional funding from the Academy of Sciences for a German edition of the map.[47]

Unlike the Academy of Sciences, which had existed in various forms for more than a century, the Russian Geographical Society was a newcomer to intellectual life in the imperial capital. The Geographical Society had been founded half a year earlier in the autumn of 1845 on the model of the similar societies established in Paris (1821), Berlin (1828), and London (1830).[48] The learned society was formally subordinate to the Ministry of Interior, allocated an annual budget from the Imperial Treasury of 10,000 roubles, and Grand Duke Konstantin Nikolaevich, the second son of Tsar Nicholas I, was appointed as the first president. On this basis, the Geographical Society has often been presented as an intelligence-gathering branch of the tsarist state designed to improve the administration of imperial territories.[49] However, like any scholarly association, the Geographical Society attracted members with diverse aims and agendas. Köppen was part of the founding group of academicians, scientists, explorers, and military personnel who wanted to promote research on the geography of the empire and enhance the prestige of Russian science internationally. Like Köppen, many of the founders were German-speakers and several had family roots in the Baltic provinces, including navy admiral and explorer Ferdinand von Wrangel and natural scientist Karl Ernst von Baer. These Baltic Germans pursued professional careers in tsarist service, contributing to exploration and the consolidation of imperial rule. We can think of them together with other social and ethnolinguistic groups of elites from the empire's border regions, such as Armenians in the South Caucasus, whom Stephan Riegg argues on account of their involvement in the military, civil service, and academia functioned as 'both agents and recipients' of imperialism.[50]

[45] SPbF ARAN f. 40, op. 1, d. 465, l. 20.
[46] Semenov, *Istoriia poluvekovoi deiatel'nosti IRGO*: 42.
[47] On Köppen's negotiations with the Academy of Sciences and Russian Geographical Society to co-finance map production, see Gibson, 'Problema sozdaniia etnograficheskogo atlasa Rossii'.
[48] Sukhova, 'Eshche raz o predystorii RGO'.
[49] Seegel, *Mapping Europe's Borderlands*: 110–11.
[50] Riegg, *Russia's Entangled Embrace*: 5.

The Geographical Society played an important role in promoting the kind of collaborative and interdisciplinary research that was crucial for ethnographic mapmaking. Prior to the 1840s, ethnographic research in the Russian Empire had primarily been conducted by individual scholars, with little sense of an overarching agenda.[51] The new Geographical Society established specialist sections for physical geography, mathematical geography, statistics, and ethnography. Ethnographic mapmaking overlapped with the interests of the Statistical Section, directed by Köppen, and the Ethnographic Section, led by Baer. Among the Society's founding members were several figures with experience in mapping, including as the military topographer and geodesist Mikhail Pavlovich Vronchenko and geologist Gregor von Helmersen, who produced the first geological map of European Russia in 1841. The Geographical Society thus provided both a physical and social meeting place to bring these different areas of expertise together (Figure 1.1). The community of the learned society fostered the kinds of knowledge exchange between various geography-adjacent disciplines that lay at the heart of ethnographic mapmaking.

Köppen made a strong case in the spring of 1846 for why both the Academy of Sciences and Geographical Society would be interested co-financing an ethnographic map of European Russia. Like Köppen, many members of the Geographical Society were also academicians of the Academy of Sciences and so the research interests of both organizations had clear overlaps. In the spring of 1846 the organizations also physically shared the same building. Prior to the Geographical Society securing its own premises in an apartment near Pevchevskii Bridge it held its meetings in the conference hall of the Academy of Sciences.[52] Yet, the two scholarly organizations also had clear differences. Despite the strong presence of members with German heritage among the founders, the Geographical Society published its proceedings in Russian, whereas in the 1840s the Academy of Sciences also published in German and French. Köppen sought to take advantage of the different languages of science communication preferred by these two organizations and asked the Academy of Sciences to contribute a further 300 roubles to the project to fund a German language edition of the ethnographic map and commentary so that his work could circulate outside of the Russian Empire. Köppen's different patronage networks—the Academy of Sciences and Geographical Society—held diverging priorities about what was valuable and useful knowledge. Köppen was aware of how the different organizations related to one another and to state policy, and Köppen had to appease them to obtain funding and credibility. It also highlights the tensions mapmakers faced, on the one hand, between restricting access to knowledge

[51] Petronis, *Constructing Lithuania*: 116.
[52] Sukhova, 'Russkoe geograficheskoe obshchestvo i Akademiia nauk': 567.

Figure 1.1. The Imperial Russian Geographical Society as a social space for mapmaking.

Source: *Niva* 21, no. 16 (1890): 433. Public domain.

and, on the other hand, the need to finance map production and disseminate knowledge by making maps available to consumers.[53]

Fuss, the Academy's Secretary, replied to Köppen that 300 roubles, in addition to the 123 roubles already promised for the atlas, was a large ask. However, the committee felt that the cost of making the work accessible in German could be substantially reduced if the Geographical Society agreed to share the engraved plates of the map with the Academy of Sciences, since the plates were one of the most expensive parts of making a map. Even then, the committee of the Historical-Philological Section argued that there was no real need to print another edition of the map in Latin script as 'Russian writing [i.e. Cyrillic] can hardly be seen as a disadvantage, since it is to be assumed that anyone who studies the ethnography of Russia as the subject of their special studies must be able to read the Russian alphabet with ease'.[54] The Academy's statement echoed Bodianskii's earlier remark about how visual cartographic communication transcended linguistic divides in the case of Šafařík's map. The Academy did however agree to fund a German translation of the map's commentary.

Work on the ethnographic atlas proceeded quickly but costs spiralled. The total expense of producing the three copies of the atlas amounted to nearly 200 roubles, almost one and a half times what the Academy originally estimated. By far the most expensive component was the cost of purchasing the printed topographic base maps from the Military Topographic Depot. The remaining money was spent paying the illuminator to hand-colour the leaves and the bookbinder to bind the pages and cover the atlas in leather.[55] Human errors and editorial changes to the maps pushed up the cost as additional base maps had to be purchased during the rounds of correction.

Köppen's efforts to repurpose topographic maps to make an ethnographic atlas resulted in a beautiful and impressive tome, but the final product was neither functional nor user-friendly as a cartographic object to work with. The large 60 × 80 cm format of the printed topographic maps meant that the atlas was oversized, heavy, and cumbersome to manoeuvre. The composite nature of the atlas's provenance also created a dissonance between the topographic base maps, which were organized by province according to an administrative spatial logic, and the ethnic groups coloured onto the maps. The sequence of map sheets in the ethnographic atlas followed the same order as the topographic series, which ran from west to east and from north to south across the western part of the Russian Empire.[56] Consequently, the order of maps bore little correspondence to the spatial distribution of ethnic groups. For instance, Estland province appeared on sheet 12 of the atlas and then the reader had to leaf through the atlas to sheet 17 covering

[53] On the similar dilemma facing the Military Topographic Depot during this period, see Jeske, 'Topographische Karten im Zarenreich'.

[54] SPbF ARAN f. 30, op. 1, d. 465, l. 22.

[55] SPbF ARAN f. 30, op. 1, d. 465, l. 13ob.

[56] The sequence of the plates is indicated on the index map, see *Sbornyi list spetsial'noi karty Zapadnoi chasti Rossii.*

northern Livland province to find the other half of the Estonian ethnolinguistic area (Plate 3). Smaller patches of Estonians also appeared on sheets 16 and 11, which covered the islands of Osel (Saaremaa) and Dagden/Dagö (Hiiumaa). Readers had to flip back and forth between different pages to get a sense of the territory inhabited by Estonians. The large scale of the topographic base maps presented an overview of the empire at a level of detail that facilitated military planning and imperial administration. As an ethnographic atlas, however, the large scale hampered learning and understanding as it made it difficult to get any overview of where different ethnic groups lived in the empire. Moreover, the different scales of the two topographic map series used as the base maps hindered comparisons of the relative size of areas inhabited by ethnographic groups living in different regions of the empire.

Despite the atlas's shortcomings, the process of repurposing existing topographic maps for ethnographic mapmaking did have its advantages in terms of production costs and time. While the ethnographic atlas was completed quickly over a period of two years, the process of making the ethnographic sheet map dragged out over five years. Köppen was able to produce an initial draft for the Geographical Society by February 1848. At this point, the four sheets were engraved and coloured but were still missing inscriptions.[57] The following year, Köppen announced that the two eastern sheets of the map had been successfully completed, but that the engraving work on the other half had slowed down. Köppen explained that the delay was due to his desire to finish the map to the high standards and the fact that he kept receiving more data, which meant that the map was continually being modified.[58] The ethnographic map was eventually published in 1851 and subsequently reissued in two further editions. The second edition alone was printed in three hundred copies, an extraordinary number for that time, and accompanied by a forty-page commentary.[59] The Academy of Sciences did not pay for a separate German edition of the map. However, German translations of the names of the ethnic groups were printed below the Russian terms in the map's legend and the Academy of Sciences published a German translation of the map commentary in 1852.[60]

Understanding the institutional contexts in which Köppen worked exposes how the economics of map production profoundly impacted approaches to ethnographic mapmaking. Money had a bearing not only on the content of Köppen's maps, but also their appearance, format, and intended audiences. Ethnographic map production in 1840s imperial Russia depended on the individual agency of mapmakers who were able persuade patrons of the value of their project and

[57] *Geograficheskie izvestiia* 1 (1848): 7.
[58] 'Otchet Russkago Geograficheskogo Obschestva za 1849 goda', *Zapiski IRGO* 5 (1851): 19–20.
[59] Keppen, *Ob etnograficheskoi karte evropeiskoi Rossii.*
[60] Köppen, 'Über die Anfertigung der ethnographischen Karte'.

tactfully negotiate funding. Köppen's interest in ethnographic mapmaking happened to align with the interests of the Academy of Sciences and Geographical Society at a particular moment in time, and indeed his insider position as a member of both these organizations meant that he also had a role in shaping their priorities in the second half of the 1840s. Yet, money for ethnographic mapmaking also came with conditions attached. The various organizations funding map production all had distinct aims and logics governing their activities. In the process of making his maps, Köppen was forced to answer to the demands of different stakeholders and funders. These differences could be potentially productive for people like Köppen who were experienced at navigating between different branches of the imperial bureaucracy and learned life in the capital, as exemplified by the way he strategized (albeit unsuccessfully) to secure funding for both a Russian and a German edition of his ethnographic map. However, as we shall see, funding bodies' internal quarrels and institutional politics could also be a source of frustration for mapmakers and, ultimately, determine what could and could not be mapped.

Empty Spaces and Ethnic Enclaves: Mapping Finno-Ugric and Baltic Peoples

Both Köppen's *Ethnographic Atlas of European Russia* (1848) and *Ethnographic Map of European Russia* (1851) (Plates 2 & 3) are unusual because of the conspicuous amounts of blank space lying at the geographical centre of both maps. Notably, in both instances Köppen chose not to shade the Russian Empire's Slavs. Instead, the use of colour drew readers' eyes to the non-Slavic ethnolinguistic groups inhabiting the imperial border regions. In Köppen's case, what was and was not depicted on the map spoke volumes about the motivations driving ethnographic map production in the Russian Empire.[61]

Scarcely populated regions and inhospitable environments have a long tradition of being portrayed on maps as empty spaces devoid of colour, markings, or symbols. As Jennifer Keating notes, discourses about 'creating and upholding both myths and realities of vacant spaces' were often used by colonial or imperial powers to justify the appropriation of these territories, which were allegedly in need or civilizing or taming. At the same time, Keating underscores how tropes of emptiness also served as reminders of the challenges of the imperial mission as 'emptiness did not automatically behove opportunity and improvement'.[62] Both of these subtexts were deeply embedded in Köppen's approach to ethnographic mapping. As he wrote in a letter to the Academy of Sciences in the spring of 1846, one of the greatest challenges facing contemporary Russian society was the lack of

[61] On the significance of empty spaces on maps, see Harley, 'Silences and Secrecy': 71.
[62] Keating, 'Amid the Horrors of Nature': 33–4.

knowledge among the intelligentsia and imperial administrators about the inhabitants of the empire, even those living on St. Petersburg's doorstep. Köppen noted that 'With the exception of individual points, a more or less dark veil still generally hovers over the distribution of the various peoples of European Russia.'[63] Köppen saw ethnographic mapping as a tool to bridge what was known and what was unknown, expanding the imperial government's knowledge and administrative reach over its dominions. At the same time, his choice of imagery hints at how lifting this 'dark veil' to peer underneath it had the potential to expose the limits of imperial power in the non-ethnically Russian borderlands.

In both the ethnographic atlas and map, Köppen chose to subvert conventional cartographic discourses of exploration and conquest which presented mapping as being about filling in blank spaces. Instead, he left what was known—in this case, Slavs—blank, signifying their default presence, and instead used ethnographic mapping to illuminate the empire's unknown "Others". The absence of Slavs on his ethnographic maps reveals how Köppen conceived of ethnographic mapmaking as a practice directed towards particular kinds of people. Köppen drew a sharp distinction between his own sense of Russian-German selfhood as a mapmaker, academician, and imperial civil servant, and the subjects of his maps, which he referred to as '*inorodtsy*'. This was a legal term used in the Russia Empire since 1822 originally to describe certain ethnic and religious minorities, often nomadic peoples, living in the east of the empire, and which was extended in 1835 to include Jews.[64] Gradually, throughout the nineteenth century the term came to be used in an unofficial sense to refer to all non-Orthodox and non-Slavophone populations, and this was how Köppen used the concept.[65] As we shall see, later approaches to ethnographic mapmaking would gradually come to collapse the distinction between 'civilised subject and savage object' with the increasing focus in Russian imperial ethnography on studying aspects of Russian nationhood (*narodnost'*).[66] However, in the mid-1840s Köppen approached the task of making an ethnographic map with a clear sense of separation between self and "Other".

Köppen's focus on using ethnographic mapping to study the Russian Empire's non-Slavic peoples was shaped by a long intellectual tradition of expeditionary ethnography. The very concept of ethnography as a research field in the Russian Empire emerged through academic expeditions to Siberia organized by the Academy of Sciences in the 1730s and 1740s and was closely connected to studies of Finno-Ugric peoples.[67] Interest in Finno-Ugric studies further increased after

[63] SPbF ARAN f. 30, op. 1, d. 465, l. 2ob. [64] SPbF ARAN f. 30, op. 1, d. 465, l. 3.

[65] Slocum, 'Who, and When, Were the Inorodtsy?'. Köppen used this term in the title of his 1861 publication to mean all non-Slavs: Keppen, *Khronologicheskii ukazatel'*.

[66] Knight, 'Seeking the Self in the Other': 117–18. Similarly, on the tendency to exclude Baltic Germans, as a group of regional socioeconomic elites, from general ethnographic and anthropological studies of 'the peoples of Russia', see Brüggemann, 'Fremde im eigenen Imperium': 56–7.

[67] On eighteenth-century Russian imperial ethnography, see Vermeulen, *Before Boas*; Slezkine, *Arctic Mirrors*; Slezkine, 'Naturalists Versus Nations'.

the Grand Duchy of Finland was ceded by Sweden to Russia in 1808–9 and in the first half of the nineteenth century a large proportion of ethnographic studies focused on the history, language, and onomastics (the study of the history and origin of proper names and place names) of the empire's Finno-Ugric peoples. In particular, the research by linguist, ethnographer, and historian Anders Johan Sjögren on Ingria and St. Petersburg province galvanized popular interest in Finno-Ugric studies in the Russian Empire, providing inspiration for Köppen's own work.[68]

Approaches to ethnographic mapmaking in the 1840s were not universalist. Whereas in Central Europe the impetus behind the surge in ethnographic mapmaking was closely linked to questions of nationhood against the political backdrop of the revolutions of 1848, in the Russian Empire early ethnographic mapping remained closely tied to Enlightenment ideas about the classification of the natural world. The Russian imperial context in which Köppen's maps were produced highlights how place 'lays down conditions for the appearance of objects of science, for their validation as real, and for the terms on which they are knowable'.[69] What passed as an ethnographic map in 1840s St. Petersburg was borne out of a long ethnographic tradition of researching the empire's Finno-Ugric peoples, which influenced practices in Russian imperial cartography and understandings of what ethnographic maps should depict.[70] The strong influence of Finno-Ugric studies on ethnographic mapmaking in the Russian Empire becomes clear if we broaden our focus out from Köppen's ethnographic atlas and map of European Russia to examine several of Köppen's other, lesser-known ethnographic mapping projects during the late 1840s and early 1850s. The process of researching and gathering data to make his two maps of European Russia raised questions which Köppen felt merited closer investigation at the regional or provincial scale.

The first ethnographic map published by Köppen was not the *Ethnographic Atlas of European Russia* (1848), as is often assumed, but rather a small fold out ethnographic map of the Grand Duchy of Finland (Plate 1).[71] Köppen published his *Ethnographic Map of Finland* (1846) in one of the specialist journals of the Academy of Sciences to accompany an article about ethnographic statistics on the inhabitants of the Grand Duchy of Finland. In the spring of 1846, Stepan Ivanovich Baranowskii, mapmaker and professor of Russian language at the University of Helsingfors (Helsinki), sent Köppen a copy of August Wilhelm Eklund's *Map of the Grand Duchy of Finland* (1840), a Swedish-language map that included some information about the geographical distribution of ethnolinguistic groups.

[68] Sjögren, *Über die finnische Bevölkerung des St. Petersburgischen Gouvernements*. On Sjögren's influence on Köppen, see SPbF ARAN f. 30, op. 1, d. 105a. On Sjögren and the Finno-Ugric emphasis in early nineteenth-century Russian imperial ethnography, see Branch, *A. J. Sjögren*.

[69] Ophir & Shapin, 'The Place of Knowledge': 15.

[70] On the relationship between ethnography and aesthetics, see Wortman, *Visual Texts*: 101–21.

[71] Köppen, 'Finnland in ethnographischer Beziehung'.

Köppen presented these materials to the Academy of Sciences and decided to make his own ethnographic map of the region. Köppen's map of the Grand Duchy of Finland depicted the territory's five main ethnolinguistic groups: Finns, Swedes, Russians, Sámi (listed as Lapps), and Germans. Köppen further subdivided Finnish-speakers into two 'main dialects' of Tawaster (Hämäläiset) and Karelian, and also included a mixed category of 'Finns and Swedes' to denote areas inhabited by both groups. Köppen further subdivided Karelians into three religious groups of Lutherans, mixed Lutherans and Orthodox, and Orthodox across a gradient from west to east, revealing how confession was perceived to be an important cultural marker of difference within ethnolinguistic groups.

Köppen presented the *Ethnographic Map of Finland* in a talk at the Academy of Sciences on 13 November 1846. While Köppen did spend a portion of his presentation describing the main empirical findings, what he was more interested in discussing were the theoretical and technical aspects of ethnographic cartography, the reliability of data, and the wider applicability of the approach. Köppen enlightened his audience on methodological challenges, such as how the ethnolinguistic border between Sámi and Karelians 'flowed into one another' to the extent that it 'cannot be specified with certainty'. He noted how the mapmaking process had been limited by the fact that the population data available to him was organized according to the administrative divisions of the Grand Duchy, as well as the Lutheran archdiocese of Åbo (Turku) and dominantly Swedish-speaking diocese of Borgå (Porvoo), which meant that the ethnographic map was bound by these same administrative structures. Moreover, the available data mostly classified the population by confession rather than language or ethnicity. Köppen nevertheless remained optimistic about the potential of ethnographic mapping, especially the ability of colour shading to convey the 'nuances' of ethnographic settlement and make the relations between the different groups 'vivid'. At the same time, he cautioned that the small format of the map meant that generalizations were inevitable; without such simplifications, a map would contain so much information that it would become impossible to read. Köppen used his talk in front of an expert audience to point out how the transformation of data from one medium to another—in this case, from ethnographic descriptions and statistics to a map—resulted in the creation of different ways of seeing the same phenomenon.[72]

During his work on the *Ethnographic Map of European Russia*, Köppen came to realize that the map would only provide an 'overall overview'. The scale was too small and the map too generalized to convey all the 'ethnographic nuances' and 'shades' of the Russian Empire's ethnic composition.[73] Based on the positive reception of his *Ethnographic Map of Finland*, Köppen applied for further

[72] Köppen, 'Finnland in ethnographischer Beziehung': 434–5.

[73] Köppen, *Über die Deutschen*: 1; Keppen, *Vod' s S. Peterburgskoi gubernii*: 1.

funding from the Academy of Sciences to prepare a detailed ethnographic map of St. Petersburg province.[74] Köppen's project to map St. Petersburg province ran in parallel with the mapping activities of the Military Topographic Depot, which had started work in 1845 on a 1-*verst* topographic map of the province. Although the topographic map was only completed in 1854, the coexistence of both mapping projects suggests the close links between the accumulation of ethnographic data, mapmaking, and administrative changes to toponyms that were implemented in the province in the late 1840s and 1850s.[75]

The draft copies of the map point to the laborious rounds of corrections Köppen undertook as he wrestled with one of the key tensions in nineteenth-century scientific image-making, as described by Lorraine Daston and Peter Galison (Plate 4).[76] Namely, on the one hand, Köppen paid fastidious attention to detail and strove towards an epistemological realism. The pencil scrawls and pen marks on the draft maps reveal how he corrected the spellings of place names as he attempted to match the Finno-Ugric and Slavic versions of toponyms.[77] Köppen endeavoured to make his map as truthful as possible, continuously adding new data and correcting minute errors based on information he received from local correspondents, who sometimes even sent him small sketches with their letters to show how the map should be changed.[78] The emphasis on exactitude was also demanded by advances in cadastral surveying in the first half of the nineteenth century, which meant that the topographic base maps that Köppen was working with were increasingly detailed and precise. As Jason Hansen notes: 'An errant line or splotch of color, therefore, would transmit incorrect meaning, and as a result such mistakes had to be avoided to the greatest extent possible.'[79]

On the other hand, Köppen readily intervened in the image-making process to compensate for discrepancies or lack of data. In the commentary accompanying the map of St. Petersburg province, he noted that while he was able to include ethnic 'enclaves', he was unable to mark every scattered ethnic group (Figure 1.2). Where Germans and Jewish communities lived in separate villages he shaded them in the corresponding colour. However, in cases where they lived with other groups 'and did not consist of the majority population, it was inconvenient to show them.'[80] It is not clear if by 'inconvenience' Köppen was referring to political or technical challenges, or a combination of both. Either way, the interpretative process of transferring data onto a map made certain groups disappear from the ethnographic landscape. Despite these compromises, Köppen was satisfied with the outcome of the ethnographic map of St. Petersburg province, which was able to present a far more detailed portrait of the ethnographic landscape of the region

[74] Köppen, *Ethnographische Karte des St. Petersburgischen Gouvernements*.
[75] Krasnikova, 'Sankt-Peterburg i guberniia.'
[76] Daston & Galison, *Objectivity*: 42.
[77] Köppen, *Über die Deutschen*: 4–5.
[78] SPbF ARAN f. 30, op. 1, d. 102, l. 48–50.
[79] Hansen, *Mapping the Germans*: 58.
[80] Keppen, *Ob etnograficheskoi karte*: 25.

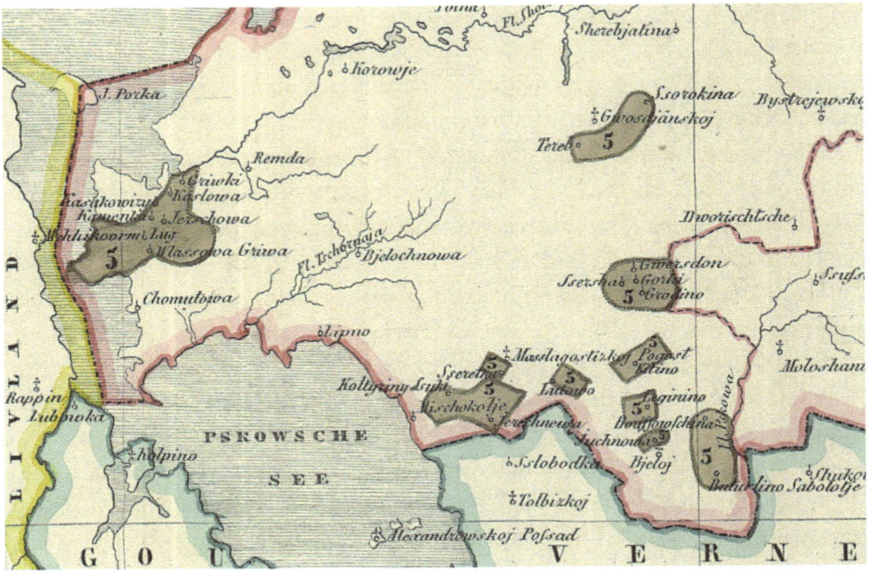

Figure 1.2. Excerpt from Peter Köppen's *Ethnographic Map of St. Petersburg Province* (1849) depicting the settlements of Estonians (no. 5 brown) on the the eastern shore of lake Pskov/Pikhva. Zooming in on a single province allowed Köppen to conduct a more detailed examination of the microgeographies of Finno-Ugric settlement. See Plate 5 for full map.

Source: Peter von Köppen, *Ethnographische Karte des St. Petersburgischen Gouvernements* (St. Petersburg: Kaiserliche Akademie der Wissenschaften, 1849). Reproduced with permission from the Russian Geographical Society.

than his earlier map of the Grand Duchy of Finland.[81] Comparing Köppen's approach to mapping the territory around St. Petersburg on the *Ethnographic Map of European Russia* (Figure 1.3) with his dedicated map of St. Petersburg province (Plate 5) reveals how he was experimenting with mapping ethnolinguistic groups at different scales. Köppen's maps invited readers to think about the scale of spatial thinking at which the subdivisions between various linguistic and dialectal communities become significant.

The Baltic provinces also featured prominently on the intellectual horizons of ethnographers in the second half of the 1840s as a field site for researching Finno-Ugric peoples. The first ethnographic expedition organized by the Russian Geographical Society in the summer of 1846 was dispatched to Livland and Kurland province to study Livs and Krevings, two small Finno-Ugric-speaking populations, as well as the Swedish-speaking coastal inhabitants.[82] The proximity

[81] Köppen, 'Finnland in ethnographischer Beziehung': 463.

[82] For the full report of the expedition, see 'Izvlechenie iz otcheta, predstavlennogo russkomy geograficheskomy obschestvy chlenom-sotrudnikom A. Shegrenom. Ob etnograficheskoi ekspeditsii v

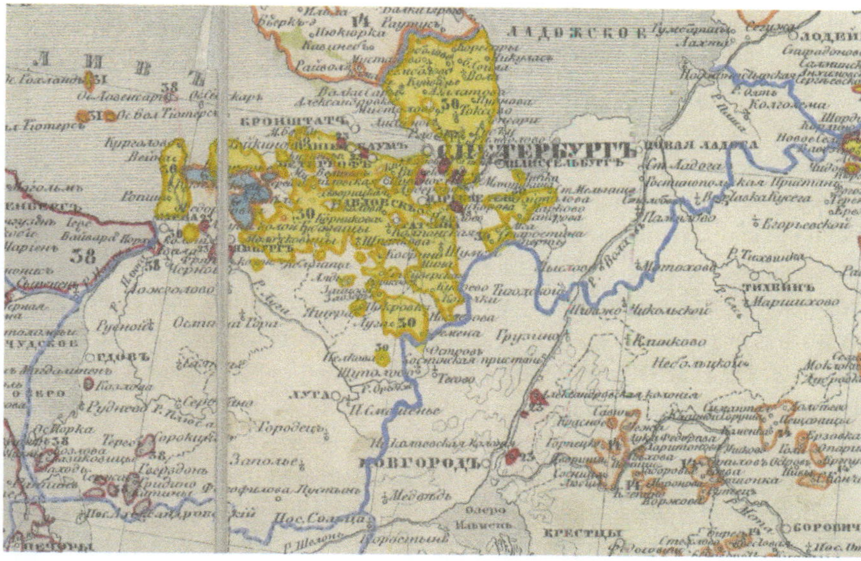

Figure 1.3. Excerpt of the area around St. Petersburg from Peter Köppen's *Ethnographic Map of European Russia* (1851). See Plate 2 for full map.

Source: Petr Keppen, *Etnograficheskaia karta Evropeiskoi Rossii* (St. Petersburg: Russkoe geograficheskoe obshchestvo, 1851). Reproduced with permission from the Russian Geographical Society.

of the Baltic provinces to St. Petersburg and the compact settlements of these populations likely had a bearing on the choice of location of this first ethnographic expedition, as it meant that the research trip was relatively short and the whole project cost just 1,200 roubles.[83] The expedition was led by Anders Johan Sjögren, the Russian Empire's most renowned expert on Finno-Ugric ethnography and linguistics at the time. August Georg Wilhelm Pezold, a Baltic German artist from Wesenberg (Rakvere) in Estland, accompanied the expedition to produce ethnographic drawings.[84] The research by Sjögren and Pezold left a lasting mark on ethnographic depictions of the region. Despite their small numbers, Livs and coastal Swedes appeared on almost all ethnographic maps of the Baltic provinces published in the Russian Empire throughout the second half of the nineteenth century. For instance, N. I. Zuev's *Detailed Atlas of the Russian Empire* (1860)

lifliandiiu i kurliandiiu', *Zapiski RGO* 1–2 (1849): 311–22. On Sjögren's research results, see Sjögren, 'Bericht über eine im Auftrage der russischen geographischen Gesellschaft'. On the expedition, see Blumberga, 'A. J. Sjögren's Expeditions to the Livs'; Cimermanis, 'The Livs of Svētciems Pagasts'.

[83] NA RGO f. 2, op. 1, d. 33, l. 35–6.

[84] On Baer's appointment of Pezold, see NA RGO f. 1–1846, op. 1. d. 4., l. 10. Pezold's original drawings have been preserved in an album by the Russian Geographical Society: NA RGO f. 116, op. 1, d. 61.

marked the location of Livs and Swedes far more prominently than the more numerous German and Jewish populations.[85]

In 1846 Köppen also began preparations to make ethnographic maps of Estland and Livland on a similar model to those of the Grand Duchy of Finland and St. Petersburg province. In 1846 he published several short articles on the ethnographic composition of the three Baltic provinces.[86] Köppen wrote to officials in Estland and Livland in 1847 to ask for more detailed and up-to-date information about the location of villages inhabited by Estonians, Latvians, and Russians with a view to preparing ethnographic maps. Yet, it appears that Köppen never got any further than the data-collecting stage and no more is mentioned of these maps after 1847. The sources here are slightly ambiguous as Köppen appealed to authorities in the Baltic provinces for information for his 'map', thus it not entirely clear which cartography project he was referring to. However, the titles of the archival collections held by the Russian Geographical Society containing Köppen's correspondence with officials in the Baltic provinces support the interpretation that Köppen planned to make separate ethnographic maps of Livland and Estland.[87]

Paying attention to truncated cartography projects that never materialized, as was the case with Köppen's ethnographic maps of Estland and Livland, reminds us of the practical and financial challenges facing cartographers. Quests for knowledge were often aspirational and mapmakers' grand visions were frequently constrained by the budgets and institutions that they had to work with. Yet, the kinds of maps of European Russia that did receive substantial investment from the Academy of Sciences and Geographical Society also tell us about the kinds of cartographic knowledge that was perceived to be valuable and useful in the 1840s and early 1850s. The focus on the empire's small ethnolinguistic groups cultivated a particular culture of ethnographic cartography in imperial Russia that provided institutional credence and financial support for mapping Finno-Ugric peoples as objects of ethnographic study.

Not all members of the Russian Geographical Society, however, agreed with this approach. The years Köppen spent working on his various ethnographic maps coincided with emerging fissures within the Geographical Society's members over the aims and methods of scientific inquiry, as well as clashing visions of Russia as a dynastic multiethnic empire or a state pertaining to its titular nation.[88] The Geographical Society's members became polarized by internal quarrels that positioned the so-called 'German' founding members against a younger group of 'Russians'. The conflicts over the mission of the Geographical

[85] 'Karta gubernii: Lifliandskoi, Estliandskoi, Kurliandskoi, Kovenskoi, Vilenskoi, Grodensskoi i Minskoi'. In Zuev, *Podrobnyi Atlas.*

[86] Köppen, 'Die Bewohner Kur- und Livlands'; Köppen, 'Die Bewohner Estlands'.

[87] 'Materialy dlia etnograficheskoi karty Lifliandii', NA RGO f. 2, op. 1, d. 202; 'Materialy dlia etnograficheskoi karty Estliandii', NA RGO f. 2, op. 1, d. 203.

[88] Knight, 'Seeking the Self in the Other': 118.

Society in the late 1840s have been extensively documented.[89] Pertinent to the present discussion is how this debate impacted ethnography cartography as represented by the viewpoints of the leading scholars of the respective camps: Karl Ernst von Baer, a Baltic German natural scientist and explorer from Estland who was a founding member of the Russian Geographical Society and the first head of its Ethnographic Section, and Nikolai Nadezhin, ethnographer, literary critic, and a former professor of aesthetics at Moscow University.[90]

Baer understood ethnography primarily as the scientific study of the empire's diverse nationalities.[91] Influenced by Enlightenment ideas about racial classifications and hierarchies, Baer regarded ethnography as part of the state's "civilizing mission" towards its native and "primitive" peoples. In Baer's view, the task of the ethnographer should be to describe and collect material artefacts about these small ethnic groups, especially those who were in danger of disappearing through assimilation due to the onset of modernity. For this reason, Baer's concept of ethnography has been compared to American anthropologist Franz Boas's practices of 'salvage ethnography'.[92] Nadezhin, on the other hand, perceived ethnography as an expression of national identity and insisted that ethnographers focus their efforts on studying the Russian nation (*narod*).[93] By 1848, the so-called Russian faction had gained dominance within the Geographical Society and Baer resigned from his post. In 1849, the changing aims of the Geographical Society were reflected with the addition of 'Imperial' to the organisation's official name.

Köppen's ethnographic maps were produced against the backdrop of these conflicting ideas about ethnography within the Geographical Society in the late 1840s. Broadly speaking, Köppen's approach to mapmaking aligned with Baer's conception of ethnography as the study of the diverse non-Russian ethnic groups or *inorodtsy* of the Russian Empire, filling in the blank spots in knowledge and making the empire knowable. As we shall see in Chapter 3, in subsequent years, the ethnographic maps published by the Geographical Society would shift away from Köppen's Baerian approach to ethnographic mapmaking as a form of 'salvage ethnography' to be replaced by a more explicit focus on promoting ethnic Russian interests.[94]

[89] Berelowitch, 'Aux origines de l'ethnographie russe'; Knight, 'Science, Empire, and Nationality': 116–22; Bradley, *Voluntary Associations in Tsarist Russia*: 82–127.

[90] Baer and Nadezhdin's respective interpretations of ethnography are outlined in: Ber, 'Ob etnograficheskikh isledovaniiakh'; Nadezhin, 'Ob etnograficheskom izychenii'.

[91] 'Otchet Russkago Geograficheskogo Obschestva', *Zapiski RGO* 2 (1847): 9–10. On Baer's views, see Tammiksaar, 'The Contribution of Karl Ernst von Baer': 146; Tammiksaar, 'Das Verhältnis von Wissenschaft und Nationalität'.

[92] Knight, 'Seeking the Self': 120.

[93] On the semantics of the term *narod*, see Miller, 'Natsiia, Narod, Narodnost''; Karjahärm, 'Terminology Pertaining to Ethnic Relations'.

[94] Knight, 'Seeking the Self': 119.

Köppen's Readers: Places of Cartographic Reception

The *Ethnographic Atlas of European Russia* (1848) languished in obscurity for most of the nineteenth and twentieth centuries.[95] Two copies were stored in the cartography department of the Russian Geographical Society and library of the Academy of Sciences, while the fate of Köppen's personal copy remains unknown. Köppen's contemporaries rarely cited the atlas in their publications. Only recently has the atlas been made publicly accessible thanks to the digitization of the atlas in 2020 by the Russian Geographical Society. The atlas's reception history, or rather lack thereof, may at first appear to fit into a familiar narrative of the long tradition of secrecy surrounding maps, especially during the Soviet period.[96] As strategic military and political assets, many maps remained closely guarded state secrets. Yet, from examining the internal correspondence of the Academy of Sciences and Geographical Society about the *Ethnographic Atlas of European Russia* a different story emerges about the consumption and reception of ethnographic maps in imperial Russia. The knowledge contained on the pages of Köppen's atlas was not restricted. The topographic maps on which it was based were widely sold in Russia and abroad and the ethnographic information was made available in digest form on the *Ethnographic Map of European Russia* (1851). Köppen was not an anonymous bureaucrat and by the 1850s he had established himself as a well-known ethnographic mapmaker among European intellectual circles. Instead, the atlas's limited audience resulted from the fact that it was conceived of as a specialist reference work, only produced in manuscript form in three copies. It was stored in the libraries of the Academy of Sciences and Russian Geographical Society and not available for loan, thus making it only accessible to a small circle of academicians in St. Petersburg.[97] The relative obscurity of the atlas thus stands as an important reminder of how, regardless of the perceived scientific value of their content, maps need to be seen, touched, and talked about to acquire meaning and have demonstrable impact.

The challenges encountered by fellow mapmakers trying to access a copy of Köppen's atlas are evident from the discussions which ensued when the Scottish geographer and cartographer Alexander Keith Johnston tried to procure a copy the atlas from the Academy of Sciences in 1853. Owner of the leading Edinburgh-based mapmaking firm W. & A. K. Johnston, Johnston was well-connected within European cartographic circles and was keen to obtain the latest works about Russian geography. However, the Academy of Sciences informed Johnston that the atlas was not for sale as the Academy's library only possessed one copy in manuscript form. If Johnston was determined to obtain a copy of the atlas, the

[95] Petronis, *Constructing Lithuania*: 185–9.
[96] Baron, 'Ot grekha podal'she'.
[97] SPbF ARAN f. 30, op. 1, d. 465, l. 4ob & 7ob.

Academy could arrange for additional copies of the base maps to be purchased and a new copy of the atlas to be especially illuminated for Johnston. However, the Academy warned that the cost would likely amount to several hundred roubles. The Academy recommended that the best option would be for Johnston to simply purchase Köppen's published commentary containing the four-sheet fold-out *Ethnographic Map of European Russia.*[98]

The exchange between Johnston and the Academy of Sciences demonstrates how factors such as format, price, and number of copies printed mattered when it came to determining whether an ethnographic map reached audiences. In contrast to the atlas, the *Ethnographic Map of European Russia* brought Köppen and the Geographical Society international fame. The portable format of the folding map on four sheets meant that Köppen was able to dispatch copies far and wide across the Russian Empire and Western Europe. Köppen personally sent copies to his network of friends and acquaintances stretching from Tifilis (Tbilisi) to Strasbourg.[99] The Geographical Society awarded Köppen the prestigious Konstantin medal and Zhkovskii prize for his achievement and the map remained a respected reference work on the ethnographic composition of European Russia for many decades.

By titling his various maps from 1846–51 as 'ethnographic maps', Köppen deliberately positioned his works in dialogue with the emerging genre of ethnographic cartography. The title 'ethnographic map' established expectations among audiences about the form and content. As Köppen noted, 'according to the colours and signs, the owners of the map will already find their way without further explanation'.[100] At the same time, the copies of the ethnographic map distributed by Köppen personally were almost always accompanied by the map commentary. For Köppen, the commentary was a crucial component of the map and he conceived of the text and map as working together. The commentary demystified the map as an object of unquestionable authority by drawing attention to the methods, sources, and design choices made. The commentary also emphasized how the map was not meant to be read as a definitive or finished work. Köppen included requests for local observers to send him information to improve the map, which he specifically addressed in a revised version of the commentary published to accompany the third edition of the map in 1855.[101] Köppen used the mapmaking process to make visible gaps in existing networks of knowledge and, in line with the founding objectives of the Geographical Society, engage educated inhabitants throughout the empire in scientific discussion. Yet, Köppen's own understanding of his maps was one thing. What his readers saw in the maps was an altogether different matter.

98 RGIA f. 735, op. 3, d. 147, l. 3–4.

99 SPbF ARAN f. 30, op. 1, d. 465, l. 20–1.

100 Köppen, *Über die Deutschen*: 8.

101 Keppen, *O tret'em izdanii Etnograficheskoi karty*: 1.

The Slavicist Izmail Sreznevskii, the aforementioned critic of Šafařík's map in the early 1840s, is a case in point of how despite Köppen's efforts to communicate his intentions in the map commentaries, readers brought their own experiences, interpretive lenses, and political viewpoints to read what they wanted to see in his maps. Freshly appointed in his post as head of the new Russian Language Department of the Academy of Sciences, Sreznevskii perceived Köppen's map as a valuable source of knowledge about the geography of the Russian language. Ironically, Sreznevskii devoted most of his review of the map published in the journal of the Imperial Russian Geographical Society to discussing the significance of the blank space in the centre of the map. Szrenevskii noted how by shading the territory inhabited by non-Slavic speakers, the map conveyed a clear sense of the territory where Russian was the dominant language. Sreznevskii felt that the map greatly improved upon earlier maps of the Slavic languages created by Šafařík and Berghaus. Yet, Sreznevskii felt that Köppen had underestimated the spread of the Russian-speaking territory as he had not accounted for ethnic groups who had assimilated: those who had 'forgotten' their native language and now used Russian as their language of everyday communication. Sreznevskii argued for an even more expansive vision of the geography of the Russian language by including settlements of Russian-speakers in Siberia and outside the borders of the Russian Empire, notably Ruthenians (today's Ukrainians) in the Habsburg Empire.[102] By reading the map against the grain, paying attention to what was not coloured on the map rather than what was, Sreznevskii found a subtext in Köppen's map that confirmed his own Russophile sympathies. The interpretivism embedded in ways of reading maps, where readers deconstructed what they saw and ascribed their own meanings to the colours and lines, foreshadowed many of the debates about ethnographic cartography in the decades that would follow.

Reviewing the map for an audience of fellow academicians with an interest in geography, Sreznevskii was also intrigued by the map from a methodological perspective. Sreznevskii was impressed by the potential of cartography as a form of data visualization to facilitate analysis and deepen understanding, since it provided a 'reading of the facts, which are not convenient to understand without a visual aid'.[103] Yet, he also anticipated that using cartographic methods would impact how ethnolinguistic groups would be studied in the future. For example, Sreznevskii noted how while it was impossible to draw 'a single completely truthful line between languages, or even between dialects', making an ethnographic map required such lines to be drawn.[104] Moreover, while linguists were aware of how languages could mutually influence one another, maps presented the relationships between languages as fixed and immobile. Sreznevskii foresaw that ethnographic mapping would shift

[102] Sreznevskii, 'Zamechaniia': 8–9.
[103] Ibid: 6.
[104] Ibid: 10.

the focus onto 'questions of the borders and space of languages and dialects', supporting Peter Haslinger's argument that by the turn of the twentieth century 'imagined territory' would become more important than 'imagined community' and that having 'one's own territory' would become a crucial attribute of any modern nation.[105] In the 1850s, there was no doubt in Sreznevskii's mind that approaches to mapping ethnolinguistic groups would impact future approaches to thinking about these populations.

The way in which context and place impinged on the reception of Köppen's *Ethnographic Map of European Russia* becomes even more apparent if we move to examine the reactions of readers outside St. Petersburg, who were inclined to read his map in the light of their local knowledge and sense of selfhood. Köppen targeted learned societies in the Baltic provinces as a key audience for his publication and sent copies to the Society for the History and Archaeology of the Baltic Sea Provinces of Russia in Riga, Estonian Literary Society in Reval (Tallinn), Estonian Learned Society in Dorpat (Tartu), and Mitau Society for Literature and Art (in today's Jelgava).[106] One of the main German-language journals in the university town of Dorpat published a full-page review of Köppen's map. The reviewer was impressed by the beauty of the map and how it drew readers' attention to curiosities in the empire's ethnographic landscape, such as the presence of small enclaves of Jews in Kherson, Germans in Volhynia, Roma (referred to by the author as Gypsies) in Bessarabia, and Tatars in Minsk province. The reviewer also pointed out that although the map text was in Russian, it was still accessible to local readers as the legend helpfully included the names of the ethnic groups in German translation.[107]

Nevertheless, the Baltic reviewer found fault with several aspects of the map based on their reading of the work from a local perspective. The reviewer questioned Köppen's decision to prioritize language and ethnic groups *(Volkstämme)* over religion as the main factor for thinking about relative similarities and differences between populations in the Baltic provinces. They found it strange that Jews were listed on the map's legend as an 'Asiatic nation' and represented as far removed from the 'blond, blue-eyed Latvians, Slavs and Germans'. Referring to a recent publication on statistical data on Christian Jews in Kurland province, the reviewer argued that by excluding religion the map obscured an important dimension of the region's cultural complexity. Another criticism concerned the methodological challenges posed by migration and multiethnic regions. Although the reader conceded that from a cartographic perspective it would be much easier 'if the dispersed people were drawn to a gathering place as much as possible', the multiethnic character of

[105] Ibid: 6; Haslinger, *Nation und Territorium*: 31. On the concept of imagined communities, see Anderson, *Imagined Communities*.

[106] SPbF ARAN f. 30, op. 1, d. 465, l. 20.

[107] 'Ethnographische Charte des Europäischen Russlands,' *Das Inland*, no. 28, 14 July 1852: 565–6. The reviewer is identified only as F. Chr. von T.

the imperial borderlands required mapmakers to add colours for mixed areas, such as 'Finns and Swedes not separated'. Based on this logic, however, the reviewer questioned why all the Baltic provinces were not marked as 'Latvians and Germans not separated' or 'Estonians and Germans not separated', since Germans were the principle landowners in the region. German-speaking readers from the Baltic provinces did not see themselves reflected in the map, nor the entire manorial economy and estate-based class system that structured society in the Baltic provinces.[108]

It also did not escape the reviewer's attention that 'by far the largest place on this colourful chart are the Slavic peoples, followed by Finnish or Chuds'. The reviewer thought that it was strange that Slavs had not been coloured and subdivided into their languages and dialects as linguists were thirsty for knowledge about the relationship between ethnolinguistic borders, such as between Ukrainian, Podolian, and Volhynian. Shading the Slavic language area and subdividing it into its constituent languages and dialects 'would have given the large white area a new look'.[109] The reviewer did not elaborate further, but the implication was that perhaps this 'new look' of mapping Slavs on an equal footing alongside the empire's *inorodtsy* might not have aligned with the priorities of those making, financing, and publishing the map in St. Petersburg.

Köppen's readers not only bought, read, and shared their thoughts about Köppen's maps, but we also have evidence of Köppen's maps being directly used as inspiration for other mapmakers. In the late 1840s in the seaside town of Hapsal (Haapsalu) in Estland province, Carl Friedrich Russwurm was busy preparing a large volume on the history, language, and folk culture of the region's coastal Swedish speakers. Born in Ratzeburg in the Duchy of (Saxe-) Lauenburg, which was granted to Sweden in 1815 at the Congress of Vienna and subsequently ruled in personal union with the Danish Crown, Russwurm moved to western Estland in 1835 to work as a tutor, teacher, and school inspector. There he became interested in the local coastal Swedish-speaking communities (who called themselves *eibefolke*, or 'island folk'), who had settled in the area between the thirteenth and fifteenth centuries. Since being incorporated into the Russian Empire, the population declined due to evictions from their farmlands, forced migration to today's southern Ukraine, and lack of access to the same rights as Estonian-peasants which had been granted the basis of the laws of 1816 abolishing serfdom.[110]

At some point in his research Russwurm came across Köppen's *Ethnographic Map of Finland*, published in one of the journals of the Academy of Sciences.

[108] Ibid.

[109] Ibid.

[110] On the Estland's Swedish-speaking population, see Põldvee, 'Russwurmist ja "Eibofolkest"'; Tøllefsen & White, 'Navigating an Orthodox Conversion'.

Figure 1.4. Carl Friedrich Russwurm's copy of Peter Köppen's *Ethnographic Map of Finland* (1846) (cf. Plate 1 for Köppen's original version).

Source: RA, EAA.854.7.14.146. Reproduced with permission from the National Archives of Estonia.

Russwurm was struck by the map and the cartographic depiction of a linguistic contact zone between Swedish and Finno-Ugric speakers. He made a detailed copy the map, possibly to keep as a reference if he did not personally own a copy of the journal or as a study exercise to practice cartographic drawing (Figure 1.4). Later, when Russwurm published his book on Estland's Swedish-speakers in 1855 with a subsidy from the Academy of Sciences he included four ethnographic maps: a

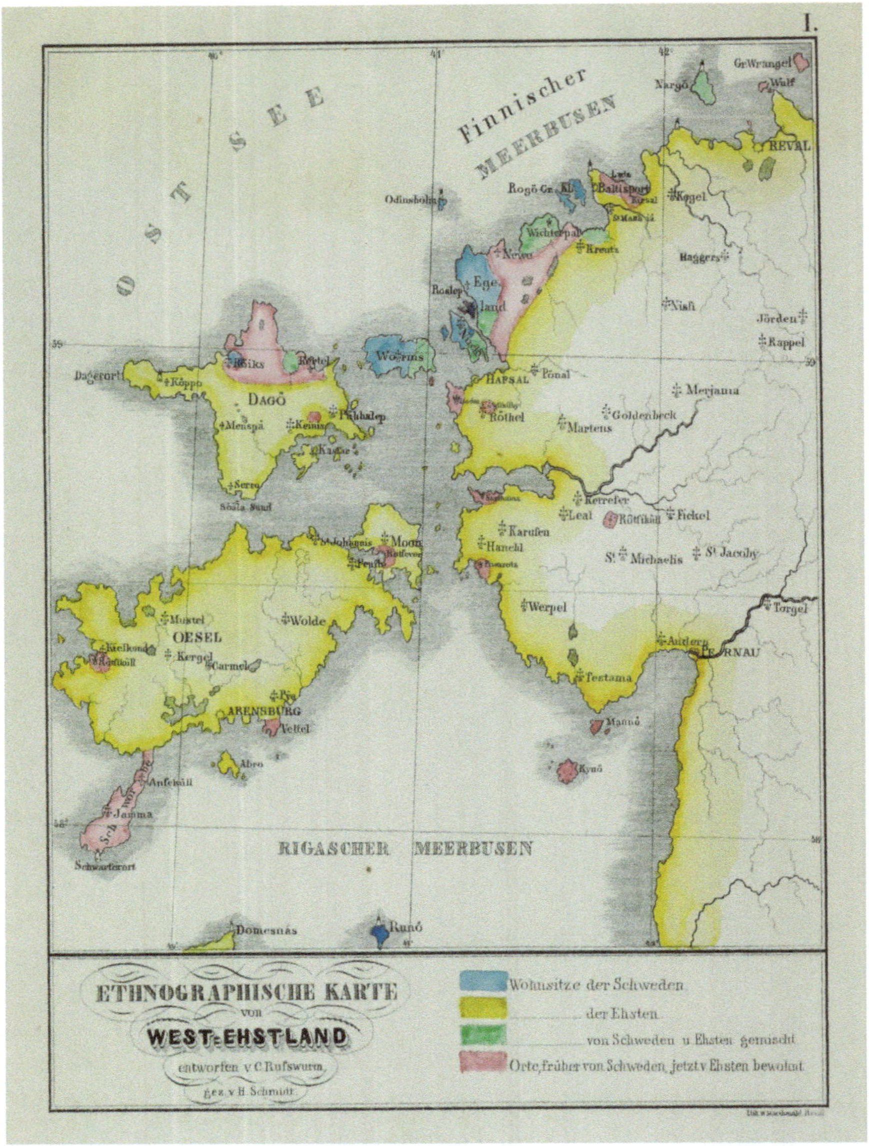

Figure 1.5. Carl Friedrich Russwurm, *Ethnographic Map of West-Estland* (1855). The colours on the map indicate the areas inhabited by Swedes (blue), Estonian (yellow), mixed Swedes and Estonians (green), and places where Swedes used to live but which are now inhabited by Estonians (red), drawing attention to the gradual decline of Estland's Swedish-speaking population.

Source: Carl Friedrich Russwurm & Johann Heinrich Schmidt, *Ethnographische Karte von West-Ehstland* (Tallinn: F. W. Macdonald, 1855). Public domain. Scan provided by the National Library of Estonia.

general ethnographic map of the Baltic coast and islands indicating the areas inhabited by Swedes and Estonians (Figure 1.5), as well as three close-up maps of the islands inhabited by Swedish-speakers.[111] Like Köppen, Russwurm included a category of mixed Swedes and Finno-Ugric peoples. Such unwritten intertextualities reveal the circulation of knowledge and approaches to ethnographic mapmaking between German-speaking scholars in the Russian Empire.

Conclusion

Köppen was but one of many mapmakers experimenting with ethnographic mapping in the mid-nineteenth century as part of a wider Pan-European interest in applying geographical and graphical methods to visualize ethnolinguistic groups. Köppen embodies how international, imperial, and local influences conjoined to shape cultures of ethnographic mapping in the Russian Empire in the late 1840s and early 1850s. The ethnographic maps of the 1840s travelled across empires and circulated in different milieus of knowledge production. They contained a high degree of intertextuality and were often founded upon the same base maps, corpuses of statistical data, and frequently copied one another in their design and aesthetics. The correspondence between Köppen and Šafařík provides us with a thread to trace how these ideas circulated. Mapmakers in imperial Russia did not simply appropriate or copy a model developed in Central Europe, but rather played a key role in the wider transnational conversation on ethnographic mapping from its early years. Critical to sustaining these conversations was German, a language which connected mapmakers living and working in the peripheries of European empires and facilitated lively discussion and knowledge exchange.

Making an ethnographic map in the 1840s was an expensive and laborious undertaking. Prospective mapmakers not only had to build networks to gather data, but prior to the founding of specialized commercial cartographic printing and publishing firms in the Russian Empire individuals had to navigate a complex world of government ministries and learned societies to raise funds and distribute their maps to prospective readers. Köppen's various mapping activities through the late 1830s and 1840s highlight how ethnographic cartography developed in the Russian Empire as a result of the aligned interests of different stakeholders, as both the tsarist state and intellectual elites turned to science to inventory and manage the natural and human resources of the empire. The complex biographies of men like Köppen who straddled the worlds of government and academia make it hard to define the line between imperial officialdom and civil society. Köppen's ventures in ethnographic mapping evolved precisely because of his ability to

[111] The maps were published in Russwurm, *Eibofolke.*

navigate between bureaucratic and scholarly institutions. The success of the *Ethnographic Map of European Russia* paved the way for the application of ethnographic cartography as a tool for visualizing ethnic statistics in many different settings in the second half of the nineteenth century. However, as will be explored in the coming chapters, ethnographic mapmaking did not remain tied to institutions and funding structures embedded in the central tsarist government and learned life of the imperial capital for long.

The reactions of Köppen's readers, both within the close-knit scholarly world of the imperial capital and in the empire's provinces, reveal how audiences engaged in critical ways with the emerging genre of ethnographic cartography. As Köppen's maps circulated among readers, they were subjected to critical scrutiny. Audiences raised astute methodological questions about the application of cartographic techniques to mapping ethnolinguistic groups. We see in these reactions to Köppen's maps how many of the core issues dominating debates on ethnographic mapping in the second half of the nineteenth and twentieth century were evident to early audiences. Readers did not take the maps at face value, and Köppen's map commentary explicitly encouraged them not to. Readers were under no illusion that the maps portrayed an accurate image of the world, which was impossible to render in all its complexity in two dimensions. There was an atmosphere of excitement surrounding these new ways of visualizing ethnolinguistic data, the kinds of new questions that maps could stimulate, and the potential of mapping to generate new knowledge. Yet, this enthusiasm for thematic mapping was tempered by an awareness of how mapping could be used to change the way people saw the world. In contrast to Köppen's contemporaries, later generations of readers often evaluated his maps as stand-alone objects and rarely read or commented on the accompanying text. Divorced from their explanatory context, his maps were interpreted as authoritative and scientifically objective representations of the world as it is, which lent a strong legitimacy to the images of the empire that they constructed.

Finally, despite his prodigious productivity, by the early 1850s Köppen's experience of becoming an ethnographic mapmaker was overshadowed by the financial burdens associated with map production. In January 1852 Köppen wrote to the Academy of Sciences to complain that he was still owed 45 roubles and 65 kopeks from making the *Ethnographic Atlas of European Russia* four years earlier.[112] Köppen also wrote numerous, increasingly frustrated letters to the Geographical Society requesting reimbursement for the expenses he had incurred while making the *Ethnographic Map of European Russia*. He complained that he had only been able to sell two copies of the map that he had been given and asked the Society to purchase back the remaining 85–90 copies from him as compensation for his thirty years collecting data and the time spent

[112] SPbF ARAN f. 30, op. 1, d. 465, l. 13.

working on the map.[113] The large number of unsold copies suggests that the Geographical Society overestimated the size of the market for the map, which remained a specialist work purchased mainly by scholars, learned societies, and libraries. Moreover, Köppen explained that he could not afford to purchase the Geographical Society's bulletin as the subscription cost more than 10 roubles a year and he asked to be sent a copy via his son in St. Petersburg so that he could read Sreznevskii's review of his map.[114]

Despite the growing interest and high regard for ethnographic mapping among the empire's scholarly community, ethnographic mapmaking in the 1840s and 1850s was not lucrative. Köppen's financial difficulties draw our attention to the material and human dimensions of what it meant to be a mapmaker in mid-nineteenth-century imperial Russia. Köppen received no renumeration for his cartographic work. Map productions costs were high and as a civil servant, Köppen could not rely on a vast personal wealth to subsidize his scientific projects. The lack of a centralized and state supported body for ethnographic mapmaking in the Russian Empire meant that mapmakers often paid upfront out of their own pockets to finance their mapping projects. Even when pledges of financial support were secured, practically speaking, it often took many years for expenses to be reimbursed. Ethnographic mapmaking for Köppen was a frustrating and financially draining undertaking. Understanding the economic and political tensions underpinning ethnographic mapping during this period adds another dimension to the question of why Köppen gradually distanced himself from the Geographical Society in the early 1850s, retired from public service in 1852, and did not make any more maps thereafter. The change of leadership and shifting research aims of the Geographical Society after 1849 towards a more Russian-orientated research agenda undoubtedly played a key role in souring Köppen's relationship with the Society he had been instrumental in founding six years earlier. Yet, his strained relationship over financial matters related to the production of ethnographic maps also left him with bitter feelings.

113 SPbF ARAN f. 30, op. 1, d. 460, l. 1.
114 SPbF ARAN f. 30, op. 1, d. 460, l. 4.

2

Map Production in the Provinces and the Rise of Cartographic Entrepreneurship

In January 1869 Aleksandr Lvovich Potapov, the newly appointed Governor-General of the Russian Empire's North-Western Territory, announced that his administration lacked a detailed cartographic overview of the provinces under his jurisdiction.[1] The region, which overlaps with today's Lithuania, Belarus, Poland, and eastern Latvia, had been the site of the Polish-Lithuanian Uprising against imperial rule several years earlier in 1863–4. In the wake of the political turbulence, the imperial administration financed a wave of ethnographic and religious mapping projects to gather intelligence about non-Orthodox and non-Russian-speaking populations in the empire's borderlands. As will be examined in the next chapter, these maps formed part of broader initiatives supported by the imperial government which aimed to diminish the perceived Catholic and Polish influence, strengthen the perception of the territory as a core part of the Orthodox and Russian-speaking lands, and legitimize Russian imperial rule in the western borderlands.[2] Still, Potapov remained dissatisfied with these previous attempts to produce ethnographic and religious maps of the North-Western Territory, arguing that they did not provide him with the depth of knowledge required to govern the territory.[3] Potapov might have also been looking to boost his reputation through the prestige associated with commissioning an ambitious cartographic project.

Potapov ordered Prince Petr Romanovich Bagration, his assistant for civil affairs and an experienced mapmaker, to begin work on a new and more ambitious regional statistical atlas. Potapov discussed the idea with the secretary of Vitebsk (Viciebsk) Provincial Statistical Committee, Aleksandr Maksimovich Sementovskii, who drew up an initial plan in February 1869 for an atlas spanning the six provinces under Potapov's oversight. Bagration and Potapov were impressed by the plan and called all the secretaries of the six provincial statistical committees in the North-Western Territory to a meeting in Vil'na (Vilnius) in early April 1869 to

[1] NHAB f. 2502, vop. 1, spr. 183, ark. 1.

[2] On the politics of ethnographic and religious mapmaking in the years after the Polish-Lithuanian Uprising, see Petronis, *Constructing Lithuania*: 194–209. On the confessional and linguistic policies of 'Russification' implemented by the imperial government in the North-Western Territory, see Staliūnas, *Making Russians*.

[3] NHAB f. 2502, vop. 1, spr. 183, ark. 1.

Geographies of Nationhood: Cartography, Science, and Society in the Russian Imperial Baltic. Catherine Gibson, Oxford University Press. © Catherine Gibson 2022. DOI: 10.1093/oso/9780192844323.003.0003

coordinate preparations for the atlas.[4] This chapter offers a detailed account of the process of making one of the maps from the atlas, the *Ethnographic Map of Vitebsk Province* (Plate 6), in the period from 1869 to 1872. Vitebsk province lay to the east of the Baltic provinces, but the population of its three western districts of Dinaburg (Daugavpils), Rezhitsa (Rēzekne), and Liutsin (Ludza) comprised a large proportion of Catholic Latgalian-speakers, a language closely related to Latvian. On this basis these districts were incorporated into the independent Republic of Latvia after the First World War. Thus, Vitebsk province constitutes a key—albeit often overlooked—part of Baltic regional history in the imperial era.[5]

The extensive archival materials documenting the cartographic activities of the Vitebsk Statistical Committee under the secretaryship of Aleksandr Sementovskii offer an unprecedented account of how provincial bureaucrats approached the task of creating an ethnographic map. As Matthew Edney argues, instead of asking what maps are or show, we should focus on mapping processes to understand how 'people variously produce, circulate, and consume maps according to their discursive conditions'.[6] The archives of the Vitebsk Statistical Committee foreground two stages in the map production process which only become apparent when taking a microhistory approach to examine the backstory of a single map, namely the collection of data and the process of printing and publishing maps.

Firstly, this chapter asks how our understanding of ethnographic mapping in the Russian Empire shifts if we relocate our gaze away from the elite members of the Academy of Sciences and Imperial Russian Geographical Society to provincial statisticians. The development of provincial infrastructures for statistics embedded ethnographic mapmaking in everyday administrative procedures and practices. While scholars have commented on the statistical foundations of ethnographic mapping, we still know little about the 'archaeology of knowledge' that lay behind the finished map, including processes of ethnographic map production and how processes of data collection, reporting, and circulation shaped what was possible in cartography in a particular time and place.[7] This chapter examines how administrative practices shaped ethnographic mapmaking processes and how, during their routine paper-shuffling, local bureaucrats exercised considerable agency over methods of counting, categorizing, and mapping imperial subjects. In doing so, the chapter argues for the importance of considering the role of places outside of imperial capital in the emergence, interaction, and reception of ideas about ethnographic mapmaking in the Russian Empire.[8]

[4] Gatsinskii, *Sbornik v pamiat'*: 34–7.

[5] On western Vitebsk province and Latgale as a linguistic borderland, see Gibson, 'The Polish Livonian Legacy'.

[6] Edney, *Cartography*: 236–7.

[7] Foucault, *The Archaeology of Knowledge*. On the statistical data underpinning ethnographic cartography in, see Hansen, *Mapping the Germans, esp. pp.20–38*; Schulten, *Mapping the Nation, esp. pp.164–95*; Labbé, *La nationalité, une histoire de chiffres*.

[8] On scientific hubs in Central Eastern Europe's 'peripheries', see Surman, 'Paris-Wien-St. Petersburg'.

Secondly, this chapter examines how ethnographic mapmaking gradually became intertwined with the nineteenth-century business history of the printing and publishing industry. The early 1870s witnessed a boom in popular print culture in imperial Russia amidst the rise of modern capitalism and a burgeoning civil society. The limited opportunities for printing ethnographic maps in the Russian Empire at this time forced provincial branches of the imperial administration to partner with commercial printing and publishing houses in St. Petersburg. This caused immense logistical challenges and placed practical constraints on mapmakers. Clashes emerged between statisticians' intellectual ambitious and the commercial interests and technical capabilities of private firms. In the 1840s–1850s, Peter Köppen's ethnographic maps were conceptualized as scholarly outputs which were heavily subsidized by elite learned establishments, who also coordinated the printing and publishing of the maps. By the early 1870s, mapmakers had to navigate an increasingly fraught relationship to ethnographic mapmaking as an activity that was, simultaneously, routine administrative work, scientific research, and a commercial endeavour.

Administrative Cartography in the Provinces

The origin story of the *Ethnographic Map of Vitebsk Province* forms part of a larger narrative about the expansion of imperial power through the systematic collection and management of knowledge. The Russian Empire, like many other imperial regimes, can be conceived of as an 'empirical empire' which came to increasingly depend on the acquisition and management of knowledge for the constitution of authority and control.[9] By the middle of the nineteenth century the Russian imperial Ministry of Interior generated more than 30 million bureaucratic documents a year.[10] Yet, knowledge produced in administrative contexts has often been demeaned as 'narrowing and humdrum', especially when contrasted to scholarly knowledge production.[11] As Robert Geraci notes, in the many studies of Russian imperial governance, 'it is often the mechanical or "nuts and bolts" aspects of the empire's administration that are least discussed', yet these are frequently the most revealing places to look to learn about how the empire actually worked.[12]

Recent literature has pointed to the importance of paying attention to 'bureaucratic knowledge' as a key part of the history of science.[13] In her work on how

[9] The phrase 'empirical empire' comes from: Brendecke, *Empirical Empire*. Key literature examining histories of statistics and knowledge management includes: Patriarca, *Numbers and Nationhood*; Bayly, *An Empire of Information*; Crooks & Parsons, eds. *Empires and Bureaucracy in World History*; Porter, *The Rise of Statistical Thinking*; Randeraad, *States and Statistics*.

[10] Lincoln, 'The Daily Life of St. Petersburg Officials': 82.

[11] Porter, 'Revenge of the Humdrum': 4.

[12] Geraci, 'On "Colonial" Forms': 180.

[13] Felten & von Oertzen, 'Bureaucracy as Knowledge': 8.

local authorities registered nationality in Prussia, Morgane Labbé argues that shifting our gaze beyond state-level institutions exposes the lack of centralization and standardization governing statistical practices.[14] Labbé's regional focus discloses the considerable influence of local epistemic practices and bureaucratic infrastructures on the data reported. For the Russian Empire, Gregory Afinogenov examines the vast network of 'anonymous or nearly anonymous officials' toiling as part of a web of intelligence-gathering operations implanted within the imperial bureaucracy to challenge the primacy of scholarly institutions as the main source of knowledge about the empire.[15]

The importance of the imperial bureaucracy as a player in the development of approaches to ethnographic mapmaking is evident in the case of Governor-General Potapov's ambitions to create a statistical atlas of the North-Western Territory. The provincial statistical committees, whom Potapov assigned to work on the atlas, formed part of the emerging provincial infrastructure for collecting statistics which had developed over the previous decades. In 1834 the Statistical Division of the Ministry of Interior in St. Petersburg established statistical committees as part of the provincial administration in the western portion of the empire to coordinate the collection of statistical data. However, most of the committees were not very active during their early years. Their activities were severely limited due to under-financing, short-staffing, and the fact that most of the local administrators had no training in statistics.[16]

In December 1860, the Ministry of Interior overhauled the organization of the statistical committees as part of Tsar Alexander II's far-ranging programme of social and economic reforms. Russia's humiliating defeat in the Crimean War of 1853–6 spurred the new tsar and his government to instigate changes to restore the empire's prosperity and military dominance. The climate of reform invigorated state-sponsored research into the socioeconomic and ethnographic character of the empire's inhabitants, which were perceived as crucial for creating a more unified and organized state. The spirit of reform reflected gradual shifts in the understanding of statistics over the previous decades under the influence of Konstantin Arsen'ev, head of the Statistical Division from 1835 to 1853. Arsen'ev advocated moving away from the eighteenth-century cameralist tradition of using statistics to extract resources from the population through taxation and military conscription to a view of statistics as a tool to study society.[17]

The provincial statistical committees were tasked with fulfilling the demands of the Ministry of the Interior for comprehensive and regular statistical information about Russia's regions. The Central Statistical Committee in St. Petersburg and the regional administration—such as the Governor-General of the North-Western

[14] Labbé, 'Institutionalizing the Statistics of Nationality'.
[15] Afinogenov, *Scholars and Spies*: 5. [16] Smith-Peter, *Imagining Russian Regions*: 90.
[17] Smith-Peter, 'Defining the Russian People': 47.

Territory—sought to hold the provincial statistical committees to uniform standards for collecting and reporting statistical data that could be compared across the whole empire. The provincial statistical committees were given more precise instructions about the scope of their activities, an annual budget of 1,500–2,000 roubles, and 750 roubles to pay the salary of the secretary to manage the committee's activities. In many parts of the empire the provincial statistical committees worked alongside the *zemstvo*, institutions of local self-government introduced in 1864, which also employed their own statisticians to carry out data collection, land surveying, and produce cadastral maps.[18] However, the *zemstvo* system was not established in the Baltic provinces nor the North-Western territory, and so the statistical committees carried out the majority of statistical work in these regions. Although the statistical committees were officially part of the provincial administration, presided over by the provincial governor, in practice many assumed the character of a learned-administrative institution of the kind that would later come to flourish in late imperial provincial life.[19] The trend to decentralize mechanisms for collecting statistical data and outsource it to local experts in the provinces during these years was mirrored in the establishment of regional branches of the Imperial Russian Geographical Society in Tiflis/Tbilisi (1851), Irkutsk (1851), Vil'na/Vilnius (1867), Orenburg (1868), Kiev/Kyiv (1873), and Omsk (1877).

The provincial statistical committees generated vast amounts of printed materials, the most famous of which were the official annual reference books (*pamiatnye knizhki* or *obzory*) presenting a statistical overview of imperial society. The committees also provided regular contributions to the unofficial part of the provincial gazettes (*gubernskie vedomosti*).[20] In the second half of the nineteenth century the scope of their activities expanded to include mapmaking, thanks in part to the allocation from 1854 of an annual sum of 1,000 roubles from the provincial typography for printing forms, plans, and maps. Sometimes the provincial statistical committees received instructions to make a map from the Central Statistical Committee or from other learned societies. For example, in the early 1890s the Ministry of Interior instructed the Livland Statistical Committee to produce a series of statistical maps depicting the percentage of Orthodox in each of the parishes in Livland and the proportion of Orthodox who, from the perspective of the Orthodox Church and imperial state, had illegally evaded or lapsed back to Lutheranism.[21]

[18] On cadastral map production, see Evtuhov, *Portrait of a Russian Province*: 165–81. On statistical work in the provinces, see Johnson, 'Liberal Professionals'; Mespoulet, *Statistique et révolution*; Stanziani, "Statisticens, zemstva et État'.

[19] *Pamiatnaia knizhki Vitebskogo gubernii* (1914): 39. On the emergence of provincial learned societies in late imperial Russia, see Bradley, *Voluntary Associations*; Loskutova, 'Uezdnye uchenye'; Kaplan, *Historians and Historical Societies*. On voluntary associations in the Baltic provinces, see Hackmann, ed. *Vereinskultur*; Hackmann, *Geselligkeit in Nordosteuropa*.

[20] Soroko, *Vitebskie Gubernskie Vedomosti*: 82–115.

[21] Gibson & Paert, 'Apostasy in the Baltic Provinces': 29–32.

In other cases, provincial statisticians made maps on their own initiative. Under the secretaryship of Alfons Baron Heyking in the early 1860s the Kurland Statistical Committee published statistical maps of land use and religious confessions.[22] Friedrich von Jung-Stilling, secretary of the Livland Provincial Statistical Committee, published a series of maps of Lutheran school statistics in 1879 and an album of statistical maps in 1881 depicting the distribution of farmland, privately owned land, forests, and distilleries.[23] In 1883 the Livland Statistical Committee mapped Riga's population density to highlight the problem of overcrowding, especially in the densely populated Old Town.[24] These mapping projects testify to the fact that interest in using cartography as a form of data visualization to inform administrative policy and governance was not limited to high-ranking imperial academicians and members of the Geographical Society in St. Petersburg, but was also flourishing at the provincial scale as well from the 1860s onwards.[25] The maps produced by the provincial statistical committees reinforced the idea of the province as spatial framework for studying social and economic life in the Russian Empire and are indicative of the increasing value placed on local knowledge.[26] In doing so, the cartographic activities of the provincial statistical committees marked an important reversal in late eighteenth- and early nineteenth-century regionalism in the Russian Empire, in which intellectuals had emphasized the universality of the empire's far-flung territories and tended to regard the provinces as 'blank and interchangeable'.[27]

Aleksandr Sementovskii is an example of one such mid-level bureaucrat and civil servant who was actively involved in provincial statistics and mapmaking. Sementovskii was born in Poltava province in 1823 and after spending some time in the military he worked his way up the ranks of the forestry department in Kiev and Podolia province, before ascending to the rank of provincial forester of Vitebsk province. From 1863 until his retirement in 1880 he served as secretary of Vitebsk Provincial Statistical Committee.[28] Under Sementovskii's leadership, Vitebsk gained a reputation among statisticians in the Russian Empire as home to a 'small but enlightened circle of people who are seriously devoted to science'.[29] Sementovskii's impact was also felt outside of Vitebsk as he worked to create a more consolidated network of provincial statisticians and through his membership in the Imperial

[22] Von Heyking, *Statistische Studien*; *Karte von Kurland mit Rücksicht auf die Verbreitung der verschiedenen Religionsbekenntnisse unter der ländlichen Bevölkerung*. Undated printed map. DSHI.190. Kurland.XI.7.

[23] Von Jung-Stilling, *Beitrag zur Statistik der evangelisch-lutherisch Landvolk-Schulen*; Von Jung-Stilling, *Ein Beitrag zur Livländischen Agrarstatistik*.

[24] Von Jung-Stilling & Anders, *Ergebnisse der baltischen Volkszahlung*.

[25] Mespoulet, *Statistique et révolution*, esp. pp.27–49.

[26] Mogilner, *Homo Imperii*: 5.

[27] Smith-Peter, 'Bringing the Provinces into Focus': 838. On a similar tendency to obscure the region in favour of national or international scales of historical analysis see Applegate, 'A Europe of Regions'.

[28] Soroko, *Vitebskie gubernskie vedomosti*: 87.

[29] 'Vitebskie Gubernskie Vedomosti za 1868.' *Vilenskii Vestnik*, no. 32 (1869).

Russian Geographical Society, Free Economic Society, and Moscow Archaeological Society. Despite this, Sementovskii has primarily been remembered in the historiography for his contributions to Belarusian archaeology and ethnography.[30]

It was against the backdrop of this burgeoning provincial infrastructure for statistics that Sementovskii was summoned in April 1869 to Vil'na by Governor-General Potapov to meet with the other secretaries of the statistical committees in the North-Western Territory to discuss plans to make a regional statistical atlas. This level of regional cooperation among the statistical committees was unusual for the time. Normally, contact between the committees was limited to the transmission of information through correspondence and sending copies of publications to one another. Even when the statistical committees in the three Baltic provinces collaborated to organize a census on the same day in December 1881, the results were published for each province separately which hindered cross-regional comparisons of the data.[31] Potapov's decision to physically bring all the secretaries together to devise guidelines and align mapmaking practices is an example of how imperial administrators attempted to confront the problem of standardizing methods of collecting data, measuring, reporting, and mapping ethnolinguistic statistics across different provinces.[32] Potapov's vision of a regional atlas involved the co-creation of cartographic knowledge. Regional cooperation was paramount to ensure the 'success' and 'uniformity' of the project.[33]

After deliberating for several days, the secretaries agreed on a plan for the atlas. It would comprise six thematic maps for each province covering administration, transport and communication routes, ethnography, religious buildings, trade and industry, and educational institutions. When it came to the ethnographic maps, the secretaries agreed that the maps would all be drawn to the same scale (1:840,000) and depict the distribution of eleven ethnolinguistic groups.[34] Sementovskii was entrusted with preparing a sample legend to create a list of uniform cartographic signs and symbols to be used across the various maps, as well as to specify the precise colours to depict each ethnic group (Figure 2.1).[35] The standardized symbology was designed to give the atlas a uniform look and make a strong statement about the common grounds for thinking about the six provinces of the North-Western Territory as a distinctive imperial region. Inspiration for this guide to cartographic symbology may have come from a similar compendium of signs and symbols to be used on boundary plans, maps, and atlases published by the Boundary Commission and the Provincial Printers in 1838.[36] Petr Bagration, who was also

[30] Bandarchyk, *Historyia Belaruskai Etnahrafii*: 122–6.

[31] Plakans & Wetherel, 'The 1881 Census': 50.

[32] On the history of units of measurement in imperial Russia, see Gordin, 'Measure of All the Russias'.

[33] NHAB f. 2502, vop. 1, spr. 183, ark. 1.

[34] NHAB f. 2502, vop. 1, spr. 183, ark. 4–6.

[35] NHAB f. 2502, vop. 1, spr. 183, ark. 6adv. & 7.

[36] *Uslovnye znaki mezhevykh planov, kart i atlasov* (1838). RA, EAA.298.2.1.1–10.

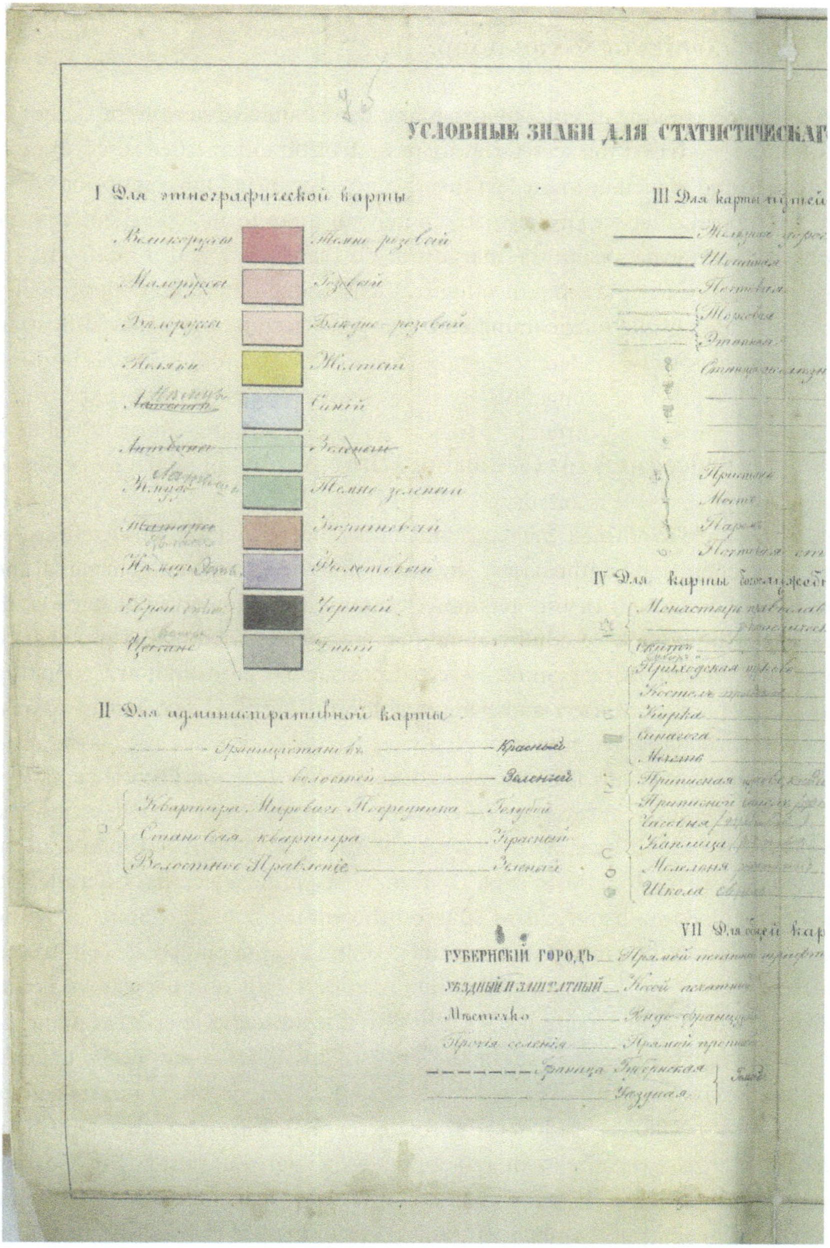

Figure 2.1. Aleksandr Sementovskii's guide to cartographic symbols for the planned statistical atlas of the North-Western Territory. The excerpt shows the sample legend for ethnographic maps. According to the original plan, the following ethnic groups were to be depicted: Russians (dark pink), Ukrainians (pink), Belarusians (pale pink), Poles (yellow), Latvians (blue), Lithuanians (green), Samogitians (dark green), Tatars (brown), Germans (purple), Jews (black), Roma (*dikii*, which literally translates as 'wild' but is interpreted here as grey). Subsequent pencil annotations to the legend, probably by Sementovskii, reveal how this plan was later modified to reflect the situation in Vitebsk province. Ukrainians, Lithuanians, Samogitians, and Tatars were scored out, Latvians, Germans, and Roma were assigned different colours, and a new group–Estonians–was added.

Source: NHAB f. 2502, vop. 1, spr. 183, ark. 7adv. Reproduced with permission from the National Historical Archives of Belarus.

present at the meeting as Potapov's assistant, contributed advice on the design of the maps. He advised the secretaries to 'pay attention to the external decoration of the map and especially to the fact that the too large text of the inscriptions does not take up much space'. He also recommended them to include the necessary explanatory information 'in the margins or on the reverse, and if necessary, on special sheets'.[37] The planning meeting in Vil'na ended on a high note of excitement and news of the forthcoming atlas was publicized by the Imperial Russian Geographical Society.[38]

Procedures for Gathering Ethnographic Statistics

Upon his return to Vitebsk, Sementovskii embarked on preparations for the atlas with great energy and enthusiasm. The question of how to gather data was at the forefront of his mind. In this section, I examine how the kinds of knowledge yielded by the process of ethnographic mapmaking was not only empirical but procedural. The statistical committee secretaries had to figure out how to organize and manage a massive data-gathering operation using their network of correspondents throughout the province.

Over the summer of 1869 Sementovskii sent printed circulars to police, school directors, and local authorities instructing them to count and describe the people living in their locality. The important role of such 'lay participants', referring to those who did not have a formal scientific role, has increasingly been recognized by historians of nineteenth-century science.[39] Some of those Sementovskii reached out to were members of the Vitebsk Statistical Committee, which in 1869 comprised twelve honorary members, four non-permanent members, and fifteen active members.[40] Orthodox, Catholic, and Jewish religious figures were a notable presence among the Vitebsk Statistical Committee's members, from archimandrites and bishops to priests and rabbis, revealing how scientific work was supported and reinforced through religious infrastructures and established practices of churchmen observing their parishioners.[41] The Russian Geographical Society relied on similar communication channels and in 1849 had praised the 'zealous and enthusiastic' efforts of Orthodox priests who had been very active in responding to the Society's requests for statistical and ethnographic

[37] Gatsinskii, *Sbornik v pamiat'*: 35.

[38] 'Statisticheskaia karta Zapadno-russkogo kraia', *Izvestiia IRGO* 5 (1869): 80–2.

[39] Hachten, 'In Service to Science and Society': 181; Vetter, 'Introduction: Lay Participation'.

[40] *Pamiatnaia knizhka Vitebskoi gubernii* (1869): 29–30.

[41] Soroko, *Vitebskie gubernskie vedomosti*: 83. On the role of religious leaders in the accumulation of scientific and statistical knowledge in the Russian Empire, see Chumakova, 'Pravoslavnye sviashchenniki-issledovateli'; Avrutin, 'The Power of Documentation'.

information.[42] Orthodox missionaries in the Volga region likewise gained a reputation during this period as 'zealous ethnographers'.[43] At the same time, the dependency of the provincial statistical committees on their members to furnish them with reports was met with mixed success. The secretaries often complained about the low-level of engagement from their members. Sementovskii's predecessor in the Vitebsk Statistical Committee, Ksenofont Antonovich Govorskii, was frustrated by the lack or 'unsatisfactory' nature of responses he received from members, writing that he 'remains as if in a desert'.[44]

The reasons for the low response rate or 'unsatisfactory' nature of the information received were varied. Sometimes the low response rate came down to a lack of time or knowledge. The experiences of the Imperial Russian Geographical Society trying to collect statistical information about Livland and Kurland reveals, for instance, how Orthodox churchmen were not dependable informants in all areas of the empire. In 1853 Archbishop Platon of the Diocese of Riga and Mitava (Jelgava) received a request from the Geographical Society to ask the Orthodox priests in the diocese to collect ethnographic data and climate observations. However, in 1855 Platon replied that the relatively limited Orthodox infrastructure in the Baltic provinces meant that the Orthodox Church had been unable to supply the required information.[45] Markian Ellinskii, the archpriest of Fellin (Viljandi) deanery in northern Livland province, explained in his letter to Platon that the priests had been unable to fulfil the request due to their lack of time and limited knowledge of agronomy.[46]

Even when the members of the Geographical Society did receive replies, they were often disparaging about the quality of the information sent to them. Statistical information was not always communicated in a way that was perceived by those on the receiving end as being 'useful knowledge' when it came to the bigger long-term goals of consolidating imperial power and contributing to the empire's socioeconomic development.[47] As one report from the Geographical Society noted:

> Some find it difficult to compile local descriptions, discerning which details should be included and which should not, and most importantly how to organize the descriptions. Often things are described that are too well-known and curious subjects, which are important for ethnography, are given without

[42] *Otchet RGO za 1849 god* (1850): 60.

[43] Kolosova, *Narodnost' and Obshchechelovechnost'*: 348.

[44] *Pamiatnaia knizhki Vitebskogo gubernii* (1864): 5.

[45] LVVA f. 7462, apr. 1, l. 531, lp. 9–13.

[46] LVVA f. 7462, apr. 1, l. 531, lp. 26.

[47] On the contributions of science and technology to the knowledge economy and the concept of 'useful knowledge', see Mokyr, *The Gifts of Athena*: 36.

> descriptions, although in the latter case the compilers of the descriptions clearly know something brief about them.[48]

This was typical of the information received by the provincial statistical committees as well. Aleksandr Gatsinskii, secretary of Nizhegorod Statistical Committee, complained that sometimes reports were received which provided a mass of information but 'often without even referring to the source from which they draw their compilations'.[49] The disparate nature of the reports, which varied widely both in terms of their content and format, made it difficult to aggregate the mass of local knowledge into a general overview for mapping purposes or carrying out cross-regional comparisons. The recipients expected knowledge to be presented in a way that made it useful for the state. For the imperial knowledge economy to function, local observations had to be reported and presented in ways that would allow knowledge to travel and be understood to distant others.

The use of pre-printed forms provided a partial solution to the problem of how to assert a measure of central control over methods of observation and reporting. Printed forms were introduced into the imperial administration during the time of Peter the Great but their use greatly expanded in the first half of the nineteenth century.[50] The forms circulated by the provincial statistical committees specified the types of data to be collected, the methods to be used, and how information was to be recorded. Form-filling, note-taking, and record-keeping thus performed important social and political functions in the making and management of ethnographic knowledge in the empire.[51]

The questions asked on this statistical paperwork also determined the kinds of answers that the Vitebsk Statistical Committee received back. In July 1869 Sementovskii issued a circular alerting the local authorities in Vitebsk province to a mistake that had been made on a form to collect data on religion. Sementovskii explained that the form had asked for information on the number of Orthodox and Catholic (Great) Russians (*Velikorusskii*), when in fact the latter column header should have read 'schismatics' (*raskol'niki*).[52] The category of 'schismatics' was part of the official vocabulary of religious dissent that had emerged in the 1860s as a way of distinguishing between 'normal' (*pravovernyi*) and other 'abnor-

[48] 'Ob'iavlenie ot IRGO', *Voronezhskie gubernskie vedomosti* 40 (3 Oct. 1853): 236. Cited in Vaskul, 'Etnograficheskaia programma RGO': 462.

[49] Gatsinskii, *Sbornik v pamiat'*: x.

[50] On the history of printed forms in imperial Russia, see Franklin, 'Printing Social Control'. On the use of forms for gathering statistics, see Mespoulet, 'Pratique de l'enquête'.

[51] For a similar discussion of bureaucratic forms in the Habsburg Empire, see Göderle, 'Administration, Science, and the State': 71–7. On methodological approaches to working with forms and lists as historical sources for the history of science, see Delbourgo & Müller-Wille, 'Listmania'; Müller-Wille & Charmantier, 'Lists as Research Technologies'.

[52] On the use term Great Russian during this period to convey they idea that Russians, Ukrainians (as Small Russians, *Malorusskie*) and Belarusians (as White Russians, *Belorusskie*) were part of a single family of Russian peoples, see Miller, 'Natsiia, Narod, Narodnost'': 387.

mal' religious practices. It was used as an umbrella term to refer to all dissenters and sectarians who deviated from the Orthodox Church over a variety of theological claims or ritual differences. Neither the Orthodox Church nor imperial state acknowledged terms such as Old Ritualists (*staroobriadtsy*) or Old Believers (*starovery*) used by the dissenters themselves.[53] Most 'schismatics' were classified on the forms collected by the Vitebsk Statistical Committee as Russians, however statistical tables sent to Sementovskii also recorded a small number of 'schismatic' Belarusians and Latvians. Whereas Belarusians, Poles, and other Slavic peoples could be either Orthodox or Catholic, the possibility of Catholic Russians was perceived by Sementovskii to be a contradiction in terms. This particular combination of religious and ethnolinguistic categorization was not even supposed to enter into the realm of possibility on the forms.[54] The politics, ideologies, and epistemic logic underlying the classification of people was embedded into the statistical paperwork from the start as the forms excluded the possibility of being a Catholic Russian.

At the same time, the use of pre-printed forms to collect statistical data did not completely remove all possibilities for respondents to draw attention to local circumstances and alternatives ways of enumerating populations. As Simon Franklin points out, there was an important 'functional distinction between routine recordkeeping and the dissemination of official pronouncements. Manuscript was the technology of record, print was the technology of dissemination.'[55] The respondents submitted their responses to Sementovskii's printed circulars and forms as handwritten reports. The flexibility afforded by free-form writing allowed local observers to inform Sementovskii of instances where their experiences and observations did not neatly fit into the rows and columns of the data tables. For instance, a report by the chairman of the Dinaburg (Daugavpils) peasant commune (*mir*) court in north-west Vitebsk province noted how the linguistic ascription 'Latvian' printed on the data table grouped together people who, in his opinion, exhibited considerable internal diversity. The chairman explained that the majority of Latvians in Dinaburg district were Catholics and were locally perceived to be distinct from Latvians living near the border with Livland province, who were predominantly Lutheran and spoke a different 'dialect' (*narechie*).[56] The report highlighted how the religious and linguistic landscape of today's Latgale region in eastern Latvia could not be straightforwardly enumerated using categories devised and prescribed by statisticians and mapmakers at the provincial (Vitebsk), regional (Vil'na), or imperial (St. Petersburg) administrative centres.[57]

[53] Paert, 'Two or Twenty Million?': 77–80.
[54] NHAB f. 2502, vop. 1, spr. 183, ark. 48.
[55] Franklin, 'Mapping the Graphosphere': 541.
[56] NHAB f. 2502, vop. 1, spr. 162, ark. 9-10adv.
[57] On the ethnoreligious particularities of Vitebsk's Latgalians, see Benz, 'Zwischen konfessioneller, regionaler und nationaler Identität'.

While these nuances might have disappeared on a small-scale map through the process of moving from a detailed local investigation to a more general overview, the provincial focus of the maps Sementovskii was making meant that some of these complexities did leave their mark on his cartographic depictions of Vitebsk province. The pencil annotations to the guide to cartographic signs and symbols reveal how Sementovskii subsequently adapted the plan for the atlas to incorporate local details (Figure 2.1). He scored out Ukrainians, Lithuanians, Samogitians, and Tatars—all groups that were not numerous in Vitebsk province—and added another category, Estonians (specifically, Lutsis named after the Liutsin/Ludza district where they lived), as a noteworthy ethnic group in the local context.[58] Sementovskii also made changes to the colours used to represent Germans (blue), Estonians (purple), and Roma (brown). Most notably, on the final map Sementovskii included three additional 'mixed' categories in the legend representing Poles, Estonians, and Jews who were 'mixed with other ethnicities' (Plate 6). In this way, the mapmaking process enabled Sementovskii to partially reclaim a sense of the specificity of place in reaction to the standardization imposed by administration of the North-Western Territory, which Sementovskii himself had at the outset been instrumental in imposing. The picture of map production that thus emerges from examining Sementovskii's activities is one of dialogue between different levels of administration in the Russian Empire.

Besides correspondence networks and pre-printed forms, several other procedural aspects of the provincial statistical committees' work come to the fore if we examine the collection of statistics on ethnolinguistic groups as a form of fieldwork. Statistics is typically conceived of as an 'armchair' discipline, involving the stay-at-home sedentary examination of data.[59] By contrast, disciplines such as ethnography, physical anthropology, and physical geography have strongly developed expeditionary cultures of fieldwork. While topographic mapmaking typically falls into the second group, as illustrated by Matthew Edney's account of the impacts of environmental and human factors on topographical mapping in British India, there has been little consideration of how the process of collecting population data in the field might have left a mark on ethnographic maps.[60] A notable exception is the work of Christine von Oertzen, who has examined the specially designed census boxes used in Prussia to sort, transport, and store individual counting cards as part of the state's 'data machinery' which began at the scale of the enumeration district.[61] While sources detailing experiences of statistical data collection during this period in the Russian Empire are few and far between, available evidence allows us to build up a picture of a range of different

[58] On Lutsis, see Jääts, 'Üks kuulus välitöö.
[59] Bridges, 'Nineteenth-Century East African Travel Records': 193–4.
[60] Edney, *Mapping an Empire*, esp. pp. 121–64.
[61] Von Oertzen, 'Prussian Census Box': 476.

environmental and social factors shaping processes of ethnographic data collection.

Environmental factors had clear a bearing on statistical work, especially weather. In the early hours of 22 October 1864 enumerators during a trial census conducted on the estate of Jensel in Livland province trudged between households in freezing temperatures. Luckily, the roads remained passable and the clear skies enabled the census-takers to read their lists. The dry weather conditions meant that the census paperwork was returned in clean, neatly bound bundles, and the information was readable.[62] The descriptions of the highly favourable weather conditions for the census suggests that this was not always the case. By contrast, during the 1881 census in Kurland province enumerators were exhausted by the end of the count after having trekked along country lanes in the bad weather for the whole day. Census forms could be easily damaged and rendered illegible in the rain.[63]

Disagreements could emerge between the organizers of mapping projects and the local officials on the ground meant to carry out instructions. In the early 1860s, mapmaker Aleksandr Fedorovich Rittikh, about whom we learn more in Chapter 3, designed routes to collect data to prepare an atlas of religion in North-Western Territory. For Minsk province, the route the counters were to follow was marked out on a large scale 3-*versts* map spread over forty-one sheets. This approach, however, proved to be very impractical as the map was too large to use as a tool for navigation in the field where there were no tables to spread it out. The local police found Rittikh's map unusable as it directed them along routes that were completely unknown to them. Whereas Rittikh had marked out routes with a view to conducting a comprehensive survey of the province, the local police were knowledgeable about the landscape and complained about being ordered to trek across swampy terrain.[64]

The cooperation of the local population was paramount for collecting individual data. The documented efforts of the statistical committees in the Baltic provinces to conduct regular censuses from 1863–81 reveal how news of an upcoming census was often met with high levels of mistrust. As I have argued elsewhere, the shift from the traditional imperial 'revisions' of 1719–1858, which had been conducted primarily to draw up tax rolls and lists for military conscription, to forms individual enumeration aroused the suspicion peasants and townsfolk in Baltic provinces. They feared that the collection of new types of personal information would be used against them by the local authorities and the Baltic German elites organising the census to either return them to serfdom, raise taxes, or resettle them. In several cases, townsfolk and villagers openly revolted against statistical

[62] *Bericht über eine am 22. Oktober 1864 auf dem Gute Jensel*: 18–20.
[63] 'Zur Volkzählung in Kurland,' *Zeitung fur Stadt und Land*, 26 January 1882: 2.
[64] Rittikh, *Atlas narodonaseleniia*: 2.

enumeration. In the winter of 1881–2 violent protests erupted against the census in Wierland (Rakvere) county in Estland and Illukst (Ilūkste) county in south-east Kurland. In the latter case, soldiers were deployed from nearby Dinaburg/Dvinsk (Daugavpils) in Vitebsk province to calm the unrest. The protests prevented the census from being carried out in this region, leaving a lacunae in Kurland's census results.[65]

Sources giving us an insight into how the arrival of a statistician, census-taker, or ethnographer was perceived by the local inhabitants are rare. Literary depictions of provincial statisticians often emphasize their incompetence and prejudices. The readiness of ethnographers to ascribe their own preconceived ideas onto the local population was satirized by the Belarusian playwright Ianka Kupala in his tragicomic drama *Tuteishiya* (1922) in a scene in which two ethnographers search for 'true Belarusian types'. The Polish-speaking scholar concludes that Belarusians are a type of Pole, and the Russian-speaking scholar thinks that Belarusians are essentially a type of Russian.[66]

Accounts of ethnographic fieldwork typically only report about incidences of discord between ethnographers and their subjects. When Anders Johan Sjögren and August Georg Wilhelm Pezold carried out their research expedition to Livland for the Russian Geographical Society in 1846, as discussed in Chapter 1, they attempted to take moulds of the face and skull of two Livs. Sjögren mentions in his report that the procedure was unsuccessful. There is no record of exactly what occurred during the incident, but—rather ominously—Sjögren wrote that they could not find anyone willing to be paid to repeat the process for any amount of money.[67] Similarly, when linguist and folklorist Oskar Kallas travelled to Vitebsk province in 1893 to study the Lutsi population, Kallas noted how the locals treated him with deep suspicion and feared he was a spy or cholera-carrier.[68]

These descriptions of environmental and human interactions sprinkled throughout reports of statistical data gathering and ethnographic fieldwork are anecdotal in nature. More research is needed to gain a deeper understanding of how fieldwork processes shaped the data which was later transformed into ethnographic maps. Nevertheless, even these brief examples impress upon us the importance of not treating maps in isolation. They remind us of the need to think critically about the circumstances in which population statistics were collected and the possible factors impacting the data before it even reached the hands of mapmakers.[69]

65 Gibson, 'Experiencing Enumeration': 9–19; Plakans & Wetherell, 'The 1881 Census': 52.

66 Per Anders Rudling discusses the significance of Kupala's play in Rudling, *The Rise and Fall*: 304–5.

67 Sjögren, 'Bericht über eine im Auftrage': 530.

68 Jääts, 'Üks kuulus välitöö': 22.

69 Gibson, 'Experiencing Enumeration': 24.

Triangulating Religion, Language, and Nationality

The process of making the *Ethnographic Map of Vitebsk Province* forced Sementovskii to confront the question of how to map nationality. Compared to the other thematic maps being prepared for the atlas, the ethnographic map presented a particular methodological challenge for the secretaries as statistical data on nationality was not collected directly, as was customary in the Russian Empire at the time. The 1897 All-Russian census included questions about religion and everyday spoken language, but the Central Statistical Committee decided that asking a question specifically about nationality in the census would not produce meaningful results among the majority of the population for whom nationhood had little meaning as a form of self-identification.[70] Borderland populations such as Vitebskians typically exhibited 'indifferent' attitudes towards national forms of identification during this period. They either identified more strongly with local, religious, or class communities, related to the idea nationality in flexible or situational ways, or proved hard to classify into national categories as a result of multilingualism and mixed marriages.[71] The Russian imperial practice of not enumerating nationality concurred with the decision reached by expert delegates at the International Statistical Congress in St. Petersburg in 1872, who decided that posing a census question on nationality would not produce meaningful results among people who did not self-identify as belonging to a national community.[72] For ethnographic mapmakers, there was a pronounced dissonance between the availability of data and the kinds of information mapmakers' needed to complete their work.

Rare occasions when a specific question about nationality was asked in statistical surveys in the Russian Empire, such as in the censuses organized by the statistical committees in the Baltic provinces from 1871 to 1881 which asked people to self-report their nationality, yielded wide-ranging and dubious results. The head of the Estland Statistical Committee, Paul Eduard Jordan, was upfront about the unreliability of the data collected on nationality during the 1881 census in Reval (Tallinn), noting that 'It is certain that the terms "*Nationalität, narodnost'*, *rahwa sugu*", were sometimes completely incomprehensible to people' as they appeared on the counting cards in German, Russian, and Estonian. To illustrate his point, Jordan highlighted cases of German-speaking Estonians who had self-reported themselves in the census as being of Russian nationality on the basis of

[70] The exception was in the Caucasus region, where local authorities requested the inclusion of a question on nationality. Cadiot, 'Searching for Nationality': 442.

[71] For a discussion of the concept of 'national indifference', see Zahra, 'Imagined Noncommunities': 106.

[72] Labbé, 'Le projet d'une statistique des nationalités'.

their Orthodox faith. Jordan also recounted the instance of a man who was undecided about whether his nationality was Estonian or Russian as he spoke both languages equally. The counter pressed him to come to a decision on his nationality by asking him which side he would usually support in a fistfight with his housemates, to which he replied 'the Estonian'.[73]

The lack of data on nationality meant that Sementovskii had to contend with using religion and language as proxies for mapping ethnic groups. When it came to the relative weight given to these factors, Sementovskii challenged earlier methodological approaches to mapping ethnicity in the North-Western Territory which leaned heavily towards religion as the determining factor. In Sementovskii's view, applying this logic to Vitebsk province had led to the misclassification of many Belarusian- and Latvian-speakers as Poles on the basis of their Catholic faith.[74] Instead, Sementovskii believed that greater weighting should be given to language as a more accurate reflection of the inhabitants' 'wishes, sympathies, traditions, origins'.[75] He also argued that a range of other ethnographic observations should be factored in, including traditional dress, handicrafts, customs, celebrations, folklore, folksongs, food, and architecture. Taking a broad perspective would help avert common problems associated with directly equating language with nationality, such as using German-language as a marker of German nationality which skewed the data as many Jews spoke German. Sementovskii's approach to ethnographic mapping illustrates how imperial statisticians 'searched for nationality' by aggregating a wide range of different linguistic, religious, and ethnographic observations.[76] In this way, 'drawing things together' to make a map became a powerful shaper for thinking about these different identity components together.[77]

Yet, the relationship between language, religion, and nationality did not always correlate in the ways that imperial bureaucrats and statisticians expected it to. Sementovskii acknowledged that using language as a proxy for nationality came with its own bundle of complexities. It was tricky to precisely delineate where people stopped speaking one language and started speaking another, especially among rural populations with low literacy rates. As Sementovskii noted, the Slavic speech of the Orthodox and Catholic inhabitants of Vitebsk province resembled 'something between' Russian, Belarusian, and Polish. A linguistic gradient that could be heard when travelling around the region: Belarusian-speakers in Vitebsk province used more Russian words, whereas those in the provinces of Minsk and Vil'na used more Polish words.[78]

[73] Jordan, *Die Resultate der ehstländischen Volkszählung*: 24.

[74] Sementovskii, *Etnograficheskii obzor*: 2.

[75] Sementovskii, *Etnograficheskii obzor*: 2.

[76] Cadiot, 'Searching for Nationality': 442.

[77] Latour, 'Drawing Things Together'.

[78] Sementovskii, *Etnograficheskii obzor*: 6.

In the western districts of Vitebsk province the local variety of Belarusian contained many influences from Latvian and Latgalian.[79]

Sementovskii also pointed out that many Vitebskians had a degree of competency in multiple languages, either as a result of mixed marriages or simply because they used, heard, or read various languages in different spheres of everyday life. We should be careful about assuming that inhabitants of multiethnic regions are necessarily multilingual and that language was an important marker of difference among local communities. Nevertheless, a study carried out by the Vitebsk Statistical Committee in 1865 revealed that proficiency in several tongues was common amongst the inhabitants of the city of Vitebsk, especially among the more educated segments of society.[80] Language-use was often situational and a resident of Vitebsk might greet his neighbour, conduct a business transaction, or read a religious text all in different languages.[81] The way in which complex intersections of socio-economic factors and gender influenced linguistic identification becomes strikingly apparent if we zoom in further to the scale of individual households. The detailed descriptions by Estland Statistical Committee secretary Paul Jordan of how families in Reval (Tallinn) answered the question on language in the 1881 census reveal how every member of a family could self-report as having a different language everyday of communication:

> A retired soldier named Wittwer, who was of Lutheran confession and Estonian descent and who indicated Estonian as his usual language, lived with three children, all three of whom, like their father, were Lutheran and Estonian by nationality. The eldest son was a crown clerk who preferably spoke Russian, the second Estonian, and the third child, a daughter, a seamstress by profession, spoke German.[82]

Finally, knowledge of a language, and the degree to which a person spoke competently and 'purely', were also highly subjective matters.[83] Most people are not equally comfortable or adept at conversing in all the languages they know. In this respect, Sementovskii noted how many Latvians in western Vitebsk province 'speak Russian badly'.[84] This statement can be read in at least two ways: as a comment on the local Latvian-speakers' limited knowledge of Russian language—they possessed only a smattering of words and simple phrases—or as a judgement on the quality of their Russian, which from Sementovskii's perspective contained many

[79] Jankowiak, 'Belarusian Dialects in Latvian Latgale'.

[80] Judson, 'Do Multiple Languages Mean a Multicultural Society?': 63.

[81] *Pamiatnaia knizhki Vitebskoi gubernii* (1865): 208–16. An elaboration of these themes in the context of the multilingual Habsburg Empire is presented in: Prokopovych et al. eds. *Language Diversity in The Late Habsburg Empire*.

[82] Jordan, *Die Resultate der ehstländischen Volkszählung*: 60–1.

[83] On the politics of defining languages and dialects, see Kamusella, *The Politics of Language*. On the question of ascribing single languages to multilingual populations, see Wolf, *The Habsburg Monarchy's Many-Languaged Soul*, esp. pp.37–41.

[84] Sementovskii, *Etnograficheskii obzor*: 33.

mistakes. Consequently, when using language as an indicator of nationality, statisticians often emphasized that they wanted information about a person's 'native national language'.[85] Native language was understood as a means of determining a person's essential national character as it was inherited from their parents, impressed on them at birth, and thus a fundamental part of their being. Knowledge of other languages, by contrast, was something that was believed to be acquired later in life and influenced by the particularities of the place and social circumstances in which the person lived.

Sementovskii's reflections reveal how the process of producing cartographic visualizations of ethnicity was an interpretative process that involved triangulating data on religion and language with ethnographic descriptions. Moreover, Sementovskii was not working on his maps of Vitebsk province in isolation but had to ensure that his methods for making judgements about ethnicity based on language, religion, and other ethnographic observations were calibrated with what his peers were doing in the other provinces in the North-Western Territory. For instance, according to Sementovskii's new calculations Poles only made up around 10 per cent of the total number of Catholics in Vitebsk province, while the rest comprised Latvians, Belarusians, Estonians, and Roma. The other secretaries seem to have generally agreed with Sementovksii's conclusions but were concerned about ensuring 'uniformity' in the way data was interpreted. As the secretary of Minsk Statistical Committee noted in a letter to Sementovskii in September 1869, the criteria for defining different ethnolinguistic groups needed to be standardized, so that the rationale for classifying someone as Polish in Minsk province would be the same as in Vitebsk province.[86] But this raised the fundamental question of whether Minsk's Poles were indeed the same as Vitebsk's? Thus, the process of creating a unified statistical atlas replicated on a smaller scale many of the same problems imposing a unified approach to mapping the Russian Empire encountered by Peter Köppen in the 1840s.

The process of triangulating nationality was clearly shaped by the political agenda of the tsarist administration in the North-Western Territory, which sought to diminish the Polish and Catholic influence by highlighting the presence of other ethnolinguistic and religious groups. At the same time, it would be an oversimplification to characterize Sementovskii's approach to ethnographic mapping as only ideological. The process of analysing data on religion and language and using it to make the ethnographic map of Vitebsk province also made Sementovskii think in new ways about the relationship between ethnicity and space. The tensions between politics and the limitations of cartographic modes of data visualization are clearly visible in the case of mapping Vitebsk's Jewish population. Whereas data tables could in theory record statistics at the precision level

[85] Sementovskii, *Etnograficheskii obzor*: 2.
[86] NHAB f. 2502, vop. 1, spr. 183, ark. 50-51adv.

of individuals, maps tended to exaggerate the presence of low-density populations spread over a large area and diminish the visual impact of high-density populations. There was a limit to the level of detail that an engraver could convey.[87] This posed an issue when it came to Vitebsk's sizeable Jewish population, which is largely absent from the *Ethnographic Map of Vitebsk Province*.

Vitebsk province lay within the Pale of Settlement, the western region of the Russian Empire in which Jews were legally allowed to permanently reside. At the end of the nineteenth century the population of the city of Vitebsk was 52.4 per cent Jewish.[88] However, Jews typically did not feature prominently on ethnographic maps as they were not sufficiently concentrated over a large enough territory to be coloured as a defined ethnographic area. The dissonance between data table and map can clearly be seen in the inset data table in the bottom left-hand corner of the *Ethnographic Map of Vitebsk Province* (Plate 6). The 87,760 Jews listed in the data table are only hinted at by the black dots on the map around cities in Vitebsk province. Another notable absence from the map are the 177 Roma listed in the table, who were split between Orthodox (151) and Catholics (26). The map contains no visibly brown areas indicating their whereabouts due to their lack of a definable ethnographic territory.[89] This dissonance between data table and map revealed how different formats of organizing and displaying information could result in strikingly different images of the world. Ethnographic mapping rested on the assumption that populations lived in permanent and fixed settlements. Populations who lived mobile lifestyles, either as seasonal labour migrants, military personnel, non-sedentary, or nomadic populations, were often rendered invisible in the process of moving between data table and map.[90]

Sementovskii was not alone in noticing this problem. His struggles with how to address questions of scale resonated with wider methodological challenges cartographers were grappling with at the time. The population of towns and cities in European Russia in the nineteenth century was often religiously and ethnolinguistically diverse, especially in the western borderlands of the empire where they usually differed in composition from the surrounding countryside. Yet the small scale of most ethnographic maps meant that information about the internal composition of a town or city was usually not depicted at all or only in a very rudimentary way. The concentrations of Germans, Russians, Poles, and Jews dotted across maps did not convey any information about the spatial relations between these groups within urban space. Moreover, these 'islands' of different minority groups

[87] Rudwick, 'The Emergence of a Visual Language': 155.

[88] Le Foll, *Vitebskaia khudozhestvannaia shkola*: 41.

[89] On ethnographic studies of Roma in imperial Russia, see O'Keeffe, 'Gypsies as a Litmus Test'.

[90] On the Russian imperial authorities' attempts to keep track of migration, see Gorshkov, 'Serfs on the Move'; Smith, 'Documents, Migration, and Governance'.

conveyed the false impression that these groups lived in isolation from one another, when in fact they were indications of the mixed ethnic character these localities.[91]

Sementovskii attempted to address the question of urban ethnographic mapping in the summer of 1869 alongside his work on the maps for the atlas. He sent out instructions to police and district officials in Vitebsk province to gather data to compile separate ethnographic plans of the towns and cities. He drew up lists of Jewish property and land ownership and instructed surveyors to shade town plans with colours denoting property ownership by different nationalities, 'or at least of Christians and Jews', to gain a more accurate picture of the distribution of nationalities within urban space.[92] Sementovskii's interest in collecting information about Jews formed part of a wave of ethnographic studies produced after 1863 which presented the Jewish population as "Others". As Claire Le Foll notes, in his ethnographic descriptions Sementovskii reduced Vitebsk's Jews to a series of stereotypes: an economically prosperous population with monopoly on trade and manufacturing, who exploited the labour of non-Jewish locals, and attracted the hostility of 'ordinary Russian people'.[93]

Correspondence between the district surveyors and Vitebsk Statistical Committee making reference to 'enclosed maps' suggest that several ethnographic plans were indeed made.[94] Sementovskii also produced a legend specifying the cartographic signs and symbols to be used and the tones used to depict different intervals of Jewish population density.[95] Unfortunately, the maps themselves have not been preserved together with the correspondence about the project in the archives of the Vitebsk Statistical Committee. Nonetheless, the discussion about the need to produce ethnographic town plans alongside the more general ethnographic map of the province indicates that Sementovskii was highly aware of the discrepancies between the demographic composition of cities and rural areas and how existing approaches to ethnographic mapping were unable to convey these differences. His proximity to the subjects of his ethnographic map and the data-collection process led him to explore alternative ways of visualizing this data in cartographic form.

Printing and Publishing Ethnographic Maps

By the spring of 1870, Sementovskii had completed drafts of three maps of Vitebsk province on the themes of trade and industry, hydrography, and ethnography. The maps were drawn with the help of forest inventory surveyor,

[91] On the contradiction of the 'island' metaphor frequently deployed by nationalists for describing multiethnic areas, see Judson, 'When Is a Diaspora Not a Diaspora?': 231.

[92] NHAB f. 2502, vop. 1, spr. 183, ark. 40.

[93] Le Foll, *La Biélorussie*: 376–7. The passage Le Foll is referring to is Sementovskii, *Etnograficheskii obzor*: 62. See also Soroko, *Vitebskie gubernskie vedomosti*: 109–10.

[94] NHAB f. 1437, vop. 1, spr. 3924, ark. 51, 76.

[95] NHAB f. 2502, vop. 1, spr. 183, ark. 43.

Konstantin Kondratovich, who was likely a contact from Sementovskii's previous job in the forestry administration. Sementovskii later informed his colleague Aleksandr Gatsinskii, secretary of Nizhegorod Statistical Committee, that he succeeded in producing all six thematic maps but only the ethnographic map was ever published. As the story goes, the remaining maps languished in manuscript form gathering dust in the office of Vitebsk Statistical Committee 'in anticipation of the best days of the activities of this institution'.[96] Unfortunately, their fate today remains unknown.

The themes of the maps—including the notable addition of a map of hydrology—indicate that for some reason Sementovskii chose to depart from the agreement reached in Vil'na the previous spring. We get no clues from the sources as to why this was. However, it is a foreboding sign of how cracks were already beginning to form in Potapov's carefully laid plans for a coordinated regional statistical atlas.[97] In the spring of 1870 Sementovskii grew increasingly frustrated with the secretaries of the other statistical committees in the North-Western Territory for failing to produce any maps. The initial enthusiasm for the project withered with the dawning realization that they had underestimated the scale of the work and money required, since all the work had to be funded from the statistical committees' own meagre budgets.[98] As Sementovskii wrote:

> The newness of the activity, the vastness and seriousness of the work adopted by the committees, the scarcity of material resources of these institutions, the numerous obligatory duties of the secretaries, who constitute almost the only moral strength of the committees, and finally, the impossibility sometimes to be content with the statistics collected by official means—all this, naturally, slowed down the progress of the matter and sometimes even complicated it.[99]

Sementovskii's account of the practical difficulties facing mapmakers highlights how individual mapmakers often had more ambitious goals than the institutions they worked for. Many secretaries held their positions for long periods spanning twenty or thirty years and, like other imperial officials, the personality and individual initiative of post-holders played a key role in determining the ebb and flow of a statistical committee's activities.[100] At the same time, for mid-level bureaucrats like the secretaries, mapmaking was first and foremost a job rather than vocation or the realization of a long-term personal ambition. Unlike Peter Köppen who could just about afford to pay upfront out of his own pocket to proceed with mapping projects while waiting for reimbursement, provincial

[96] Sementovskii, *Sbornik v pamiat'*: 40.
[97] NHAB f. 2502, vop. 1, spr. 198, ark. 20, 28-68adv.
[98] NHAB f. 2502, vop. 1, spr. 198, ark. 20.
[99] Sementovskii, *Otchet o deistviiakh*: 20.
[100] Burbank, 'Supervising the Supervisors': 19.

statisticians engaged in mapmaking as part of a civil service job. Their commitment to cartography was often determined more by their professional context rather than personal interest. Mapmaking was interrupted by other life events or sidelined when other aspects of their job were prioritized. Reflecting on the struggles of being a provincial statistician, Sementovskii's colleague Aleksandr Gatsinskii in Nizhegorod noted how 'Time takes its toll and from a small family of statistical workers, the best figures drop out of it one after another: some are kidnapped by death, others are carried away by concern for their daily bread to work of a different kind'.[101] Gatsinskii's statement reveals how ethnographic mapmaking did not necessarily mean the same thing to provincial bureaucrats as it did to academicians and central government employees such as Peter Köppen. Whereas Köppen dedicated himself to mapmaking as personal passion project, for many others it was simply part of a job.

By the spring of 1870 enthusiasm for the statistical atlas project had waned among the other statistical committee secretaries in the North-Western Territory. Undiscouraged, Sementovskii continued to push on and explore other avenues for publishing his maps. He considered publishing the maps as a separate statistical atlas of Vitebsk province, but quickly abandoned this plan due to a lack of funds. However, a new opportunity arose in the early summer of 1870 when Sementovskii, along with seventy-four other secretaries of provincial statistical committees and members of the Central Statistical Committee, convened in St. Petersburg for the first general meeting of statisticians in the Russian Empire. At the closing dinner, Sementovskii proposed that the participants publish a collection of articles to commemorate the meeting and showcase the breadth and quality of statistical work being done in the provinces.[102] Sementovskii hoped that by soliciting contributions from across the empire the publication's broad geographical scope would attract a wider audience and sell more copies. This would help make the publication financially viable and raise the profile of the statistical work being done in the provinces.[103] The project received the go-ahead from the participants. However, a year later Sementovskii again grew frustrated when many of the promised contributions failed to materialize. Sementovskii ended up writing half of the articles in himself, including an article giving an ethnographic overview of Vitebsk province accompanied by several ethnographic drawings and the ethnographic map originally created for the atlas. Thus, although it was a collective work, Sementovskii's presence was keenly felt throughout the volume.

The challenges Sementovskii faced trying to get his map out into the world did not end there. Next, he had to turn to the matter of printing and publishing the edited volume and map. Back in the spring of 1869 when Sementovskii was

[101] Gatsinskii, *Sbornik v pamiat'*: 85.
[102] Gatsinskii, *Sbornik v pamiat'*: xxxii, 1.
[103] NHAB f. 2502, vop. 1, spr. 198, ark. 20adv.

beginning work on the maps of Vitebsk province for the planned statistical atlas of the North-Western Territory, he had started enquiring about a printer and publisher for the atlas. Printing coloured statistical maps was a complex and specialist process. First, a greasy image of the base map had to be engraved onto a limestone slab or a copper plate, which would chemically bond to repel water. The base map was then lithographed in black ink. In the middle of the nineteenth century, this process was done using a hand press and the map was shaded afterwards by hand in a watercolour wash, as had been the case of Köppen *Ethnographic Map of European Russia* (1851).[104] By the 1870s, printing firms in the Russian Empire began investing in mechanical fast-printing machines, usually imported from Germany, which gradually replaced hand power with steam.[105]

Despite these technological advances, thematic maps remained particularly complicated to print as they contained many colours. Chromolithography—lithographic printing in multiple colours—was also trickier to mechanize. The map design had to be separated into different layers, each layer was traced and shaded with a lithographic crayon or pen onto a separate lithographic plate, and then each colour was separately printed onto the base map.[106] The greater the number of colours, and thus lithographic stones required, the greater the cost. Consequently, hatching and dot patterns created using a ruling machine were popular ways of adding thematic layers to a map without requiring additional lithographic plates.[107] Despite the gradual changes in methods for illuminating ethnographic maps from hand colouring to mechanized chromolithographic printing, the process was still very labour-intensive and required highly skilled workers. As one description of the chromolithography process from early twentieth-century Moscow noted, 'looking at a picture, printed, for example, in 15 colours, we cannot imagine the amount of work that is spent in the lithography to reproduce it'.[108] Moreover, the mechanization of map production did not necessarily mean that maps became more precise. Whereas copies of hand-coloured maps varied depending on the colourists' skills in following the lines, in the case of chromolithographic printing each layer had to be precisely overlaid and slight misalignments were common.

These developments in printing technologies and techniques went hand in hand with the rapid expansion of the commercial printing industry in the Russian Empire in the second half of the nineteenth century. Small private printing enterprises proliferated in St. Petersburg and Moscow in the 1860s and 1870s thanks to the growing economy, burgeoning literary culture among educated elites, and the new censorship statute of 1865 which removed preliminary

[104] On map production processes, see Cook, 'From False Starts to Firm Beginnings': 158; Pearson, 'The Nineteenth-Century Colour Revolution'.
[105] Steinberg, *Moral Communities*: 11.
[106] Ruud, *Russian Entrepreneur*: 21.
[107] Cook, 'From False Starts to Firm Beginnings': 164.
[108] *T-vo skoropechatni A. A. Levenson*: 39.

censorship.[109] The printing sector was also developing at the provincial scale. However, in the 1870s provincial capitals such as Vitebsk did not yet have printing houses equipped with the skilled personnel or machinery to print complex chromolithographed maps such as those planned for the statistical atlas. From 1864 to 1867 the annual memorial books (*pamiatnye knizhki*) compiled by the Vitebsk Statistical Committee were published in St. Petersburg.[110] After 1867, the printing process was transferred to the Vitebsk Provincial Typography, however volumes containing complex lithographic printing continued to be outsourced to printers in St. Petersburg. Alternatively, statistical committees in the Baltic provinces preferred to orientate their gaze westward, as was the case for the Kurland Statistical Committee in the early 1860s which outsourced the printing of its German-language statistical maps to the F. A. Brockhaus publishing firm in Leipzig.[111] It was not until several decades later in the 1890s that the Vitebsk Provincial Typography published its first chromolithographed maps in shades of one colour.[112] Another factor limiting the development of map printing in the empire's provinces was the dire economic situation of many of the provincial typographies. For instance, in 1872 the activities of the Estland Provincial Typography were paralysed as it was 7,500 roubles in debt; the government subsidy it received was not enough to make up for the low subscription rates to its publications.[113]

The limited local infrastructure for cartographic printing led Sementovskii in the spring of 1869 to write to A. Il'in's Cartographical Establishment in St. Petersburg to ask for a quote to print the prospective statistical atlas of the North-Western Territory. At the time, Il'in's business was already building a reputation as the leading commercial map publisher in the empire. The gradual development of the commercial cartographic printing and publishing sector in the Russian Empire in the second half of the nineteenth century was made possible by more liberal laws on the publication of maps and advances in printing technology. In the first half of the nineteenth century the Military Topographic Depot had closely regulated the production, sale, and circulation of maps.[114] However, imperial laws regulating the production and sale of maps were relaxed from 1857 allowing geographical maps to be printed after a preliminary review by the Military Topographic Depot.[115] As part of the drive to reform Russian society following defeat in the Crimea War, Tsar Alexander II granted permission in 1859 to Colonel Vladimir Aleksandrovich Poltoratskii and Captain Aleksei Afinogenovich Il'in to establish a cartographic printing and publishing firm in the imperial capital, Chromolithography of Poltoratskii, Il'in & Co. It was the first private cartographic company allowed to operate since the V. Kiprianov firm had been closed down in

109 Steinberg, *Moral Communities*: 7; Ruud, *Fighting Words*: 137–49.
110 Soroko, *Vitebskie gubernskie vedomosti*: 84.
111 Heyking, *Statistische Studien*.
112 Dolgorukova, *Vitebskaia guberniia*.
113 RA, EAA.291.1.12790.1–9.
114 Kil'dushevskaia, 'Izdatel'skaia deiatel'nost'': 97.
115 Jeske, 'Topographische Karten im Zarenreich'.

Figure 2.2. The map engraving department in A. Il'in's Cartographical Establishment in St. Petersburg in the early 1900s.

Figure 2.3. Print workers in the room for binding and gluing maps in A. Il'in's Cartographical Establishment.

Source for Figures 2.2 and 2.3: Piatidesiatiletie Kartograficheskogo zavedeniia A. Il'ina, 1859–1909 (St. Petersburg: Tipografiia Pervoi Sb. Trudovoi Arteli, 1909). Public domain.

1723.[116] In 1864 Poltoratskii left the firm due to military service and left full ownership of the company to his partner Il'in.

Over the course of the second half of the nineteenth century, Il'in and his sons Aleksei Alekseevich and Afinogen Alekseevich (who took over the company on their father's death in 1889) developed the firm into a successful business with de facto monopoly on printing and publishing civilian maps in the empire.[117] In 1868 the A. Il'in Cartographical Establishment acquired their first engine press from the factory Koenig & Bauer in Würzburg and by 1884 the firm employed over one hundred workers, including thirty-five engravers and apprentices, twenty printers, and forty-five assistants (Figures 2.2 and 2.3). Among these cartographic print workers were a significant number of women who worked in the departments for quality inspection, binding, and gluing.[118] By the end of the nineteenth century, the A. Il'in firm was responsible for printing 90 per cent of all civilian cartographic materials within the Russian Empire. The firm not only printed the materials of the state ministries and departments, Geographical Society, Central Statistical Committee, and Academy of Sciences, but also profited from the increasing demand for popular cartographic print culture, including school wall maps, atlases, and illustrated journals. In this way, the firm functioned as the Russian equivalent of Justus Perthes Geographical-Cartographical publishing house in Gotha and was emblematic of the emerging technical elite of professional experts in map production in the second half of the nineteenth century. The firm was nationalized in 1918 following the revolution.

In the spring of 1869 Il'in senior provided Sementovskii with a quote for 175 roubles to engrave and print six statistical maps of Vitebsk province in twenty-five copies. This price included the cost of paper, engraving, printing, and transportation to and from St. Petersburg. The engraving process consumed the majority of the costs at 25 roubles per map, which Il'in anticipated would take at least one month to complete and require careful proofreading and corrections.[119] Two years later, after abandoning the larger atlas project and deciding to publish his ethnographic map in the edited conference proceedings, Sementovskii again sent out letters to various printing and publishing houses to ask for quotes to print the volume of articles with the ethnographic map inserted as a fold out sheet. In June 1871 Il'in's firm quoted Sementovskii 300 roubles to print 825 copies of the map.[120] Meanwhile, Al'vin Andreevich Kaspari's printing and

[116] Kil'dushevskaia, 'Izdatel'skaia deiatel'nost'': 97.

[117] Seegel, *Mapping Europe's Borderlands*: 159.

[118] *Dvatsatipiatiletie Kartograficheskogo zavedaniia A. Il'ina*; *Piatidesiatiletie Kartograficheskogo zavedeniia A. Il'ina*; Gibson, 'Mapmaking in the Home and Printing House': 78–80.

[119] NHAB f. 2502, vop. 1, spr. 183, ark. 10-10adv.

[120] NHAB f. 2502, vop. 1, spr. 198, ark. 23adv.

lithographic firm in St. Petersburg quoted Sementovskii 600 roubles to print the whole edited volume, which included 825 copies of the book with five ethnographic drawings and the map. Kaspari specified that the map component would cost 180 roubles, including the cost of the paper.[121] The Vitebsk Statistical Committee's budget was limited so Sementovskii decided to go with Kaspari's general printing and lithographic firm rather than the specialist, but more expensive, cartographic printing service provided by Il'in.

Logistical and Social Dimensions of Map Production

Kaspari's printing and lithographic firm would later become one the major printing houses in late imperial Russia. Yet in the early 1870s Kaspari had only recently left his position as managing director of the K. I. Wolfe printing house to branch out on his own. This was an era when printing houses were multiplying severalfold in the capital as entrepreneurs sought to take advantage of the booming consumer market for printed materials.[122] Unfortunately, Sementovskii's gamble to cut costs and work with a new printing house did not pay off. The events are described in an extraordinary series of letters from the publisher Manuil Alekseevich Khan (who signed himself off when writing in German as Hahn), a third key player in the production process, to Sementovskii. Reconstructing the timeline gives us a vivid insight into social relations of map production and the complex logistics of cartographic printing and publishing at the time.[123] Printmaking was a collaborative endeavour, yet as Karen F. Beall reminds us, 'to a great extent it is [also] a business story'.[124]

Sementovskii sent the finished manuscript of the edited volume and ethnographic map to Kaspari in June 1871 to begin production. In late November Kaspari subcontracted the printing of the manuscript to Khan, a medical doctor who ran a publishing business on the side, and Kaspari paid Khan a deposit of 300 roubles for the work.[125] However, in March 1872 Khan wrote to Sementovskii to complain that he had still not received the drawings or paper from Kaspari, despite sending reminders every two days. Frustration mounting, Khan even tried sending a letter at 7am when he expected Kaspari to be in his bed, but he heard from the messenger that Kaspari was very rarely home. Unbeknown to

[121] NHAB f. 2502, vop. 1, spr. 198, ark. 22-22adv.

[122] On consumer culture in late imperial Russia, see Rowley, *Open Letters*; Marjorie L. Hilton, *Selling to the Masses*.

[123] Withers, 'On Trial—Social Relations of Map Production'.

[124] Beall, 'The Interdependence of Printer & Printmaker': 195.

[125] NHAB f. 2502, vop. 1, spr. 198, ark. 28.

Khan and Sementovskii, Kaspari had declared bankruptcy. Kaspari's financial troubles were not unusual in the context of the competitive marketplace of the printing business in St. Petersburg at the time, where printing houses operated with low profit margins and firms were frequently short-lived.[126] However, Khan was concerned about whether the drawings, map, book covers, and paper could be recovered from the bust printing house.[127] Khan asked the Vitebsk Statistical Committee to send 230 roubles to allow him to buy more paper and to complete the engraving work, promising to finish within the next twenty days.[128]

On 4 April, the contents of Kaspari's workshop in St. Petersburg were auctioned off to pay his creditors. Khan went along to the auction in the hope of buying the engraved lithographic stone for the ethnographic map. Kaspari was there and attempted to hurry away to his carriage when he saw Khan, but Khan managed to confront him. Kaspari promised to send him the already printed black and white sheets of the base maps, and reassured Khan that the lithographic stone for the map had not gone to the auction. Kaspari reassured Khan that as his printing press was expensive and as nobody had bought it at the auction, he could finish printing the map. Khan remained sceptical about the veracity of Khan's promises since he doubted that the creditors would have allowed Kaspari to keep his valuable printing press and Kaspari was still hiding from the police and the people he owed money to.[129] To complicate matters further, Khan's business affairs were also unstable and on 18 April Khan informed Sementovskii that he had sold his publishing business. However, Khan appears to have continued to manage the project, asking Sementovskii to address all future correspondence him personally as Doctor M. Khan rather than at his business address.[130] Khan was probably eager to ensure that the project was completed since he was still owed the remaining balance for his work and would not be paid until the final product was delivered.

By early May 1872 all Kaspari's property had been sold off to pay his creditors and his shop lay empty and shuttered. Khan managed to get hold of the lithographic stone and the black and white printed maps.[131] However, without the original manuscript of the ethnographic map he was unsure how to proceed. Kaspari himself was nowhere to be found. Meanwhile, Sementovskii wrote to the Ministry of Interior to ask the St. Petersburg police to retrieve the invoice for the work completed by Kaspari.[132] To everyone's surprise, Kaspari appeared on 11 May to hand over the original ethnographic map to Khan. Together, Khan and Kaspari then went to pay a visit to Arnold, a Viennese printer who had set up

[126] Steinberg, *Moral Communities*: 20.
[127] NHAB f. 2502, vop. 1, spr. 198, ark. 37-38adv, 40-40adv.
[128] NHAB f. 2502, vop. 1, spr. 198, ark. 41.
[129] NHAB f. 2502, vop. 1, spr. 198, ark. 42-44adv.
[130] NHAB f. 2502, vop. 1, spr. 198, ark. 46adv.
[131] NHAB f. 2502, vop. 1, spr. 198, ark. 51–53adv.
[132] NHAB f. 2502, vop. 1, spr. 198, ark. 48.

shop in St. Petersburg, to see about chromolithographing the ethnographic map. Arnold quoted a price per colour of 17 roubles, meaning the total cost of printing the ethnographic map with seven colours, including paper, would amount to 129 roubles. However, Arnold proposed that the price could be reduced to 95 roubles by limiting the number of tones to five.[133] Moreover, on 13 May 1872 the St. Petersburg police sent the Vitebsk Statistical Committee copies of Khan's receipt and Kaspari's invoice for the work completed before the firm went bankrupt, revealing that more than half of the budget had already been spent.[134]

By the second half of May work on the publication had resumed but the ethnographic map in particular continued to give Khan troubles. Arnold was busy and estimated that the job would take six to eight weeks. Khan visited several other lithographers, but he could not find anyone who could do the work in less than two months.[135] Khan also raised concerns about the level of knowledge among the city's many lithographers of how to print a chromolithographed map, noting that 'their carelessness gave me doubts about their thorough knowledge of the matter. I am afraid that they will print the colours inaccurately and thus spoil the whole map.'[136] Khan's concerns were valid as printing houses habitually cut costs by using cheaper materials, the quality of the work was highly variable, and not every lithographic firm had the skill and experience necessary to take on such a complex printing job.[137]

Khan reported to Sementovskii that he was dissatisfied with Arnold's work and that he had found another business, the lithographic firm of R. Cillis, who promised to print the map with seven colours within a month for 170 roubles. However, like Arnold, Cillis was hesitant about taking on a map with seven colours and warned that it will be 'difficult and dangerous' to chromolithograph and he 'cannot guarantee accuracy'.[138] Moreover, Cillis pointed out that the black and white printed base maps prepared by Kaspari were not suitable for chromolithographing as they did not contain the dot markers required to line up the lithographic stones for accurate overlaying during the printing process. Khan's worries about the poor quality of the lithography work were confirmed when Cillis advised that manually aligning each map would be a 'very laborious process' and in his professional opinion the engraving error could not be easily corrected on the base maps.[139] Instead, Cillis proposed to add the alignment points to the lithographic stone of the base map and print the entire map again, arguing that it would be wiser to pay more now to have the mistake corrected than 'risk an incorrect map'.[140]

In early June, Khan confirmed that Cillis was 'working diligently' and was on track to complete the maps by 15 June. Khan arranged for the bookbinder to also

133 NHAB f. 2502, vop. 1, spr. 198, ark. 54.
134 NHAB f. 2502, vop. 1, spr. 198, ark. 49–50.
135 NHAB f. 2502, vop. 1, spr. 198, ark. 59adv.
136 NHAB f. 2502, vop. 1, spr. 198, ark. 60.
137 Steinberg, *Moral Communities*: 20.
138 NHAB f. 2502, vop. 1, spr. 198, ark. 60.
139 NHAB f. 2502, vop. 1, spr. 198. ark. 57.
140 NHAB f. 2502, vop. 1, spr. 198. ark. 60adv.

finish by 15 June 'so all that will be required is to paste the maps into the books'. He asked Sementovskii to send him 1,191 roubles to settle the account.[141] In mid-June Khan sent the proofs of the book and a colour copy of the ethnographic map to Sementovskii. Despite earlier reservations about the number of colours, Khan was pleased with the result of the map and noted that 'It seems very well done'.[142] However, he still had not received payment and asked Sementovskii to urgently send the money to pay Cillis. The final bill for the project came to a staggering 1,421 roubles, equivalent to twice Sementovskii's annual salary or almost one and a half times the annual publication budget of the Vitebsk Statistical Committee's activities. On 11 November, Khan complained that he had not received payment and threatened to write to the governor of Vitebsk province and, failing that, to the Ministry of Interior.[143]

The correspondence between Sementovskii and the St. Petersburg printers dries up at this point. However, the volume of articles was eventually published in on 25 September 1872 with the appended ethnographic map of Vitebsk province.[144] These altercations between authors, publishers, and printers seem to have been fairly typical of publishing processes at the time. The debacle with Kaspari's bankruptcy, encounters with lithographic printing firms of dubious quality and repute, a missing manuscript, incorrect lithographic stones, and the prolonged back and forth between multiple private printing enterprises and craftsmen did not dissuade the conference attendees from compiling a second volume in September 1872, which was edited by Sementovskii's colleague, Aleksandr Gatsinskii, secretary of Nizhegorod Statistical Committee.

The process of printing the *Ethnographic Map of Vitebsk Province* took place against the backdrop of boom and bust on the St. Petersburg printing scene. Small privately owned printing enterprises sprung up across the city and jostled for business. As their names suggest, Khan (Hahn), Kaspari, Arnold, and Cillis were all non-Russians, either foreign-trained or children of immigrants, who together owned a disproportionate swathe of the printing industry. In 1875, 23 per cent of printing businesses in St. Petersburg were owned by foreign nationals and both Kaspari and Cillis corresponded with Khan in German.[145] Although Sementovskii's name appears on the cartouche and the name of Cillis's business is listed in the map's bottom righthand corner, the map passed through the hands of many others at various stages. The process of printing a map did not simply involving reproducing a manuscript in print form but was a collaborative endeavour. Printers played a key role in shaping the outcome of the final product.[146] Cillis's

[141] NHAB f. 2502, vop. 1, spr. 198, ark. 65.
[142] NHAB f. 2502, vop. 1, spr. 198, ark. 68.
[143] NHAB f. 2502, vop. 1, spr. 198, ark. 31.
[144] Sementovskii, *Sbornik v pamiat'*: 297–72. The chapter on ethnography and the ethnographic map were also published as a separate booklet: Sementovskii, *Etnograficheskii obzor*.
[145] Steinberg, *Moral Communities*: 13.
[146] Beall, 'The Interdependence of Printer & Printmaker'.

intervention to correct Kaspari's error with the engraving of the lithographic stone likely impacted the precision of the colour alignment of the different ethnographic groups depicted on the map. Arnold and Cillis both questioned Sementovskii's decision to commission a seven-colour map, cautioning that from a technical and financial point of view a map with five colours would be better. Here were see the diverging viewpoints between statisticians and bureaucrats, on the one hand, and printers, on the other hand, about what makes a "good" map. The story of the map production process also tells us about how the printing process shaped the materiality of the final published map. The ethnographic map had to be printed separately from the text and on a different size and type of paper. It was then bound at the end of the book, a practice which was 'suited to convenience of the binder rather than the needs of the author'.[147] Throughout the process Sementovskii relied on his brother who lived in St. Petersburg to inspect the work being done in the printing shops and act as his proxy when decisions had to be made on the spot.

Conclusion

Ultimately, only one of the thirty-six maps planned for the atlas of the North-Western Territory was ever printed and published. This brings into perspective the gulf which existed between cartographic aspirations and the practical realities of mapmaking. Considering the multiple stumbling blocks encountered by Sementovskii, it is remarkable that the *Ethnographic Map of Vitebsk Province* was eventually published and in a staggering 940 copies. As one Vil'na newspaper remarked in praise of Sementovskii's work: 'His activities with the most meagre means, fruits abundant in quantity and quality, are gratifying and instructive, especially since the Vitebsk circle (*kruzhok*) receives neither salaries in the thousands, nor special sums nor grants for publication, as is seen in some scientific institutions.'[148] Paying attention to these often seemingly mundane bureaucratic activities of the provincial statistical committees serves as an entry-point to push our understanding of ethnographic mapmaking further. The story of the *Ethnographic of Map of Vitebsk Province* opens a window onto several dimensions of the process of making ethnographic maps which are worth reflecting on.

Firstly, the mapmaking activities of provincial statistical committees challenge the notion that there was a well-defined centre of ethnographic cartographic production in the Russian Empire. The process of making ethnographic maps not only produced geographical knowledge about the empire but was itself also a mobile enterprise that spanned different parts of the imperial space. The

[147] Rudwick, 'The Emergence of a Visual Language': 155.
[148] 'Vitebskie Gubernskie Vedomosti za 1868.' *Vilenskii Vestnik*, no. 32 (1869).

secretaries of the provincial statistical committees sent out circulars to gather information from local correspondents. Letters, manuscripts, payments, and proofs chuntered back and forth along the newly built railway line connecting Vitebsk to St. Petersburg via Dinaburg/Dvinsk (Daugavpils). Examining mapmaking at the provincial scale also draws our attention to the uneven development of different stages in the mapmaking process at the provincial scale. While the statistical committees were well-positioned to gather data, interpret its local nuances, and use the services of local draughtsmen to produce manuscript maps, the provincial infrastructure for cartographic printing and publishing lagged behind. The dependence on cartographic printing firms in the imperial capital added an additional dimension of contingency to the map production process.

Secondly, mechanisms of knowledge-gathering in the Russian Empire were less coherent and vertical than has often been suggested. Despite coordinated efforts to standardize ethnographic methods, terminology, recording processes, and the cartographic symbology for mapping people in different provinces at both the imperial and regional scale, the activities of the administrative provinces continued to be an important frame of reference for thinking about ethnographic groups. The provincial statistical committees were highly adaptable to local circumstances and their work reinforced the idea of the province as a spatial container for thinking about the relationship between ethnic groups and space. The combination of regional expertise and close collaboration with local informants enabled the statistical committees to draw attention to specific characteristics of the population in their province and reflect on them in the light of the categories devised at regional or imperial centres.

Thirdly, ethnographic map production at the provincial scale emerged through a combination of state patronage from the imperial government and regional administration together with the free enterprise of private printing firms. The activities of the cash-strapped provincial statistical committees were shaped by funding prerogatives and the instructions they received from the Central Statistical Committee in St. Petersburg or the Governor-General, which drove them towards certain topics and determined publications formats. At the same time, the low success rates of cartography projects which were rarely seen through to completion afforded ambitious provincial bureaucrats such as Sementovskii opportunities to pick up the loose strings and pursue their own interests and research agendas.

Finally, tracing the trajectory of map production processes from proposal to printed material object reveals just how much the nature of an ethnographic cartography project could evolve over the years. Initially conceived as a part of a large thematic regional atlas initiated by the Governor-General of the North-Western Territory in March 1869, the final published version of *The Ethnographic Map of Vitebsk Province* in late 1872 ended up looking very different as a single fold-out map in an edited collection aimed at a wider audience of statisticians and imperial learned society. Wavering commitment from contributors and financial constraints meant that the map underwent a radical reframing, transforming it from part of a project closely aligned with imperial policy and regional governance in the borderlands to an instrument to showcase the rich statistical activity taking place in the provinces. The map became detached from the context of the atlas in which it was originally conceived. The ethnographic profile of Vitebsk was not presented in comparative gaze alongside the maps of the other provinces in the empire's North-Western Territory, thereby narrowing the scope from the regional to the provincial. At the same time, the detailed commentary on the map in the form of Sementovskii's article provided much more information about the context and data than if the map had ended up as planned in the atlas. Publishing the map as part of a conference proceedings also meant that the ethnographic map was able to reach a much broader audience and was sold as a more widely distributed commercial publication. Sementovskii anticipated that 'the book will find itself not only on the table of a scientist, but also in the library of every educated person, everyone who sympathizes with science in the fatherland'.[149]

The bureaucratic context of ethnographic map production in the imperial Russian provinces, which privileged the administrative province as a spatial unit for thinking about ethnographic groups, did not go unchallenged. In the second half of the nineteenth century, various mapmakers began to counter traditional imperial administrative divisions of territory and experimented with ethnographic mapping across different scales. These developments prompted shifts in trends in ethnographic mapping in the Baltic provinces, as mapmakers gradually consolidated the idea of Estonian-speakers in Estland and northern Livland as a single collective, and of Latvian-speakers in southern Livland, Kurland, and sometimes also in western Vitebsk as a single Latvian ethnolinguistic group. This transformation in ethnographic mapping was accompanied by changing audiences for maps and the commercial demands of advertising and selling maps to these new consumers, developments to which we now turn to in the next two chapters.

[149] NHAB f. 2502, vop. 1, spr. 198, ark. 24adv.

3

The Baltic Question in Cartographic Imagination

'Livland and Afghanistan! Were there ever two less comparable lands? What is common to both the Hindu Kush and the Peipus [lake], or to both the horrors of the Kyber Pass and the steamboat-furrowed entry to the mouth of the [river] Düna?' The Baltic German lawyer and publicist Waldemar von Bock posed these thought-provoking questions to readers in the opening of his book, *The German-Russian Conflict on the Baltic Sea* (1869). Bock went on to explain the logic of his comparison:

> And yet the point of comparison is not quite as far away as it should seem at first sight! Because, in light of comparative geography, Afghanistan, at the gates of India, and Livland, in front of the gates of Germany, are considered typically one and the same; a historically significant point of encounter between the Russian peoples and the Germanic tribes.[1]

For Bock, the ongoing rivalry between the Russian and British Empires over influence in Central Asia meant that the Baltic provinces and Afghanistan were both spaces of encounter and conflict between Slavic and Germanic peoples.[2] 1869 was an important year for thinking about the Baltic provinces in terms of a more expansive 'comparative geography', as advocated by Bock. Pan-Slavicist Nikolai Danilevskii published his milestone work on *Russia and Europe* (1869), which presented the Germanic and Slavic peoples as representatives of two distinct civilizations in political competition for territorial control and dominance over other groups in the east and west.[3] That same year, debates about Russian imperial rule in the Baltic provinces and the role of the region's Baltic German elites also came to a head in the heated polemics between Slavophile Iurii Samarin and Baltic German academic Carl Schirren.[4]

[1] Von Bock, *Der deutsch-russische Konflikt*: 19.

[2] On rival Russian and British imperial ambitions in Central Asia, see Morrison, *The Russian Conquest of Central Asia*: 10–13.

[3] Danilevskii's text was serialised in the journal Zaria before being published as a book in 1871, see Danilevskii, *Rossiia i Evropa*. On Danilevskii, see Sdvižkov, 'Russian and German Ideas of the West': 101.

[4] Schirren, *Livländische Antwort*; Samarin, *Otvet G. G. f. Bokku i Shirrenu*.

Geographies of Nationhood: Cartography, Science, and Society in the Russian Imperial Baltic. Catherine Gibson, Oxford University Press. © Catherine Gibson 2022. DOI: 10.1093/oso/9780192844323.003.0004

These discussions about the Baltic provinces as a space of overlapping Pan-Slavic and Pan-German influence formed part of a range of social, political, economic, and national 'questions' that crystallized in intellectual consciousness and gained prominence in the second half of the nineteenth century. The Polish Question, Muslim Question, Jewish Question, Armenian Question, Peasant Question, Woman Question, and many others, all articulated problems of imagining and governing diverse groups of peoples and territories in the context of a modernizing imperial society.[5] The Baltic Question was one of the early questions to gain traction in the Russian Empire in the mid-nineteenth century in conjunction with discussions among imperial administrators and intellectuals about the degree of regional self-governance that should be permitted, the best way to modernize the empire and abolish serfdom, and the privileges and influence of the Baltic German elites. Over the course of the second half of the nineteenth the Baltic Question ballooned to encompass far-ranging social and political issues concerning the cohesiveness of the empire, the loyalty of transimperial borderland populations, and Germanic and Slavic spheres of influence.

Intellectuals in the Russian Empire were by no means unique in formulating critiques of modern society as questions. As Holly Case argues, the nineteenth century was an 'age of questions' in which publicists engaged in a similar pattern of thinking that framed contemporary issues as problems to be solved. As Case notes, 'the nineteenth-century drive to settle or solve these questions reveals something essential about them: they were construed as problems. The "question" had become an instrument of thought with special potency, structuring ideas about society, politics, and states, and influencing the range of actions considered possible and desirable.'[6] By formulating critiques of social reality as questions, the rhetorical floodgates were opened for people to respond with answers. Yet, as the prominent Slavophile publicist Ivan Aksakov observed, it is much easier to ask questions than it is to propose solutions: 'We are all very prone to raising questions... we have a whole forest of questions, so you can get confused. And answers?... There are almost none, or very few.'[7]

Controversies over questions not only filled the pages of newspapers, books, and pamphlets, but also fuelled mapmaking. Mapmakers weighed in on these debates, contributing spatial insights to these contemporary issues and experimenting with different methods of data visualization to communicate possible solutions. Compared to other discursive mediums used to debate the Baltic Question, mapping brought spatial perspectives to these polemics and gave visual

[5] On these various questions in imperial Russia, see Stites, *The Women's Liberation Movement in Russia*; Klier, *Imperial Russia's Jewish Question*; Campbell, *The Muslim Question*; Riegg, *Russia's Entangled Embrace*.

[6] Case, *The Age of Questions*: xiv.

[7] This quotation comes from the essay entitled 'O bratstvakh v Zapadnom krae' in Aksakov, *Sochineniia*: 333–4.

and material expression to territorial claims. Thematic cartography, especially ethnographic mapping, was an especially effective medium for debating questions. The explosion of thematic mapping in the second half of the nineteenth century paved the way for what Susan Schulten describes as 'a more argumentative concept of cartography'.[8] As Schulten notes: 'topography and borders were secondary to hypotheses, problems, or themes.... If traditional topographic maps were akin to description, thematic maps functioned more like an argument, and this made them particularly relevant for science, social science, education, and governance.'[9]

This chapter traces how ethnographic mapmaking in the age of questioning led to several key shifts in how mapmakers conceived of their work. Firstly, it demonstrates that whereas earlier generations of mapmakers, such as Peter Köppen, used ethnographic maps to create visual representations of the sum of all accumulated knowledge, by the 1870s mapmaking was more explicitly understood as a process of communicating data-driven arguments and—crucially—of using visual analysis to inform future policies towards these issues. In the context of this interrogative turn in ethnographic cartography, mapmakers became visual 'querists' and used their maps to construct compelling cartographic narratives to persuade their audiences of how to see, think—and ultimately, act—with respect to the Baltic provinces.[10] Secondly, examining this age of questioning through the prism of cartographic visual and material culture draws our attention to how space and territory emerged as key concepts in the history of questioning. In an era described by Deborah Coen as a time of 'restless' scalar imagination, mapmakers put pen to paper to interrogate 'relations between the individual and the state, between nation and empire, between the small scale and the large'.[11] Approaching these big questions of the day from the perspective of cartography shaped how these issues were understood and the nature of the conclusions reached. Notably, mapmakers often looked for answers by examining the issue at stake across different scales.

As scholarship in the field of critical cartography has demonstrated, much of the persuasive power of maps comes from the crafting of the map design and cartographic aesthetics.[12] Mapmakers decide how to make space legible, how to direct audiences' attention to certain features of the map to drive home a key message, and how to make their arguments resonate affectively with viewers. By changing symbology, colour, toponymy, and scale the same data could be

[8] Schulten, *Mapping the Nation*: 8. [9] Schulten, *Mapping the Nation*: 2.

[10] I follow Case in thinking about the mapmakers who contributors to these debates as 'querists': Case, *The Age of Questions*: 4.

[11] Coen, 'Big Is a Thing of the Past': 314–15.

[12] Classic works on how maps are social constructions include Monmonier, *How to Lie with Maps*; Wood, *Rethinking the Power of Maps*; Harley, *The New Nature of Maps*; Crampton, 'Maps as Social Constructions'; Pickles, *A History of Spaces*.

transformed into radically different images of the world. This chapter pays particular attention to how mapmakers in the second half of the nineteenth century experimented with various elements of cartographic design to tell different stories about the Baltic provinces. It departs from the focus on the lives of individual mapmakers and processes of map production examined in the previous two chapters to offer a more synoptic overview of diverse approaches within this argumentative and analytical turn in ethnographic mapping in response to the big questions of the age.

The Origins of the Baltic Question

The Baltic Question emerged in the middle of the nineteenth century in conjunction with discussions about modern society in imperial Russia. In the wake of the failed Decembrist uprising of 1825, the empire's intelligentsia began to evaluate the legacy of Peter the Great's 'Westernizing' reforms of the early eighteenth century and their impact on Russian identity and Russia's place in the world. The particularities of imperial rule in the Baltic provided intellectuals in the Russian Empire with an arena to engage with broader debates about the essential character of Russia and the principles by which the empire should be governed. On the one hand, the influence of Western Romanticism created a 'fascination with foreignness' among the Russian intelligentsia.[13] The Baltic provinces were portrayed in political discourse, travelogues, and literature as a German, and thus very European, part of the empire. Intellectuals studied the Baltic provinces as a site of social and political experimentation following the abolition of serfdom in the Baltic provinces under Tsar Alexander I. Writers and tourists wrote lavishly about the beautiful landscapes, architecture, and their encounters with the culturally foreign Baltic German society and traditions. The perception of the Baltic provinces as part of the West was used to present Russia as a truly European empire. On the other hand, Romanticism and growing national sentiment led a group of thinkers known as Slavophiles to formulate a new idea of Russia as a unique national culture. They argued that Russian society and values were fundamentally opposed to the West and that the Petrine reforms were the point at which Russia had been led astray.[14] The Baltic provinces reaffirmed anxieties about the Russian Empire's ethnocultural diversity as a threat to the Russian nation and was increasingly presented as an "Other".[15] The official use of German as the language of administration, strong presence of the Lutheran faith, special code of laws, and the separation

[13] Brüggemann, 'The Baltic Provinces': 118. On the fascination with the Baltic provinces as a culturally foreign region in the Russian Empire, see pp. 116–24.

[14] On Slavophile thought, see Aizelwood, 'Revisiting Russian Identity'; Rabow-Edling, *Slavophile Thought*.

[15] Brüggemann, 'The Baltic Provinces': 124–33.

of the four noble corporations of Kurland, Livland, Estland, and Oesel (Saaremaa) from the All-Russian Table of Ranks, were all perceived as extensions of 'the West' into the borders of the Russian Empire. The particularities of the Baltic provinces forced the empire's intelligentsia to confront the question of how to justify Russian rule over non-Russian populations. Thus, as Karsten Brüggemann argues, the Baltic provinces presented a key dilemma for Russian intellectuals: 'should the Baltic be loved as "our West" or hated as a potential threat?'[16]

The mass conversions of Livlanders to Orthodoxy in the 1840s both galvanized perceptions of the Baltic provinces as a fundamentally Russian territory and contributed to a growing 'Baltic German Russophobia'.[17] Prior to the 1840s, the Orthodox presence in the Baltic province had been weak, limited to provincial capitals and main towns where civil servants, soldiers, and merchants lived. However, in the second half of the 1840s around 110,000 Estonian and Latvian peasants in Livland voluntarily converted from Lutheranism to Orthodoxy, bringing the percentage of Orthodox up to 17 per cent of the province's population.[18] In subsequent decades, there were several smaller waves of conversion to Orthodoxy in Livland between 1880–2, Kurland in 1883–6, and north-west Estland from 1883–5.[19] Historically, these mass conversions were attributed to the peasants' efforts to obtain material and political benefits from the imperial government by adopting the 'tsar's faith', as well as religious dissatisfaction with the Lutheran Church.[20] Contemporary historians tend to see the conversions as arising from a combination of factors. Although material and religious concerns may have provided initial motivation, conversion to Orthodoxy was increasingly used by peasants as a form of social protest against their Lutheran landlords and as a means to explore alternatives to the dominant German cultural and socioeconomic influence in the region.[21]

The conversions to Orthodoxy were a gamechanger for Russian perceptions of the Baltic provinces. For Slavophiles, the conversions to Orthodoxy were a sign of the supposedly natural and deep-rooted historical links between the Baltic littoral and the Russian world. The Estonian and Latvian populations were perceived as have a close cultural affinity with their fellow Orthodox Slavs. This viewpoint was encapsulated in the polemical writings of Iurii Fedorovich Samarin, an incendiary young conservative aristocrat and low-ranking official of the Ministry of Interior who played a key role in conceptualizing the Baltic Question within the wider Slavophile debate.[22] Samarin's first-hand experiences of working as an

[16] Brüggemann, 'The Baltic Provinces': 122.

[17] Von Pistohlkors, 'Russifizierung'.

[18] Bertash, 'Pravoslavnoe tserkovnoe stroitel'stvo': 38.

[19] Rebane, *Usuwahetuslik liikumine Põhja Eesti*; Ante, 'Popytki ukoreneniia pravoslaviia'.

[20] Kruus, *Talurahva käärimine*; Von Jannau, *Vene-õigeusu tulekust*; Gavrilin, *Ocherki istorii Rizhskoi Eparkhii*: 73–182.

[21] Ryan, *The Tsar's Faith*, 144. For scholarship building on Ryan's arguments, see Brüggemann, 'Imperiale und lokale Loyalitäten'; Tøllefsen & White, 'Navigating an Orthodox Conversion'.

[22] For an account of Samarin's life, see Nol'de, *Iurii Samarin*.

imperial official in Riga during the time of the conversions, coupled with the appointment in 1848 of a pro-German Governor-General of the three Baltic provinces, Prince Aleksandr Suvorov, incited Samarin into writing a polemical book-length manuscript, *Letters from Riga* (1848).[23] In the letters, Samarin attacked the privileges of the Baltic German nobles, arguing that the medieval Teutonic Order had taken the Baltic lands from the Russians by military conquest, converted the population from Orthodoxy to Catholicism, and subjugated the local inhabitants. For Samarin, Russian rule in the Baltic was natural, legitimate, and based on centuries of harmonious close contact and cultural-religious kinship with the native Estonian and Latvian populations. German rule, by contrast, was artificial and colonial. Samarin criticized the Baltic Germans for being self-serving and disloyal imperial subjects for their refusal to recognize full imperial sovereignty and use Russian language. Samarin concluded with a strong statement about the need for the imperial government to bring the Baltic provinces closer to the other parts of the Russian Empire with a view to its eventual 'merging' (*sliianie*) with Russia proper.[24]

Despite the censorship restrictions, which meant that Samarin's letters circulated in manuscript form as a kind of imperial samizdat, they were widely read by high-ranking officials and educated elites in Moscow and St. Petersburg in 1848–9. However, Samarin's recommendations for a tougher imperial policy towards the Baltic provinces largely fell on deaf ears. While Tsar Nicholas I endorsed the ideology known as 'Official Nationality', derived from the formula of 'Orthodoxy, Autocracy, Nationality' developed by the Minister of Public Education Count Sergei Uvarov, Nicholas perceived the Slavophile brand of cultural nationalism as a threat to the stability of imperial authority in the Baltic.[25] Moreover, in the context of the recent revolutionary uprisings that spread across Europe in 1848 and the involvement of the Russian imperial army in stamping them out in nearby territories, especially in Prussian Poland and Habsburg Hungary, Nicholas was wary of threats to his rule. In March 1849, Samarin was arrested and personally reprimanded by the tsar for fanning the flames of the contentious question of the relationship between Russia and the Baltic.[26]

In the 1860s, the Baltic Question returned to the spotlight once again in the context of the liberal transformations in the era of the Great Reforms. Russia's defeat in the Crimean War (1853–6) and the ascension of the more liberal Tsar Alexander II to the throne in 1855 prompted a discussion about the gradual

[23] The letters were later published in 1889 as part of a series of volumes of Samarin's works: Samarin, *Sochineniia, vol. 7.*

[24] For a more in-depth discussion of Samarin's *Letters from Riga* as a key text for the development of Russian discourses on the Baltic provinces, see Brüggemann, *Licht und Luft*: 199–207.

[25] On the imperfect translation of *narodnost'* as 'nationality', see Miller, 'Natsiia, Narod, Narodnost'': 154.

[26] Thaden, 'Iurii Fedorovich Samarin'; Miller, 'Official Nationality': 382–4.

introduction of reforms to address social and economic problems in the empire, including the abolishment of serfdom in 1861. Although the Great Reforms of the 1860s were less of a watershed in the Baltic provinces as serfdom had already been abolished between 1816 and 1819 during the reign of Tsar Alexander I, the Baltic garnered attention in the context of debates about peasant and socioeconomic questions. The early 1860s witnessed a flood of newspaper articles, brochures, and pamphlets published on the Baltic Question in the Russian Empire's emerging public sphere.[27] Nationalist newspapers such as *Moskovskie Vedomosti* and *Den'* regularly featured articles that condemned the Baltic Germans for exploiting the Baltic peasantry for their own economic gain as the manumitted peasants had been given neither land nor the right to mobility, which had given the peasants little opportunity to engage in independent agriculture.[28] The press critiqued the dire conditions of the Baltic peasantry, pointing to the multiple harvest failures and famines that occurred in all three Baltic provinces in 1841–7 and in Estland in 1868–9.[29]

The debate peaked in 1868–9 when Samarin returned to the public stage and published a six-volume work entitled *The Borderlands of Russia* (1868). The first volume was devoted to the Baltic provinces or, as Samarin termed them, the 'Russian Baltic seaboard (*Russkoe Baltiiskoe pomorie*)'.[30] Samarin shifted away from philosophical arguments about the nature of the Russian state that characterized his early writings to engage directly with pressing political issues. The Uprising of 1863–4 in the North-Western Territory against Russian imperial rule and the prospect of German unification brought international dimensions to the debates. Naval activities in the Baltic Sea during the Crimean War had exposed the military-strategic vulnerability of the Baltic provinces, which came to be perceived as a weak spot in the Russian Empire and susceptible to German conspiracy and foreign intervention.[31]

Samarin's arguments against the degree of self-rule in the Baltic provinces and the privileges of the Baltic German nobility did not pass without criticism. In 1869 Carl Schirren, a Baltic German Professor of Russian History at the University of Dorpat (Tartu), published a polemical response to Samarin.[32] The text, published in Leipzig, was directed against the actions of the Russian imperial government in the Baltic provinces, the Slavophiles, and Samarin personally. Schirren drew on historical sources to argue that the Baltic provinces had not been conquered by the Russian Empire in the early eighteenth century during the Great Northern

[27] On the discourse on the Baltic Question in the press, see Isakov, *Ostzeiskii vopros*; Majorova, 'Die Schlüsselrolle der "deutschen Frage"'; Andreeva, 'Die "baltische Frage"'; Karjahärm, 'Ostzeitsy i baltiitsy'.

[28] On the situation of the Baltic peasantry in the wake of the abolition of serfdom, see Lust, 'The Impact of the Baltic Emancipation'.

[29] Aksakov, *Polnoe sobranie sochinenii*: 3–167. On the famines, see Lust, 'Feeding the Landless'.

[30] Samarin, *Sochineniia: Okrainy Rossii*.

[31] Martin, 'The German Question in Russia': 392–3; Thaden, 'Samarin's "Okrainy Rossii"': 411.

[32] Schirren, *Livländische Antwort*.

War with Sweden but had voluntarily allowed themselves to be annexed based on legal agreements drawn up between the German-speaking nobility and Peter I guaranteeing their rights in perpetuity. Schirren also sought to demonstrate that German culture was superior to Russian culture and, consequently, that Russians did not deserve to rule over Germans. Samarin responded by openly refuting Schirren's claims and expressing his incredulity that such an outspoken pro-German and anti-Russian scholar could hold a university chair in Russian History financed by the imperial government.[33] As a result of the polemic, Schirren was dismissed from his post and left the Russian Empire to take up a professorship at the University of Kiel. The fiery back and forth between Samarin and Schirren encapsulated the volatility of the Baltic Question and the high stakes of such debates about how the Russian Empire should exercise imperial power in its multiethnic borderlands.

Baltic Russians and Baltic Orthodoxy: Mapping Minorities as a Strategy of Imperial Legitimation

Coincidental with the heated polemics in the literary sphere, mapmakers reflected on the Baltic Question through the visual medium of cartography. Aleksandr Fedorovich Rittikh (1831–1914), born Alexander Friedrich Rittich, played a particularly influential role in using ethnographic cartography to legitimate Russian imperial rule in the Baltic. Together with Peter Köppen, Rittikh is remembered as one of the most well-known and prolific nineteenth-century ethnographic mapmakers in the Russian Empire. The son of a doctor from Riga, Rittikh was educated in St. Petersburg at the Nikolaevskaia Engineering Academy of the General Staff. As a young engineer, Rittikh worked from 1862 to 1864 in Minsk province to oversee the construction and repair of Orthodox churches and schools. Later, Rittikh served in Kazan province and in the Balkans and Caucasus during the Russo-Ottoman War (1877–8). By the time he retired in 1894, Rittikh had ascended to the rank of Lieutenant-General. Alongside his military career, he was a member of the Imperial Russian Geographical Society and developed a keen interest in ethnographic mapmaking. Like his ethnographic mapmaking predecessor, Köppen, Rittikh was another of the Russian Empire's 'Russian-Germans'. Yet, unlike Köppen who harmonized his sense of belonging to German culture with loyalty and service to the empire, Rittikh's experiences of serving in the imperial army and administration led him to become fully assimilated into Russian culture. In his adult life, Rittikh's ties to his Baltic German ancestry were in name only. Moreover, whereas Köppen published in German, French, and Russian,

[33] Samarin, *Otvet G. G. f. Bokku i Shirrenu.*

Rittikh published almost exclusively in Russian. Hence, I choose to refer to Rittikh using the Russian version of his name that he used during his lifetime.

Rittikh's efforts to map various imperial border regions reveal how ethnographic cartography was used by conservative Russian nationalists as an instrument of 'internal colonization' to consolidate Russian imperial rule over its own territorial possessions and peoples.[34] Rittikh's attitude towards the Baltic provinces was heavily influenced by his experiences of ethnographic mapping in the North-Western Territory. In the wake of the 1863–4 Polish-Lithuanian uprising, ethnographic maps accrued value in the eyes of imperial administrators as a tool for gathering knowledge about potentially seditious borderland populations. As mentioned in Chapter 2, the imperial administration used maps to inform policies, often perceived as 'Russifying' endeavours, aimed at eroding Polish influence in the region and strengthening the Russian rule.[35] Rittikh was a key player in these government-sponsored mapping initiatives through his work on the *Atlas of the Population of the West-Russian Territory According to Confession* (1864).[36] The atlas depicted the population of this region as mostly Orthodox, a judgement arrived at by mapping Old Ritualists (*staroobriadtsy*) and former Greek Catholics (Uniates) as a 'single tribe with the general mass of Orthodox Russians', despite the fact that these faiths were condemned by the Orthodox Church and discriminated against by the Russian imperial state.[37] By contrast, Catholics were limited to Kovna, western Vil'na, Vitebsk, and Grodno provinces. Thus, the atlas made a strong statement about the religious kinship of the inhabitants of the North-Western Territory with the predominantly Orthodox Russian-speaking territories to the east. As Rittikh noted in the preface, the aim of the atlas was to counter foreign false claims with the 'historical truth' about the empire's western provinces as 'an organic and inalienable part of the Russian state'.[38]

Over the coming years, Rittikh expanded his focus and compiled a series of ethnographic studies and maps of different multiethnic and multiconfessional imperial regions, including the Kingdom of Poland, Lublin and Avgustov provinces (1864), Kazan province (1870), and the Caucasus (1874).[39] It was within the scope of this series that Rittikh presented his proposal in April 1872 to the Ethnographic Section of the Imperial Russian Geographical Society to fund a ethnographic study of the Baltic provinces. Rittikh's proposal included plans for two thematic maps depicting the distribution of ethnographic and religious

[34] On Russia as a 'self-colonizing' empire, see Etkind, *Internal Colonization*.

[35] Staliūnas, *Making Russians*: 88. See also Dolbilov & Staliūnas, 'Vvedenie k forumu'; Dolbilov, *Russkii krai, chuzhaia vera*; Weeks, 'Defining Us and Them'.

[36] Rittikh & Batiushkov, *Atlas Narodonaseleniia*. For a more detailed analysis of the atlas and the political circumstances underpinning its production, see Petronis, *Constructing* Lithuania: 195–9; Psianchin, *Istoriia etnicheskoi kartografii*: 127–30.

[37] Rittikh, *Atlas Narodonaseleniia*, preface.

[38] Rittikh, *Atlas Narodonaseleniia*, preface.

[39] Rittikh, *Prilozhenie k materialam dlia etnografii Tsarstva Pol'skogo*; Rittikh, *Materialy dlia etnografii Rossii: Kazanskaia*. For a discussion of these maps, see Psianchin, *Istoriia etnicheskoi kartografii*: 130–6.

groups in the Baltic provinces. Drawing on his experiences of mapping the North-Western Territory, Rittikh argued that the polemics surrounding the Baltic Question had rested on misconceptions and false information about the region. According to Rittikh, it was imperative that the Geographical Society conduct research on the Baltic provinces to provide scientifically informed answers to the important questions of the time.[40]

In Chapter 1 we saw how the Geographical Society paid considerable attention to the Baltic provinces in the second half of the 1840s, organizing ethnographic expeditions and cartography projects as part of a broader programme of research on the empire's Finno-Ugric peoples and their neighbours. However, in the intervening decades the Geographical Society's attention had gravitated towards other regions in the western part of the empire, especially the North-Western Territory. The shift in research focus was also reflected in the changes to the Society's institutional structure with the opening of the Vil'na branch in 1867.[41] Nevertheless, Rittikh argued that it was important not to neglect the Baltic as a key area for geographical research, especially since the region's Russian-speaking and Orthodox inhabitants had been overlooked in many German-language ethnographic descriptions of the region. Against the backdrop of the lively debates about the Baltic Question in the press, Rittikh convinced the Ethnographic Section of the contemporary relevance of the research and they awarded 300 roubles in funding for the project.[42]

The results of Rittikh's study appeared in a publication entitled *Material for the Ethnography of Russia: The Baltic Territory* (1873). The study included two fold-out maps: a larger map depicting the distribution of different ethnic groups (Plate 7) and a smaller map of confessional groups (Plate 8).[43] The remarkable speed of the project, completed within six months of receiving funding and published within a year, hints at how Rittikh's approach to mapmaking was neither as thorough nor as concerned with conforming to contemporary ideas about scientific cartographic methods as Peter Köppen's work for the Geographical Society in the 1840s. Whereas Köppen's spent years verifying data and subjecting maps to multiple rounds of corrections as part of a process of striving for an 'objective view' of the geography of the empire's ethnolinguistic groups, Rittikh's publication was rushed out to weigh in on contemporary debates about the Baltic Question.[44] Both mapmakers titled their works 'ethnographic maps', yet the rationales, motivations, and processes of map production underpinning both works contrasted starkly.

[40] 'Zhurnal Zasedaniia Otdeleniia Ethnografii, 8-go ianvaria 1872 g.', *Izvestiia IRGO* 8, sec. 2 (1872): 201–9.

[41] Petronis, *Constructing Lithuania*: 145–53.

[42] 'Zhurnal Zasedaniia Otdeleniia Ethnografii, 8-go ianvaria 1872 g.' *Izvestiia IRGO* 8, sec. 2 (1872): 209.

[43] Rittikh, *Materialy dlia ethnografii Rossii: Pribaltiiskii krai.*

[44] Daston & Galison, *Objectivity*, 17.

Köppen's and Rittikh's diverging approaches to ethnographic mapmaking reflected the dual conceptions of the Baltic provinces within imperial imagination.[45] For Köppen ethnic diversity was a symbol of imperial power, demonstrating the ability of the Russian Empire to unite and govern such diverse populations. Köppen approached the process of making an ethnographic map in the 1840s as a way to gather and communicate knowledge about the diverse, small, non-Russian ethnolinguistic groups (*inorodtsy*) inhabiting imperial Russia's border regions. Köppen rendered the Baltic provinces on his maps as a territory divided into distinct areas inhabited by Estonians, Latvians, Livonians, and coastal Swedes (Plates 2 and 3). The Baltic German and Russian populations were barely visible at all. By contrast, Rittikh's approach to ethnographic mapmaking operated on the reverse logic, viewing ethnic diversity as a threat to imperial power. For Rittikh, mapping was about pinpointing the region's various ethnic and religious minorities to draw wider conclusions about the cultural, political, and socioeconomic forces vying for influence in the Baltic. In Rittikh's hands, ethnographic maps became a tool for assessing the relative loyalty or alleged threat that certain groups and territories posed to imperial governance in the non-Russian borderlands.

The design and aesthetics of Rittikh's ethnographic map (Plate 7) constructed a visual hierarchy of ethnic groups in the Baltic provinces, drawing readers' attention to certain segments of society more than others. The washed-out tones and pale earthy shades of orange, brown, and beige depicting Lithuanians, Samogitians, Latvians, Finns, Livs, and Estonians make these groups visually retreat into the background, as if part of the natural landscape. The lack of contrasting colours makes it difficult to distinguish between these groups. By suggesting that these groups were not especially noteworthy, Rittikh gave visual expression to the views of Samarin and other Slavophiles that Estonians and Latvians were small nationalities who 'obviously were not intended for an independent political development'.[46] As Rittikh pointed out in the text accompanying the maps, the multiple rulers over this territory—Germans, Poles, Swedes, and now Russians—attested to the fact that these small ethnolinguistic groups could not self-govern.

By contrast, Rittikh used bold colours to highlight the presence of groups he considered significant political and socioeconomic players in the region: Germanic peoples (German- and Swedish-speakers), Russians (both Great Russians and Belarusians), Poles, and Jews, who were foregrounded in bright blue, green, pink, and yellow, respectively. Notably, Rittikh marked every place inhabited by Germans as also having a sizeable Russian presence, thereby presenting the Baltic as a region that was inhabited by Germans and Russians in equal measure. The text accompanying the maps reinforced Rittikh's message

[45] On the duality of Russian perceptions of the Baltic, see Brüggemann, 'The Baltic Provinces'. For a further comparison of Köppen's and Rittikh's ethnographic maps, see Gibson, 'Shading, Lines, Colors'.

[46] Samarin, *Sochineniia*, vol. 8: 304–5. Cited in Thaden, 'Samarin's "Okrainy Rossii"': 411.

about the strong Slavic presence in the Baltic littoral. Rittikh cited archaeological evidence that Slavs had inhabited the Baltic littoral for many centuries before the arrival of German 'colonizers'. He pointed to scholarship demonstrating the close linguistic contacts, and thus historical kinship, between Latvians, Lithuanians, and Slavs. As further proof of the 'natural' Slavic orientation of the Baltic provinces, Rittikh highlighted the growth of industry and expansion of the ports which had attracted many Russians for work, as well as the migration of Belarusians to eastern Kurland.[47]

In a similar vein, Rittikh's map of religion (Plate 8) presented evidence of the strong Orthodox presence in rural Livland and Kurland. Lutheranism was the majority religion in the Baltic provinces, yet this religious group was depicted in a pale green cross-hatching that made it resemble the landscape of fields and forests covering the region. Instead, Rittikh directed viewers' attention to the places inhabited by Orthodox and Old Believers (shaded in same dark green colour, thereby making them appear more numerous), Catholics (pink), and Jews (yellow). Following Samarin, Rittikh interpreted the mass conversion movement of Livland peasants from Lutheranism to Orthodoxy in the 1840s as an indicator of the Baltic's cultural and religious orientation towards the Slavic east. He prominently marked the location of Orthodox parishes in Livland and Kurland on the map. In his efforts to augment the Orthodox presence in the region, Rittikh greatly exaggerated the size of these parishes, ignored how these areas were inhabited by a mix of Lutherans and Orthodox, and neglected to account for the fact that by the 1870s many of the converts has subsequently 'lapsed' back to Lutheranism.[48] In the context of the debates between Samarin and Schirren just four years earlier, we can clearly see how Rittikh used ethnographic and religious cartography to exaggerate both the Russian and Orthodox influence to provide additional support for Samarin's views.

Rittikh's ideas about the deep-rooted cultural and historical affinity of the Baltic with the Slavic and Orthodox east also inconspicuously lurked in the maps' toponyms. Rittikh deliberately used Russian-language versions of place names where they existed, such as Libava rather than Libau (today's Liepāja), Mitava rather than Mitau (today's Jelgava), and Revel' rather than Reval (today's Tallinn). The name of present-day Tartu was printed as Derpt, the low German version of Dorpat. However, the name Iur'ev was also added in parenthesis, the ancient Russian name for the city that was only officially reinstated in 1893.

An even more significant renaming process was underway in the maps' titles. Rather than referring to the three provinces collectively as the *Ostzeiskie gubernii*, the common Russian language name used at the time, Rittikh termed the

[47] Rittikh, 'Pribaltiiskii krai i ego naselenie'; Rittikh, *Materialy dlia ethnografii Rossii: Pribaltiiskii krai*: 8–11.

[48] Weber 'Religion and Law'; Gibson & Paert, 'Apostasy in the Baltic Provinces'.

region *Priibaltiiskii krai*. Whereas the former was a Russified version of the German name, denoting the provinces on the 'Eastern Sea' (*Ostsee*) from the perspective of the German-speaking Central Europe, the latter suggested that the region constituted the border territories (*krai*) from the perspective of St. Petersburg and Moscow, on the edge of or near (from the prefix *pri-*) the Baltic Sea. Rittikh's approach was similar to Samarin's use of the term, 'the Russian Baltic seaboard' (*Russkoe Baltiiskoe pomorie*) as the subtitle of the first volume of his book series on Russia's borderlands.[49] Rittikh appears to have been one of the frontrunners in promoting *Pribaltiiskii krai* as a Russified name for the region; although the term *Pribaltiiskii* appeared as early as 1859, it took more than a decade to enter widespread usage.[50]

As Karsten Brüggemann argues, these renaming efforts contributed to the 'mental appropriation of the culturally and ethnically non-Russian region as an integral, if not natural, part of the Russian state'.[51] Rittikh's maps also constitute further evidence that the mental process of discursively Russifying the Baltic provinces was already well underway prior to the 1880s, traditionally characterized as the beginning of policies under Tsar Alexander III aimed to integrate the region more closely with the rest of the empire.[52] At the same time, these efforts to toponymically Russify the Baltic provinces remained at the level of symbolic geographies. The Baltic provinces were never officially designated as an imperial *krai*, a term which was officially applied to the other administrative borderlands of the Russian Empire, such as the North-Western and South-Western regions, Vistula, Caucasus, Turkestan, and *Primorskii krai* in the Far East. Nonetheless, Rittikh's use of the term *Priibaltiiskii krai* in the context of his wider oeuvre of maps would have evoked comparisons among his readers with these other imperial border regions.

In the context of public debates among intellectuals in the Russian Empire about the Baltic Question, Rittikh used ethnographic cartography as an 'evidence-generating tool' to advance his argument about the need to bring the Baltic provinces closer into the fold of the Russian Empire.[53] The ideas communicated in his maps went far beyond the ethnographic and religious themes implied by their titles. Rittikh used ethnicity and religion as proxies for examining questions of shared history, deep-rooted cultural kinship, and imperial loyalty. Rittikh capitalized on long-standing Slavophile anxieties about imperial governance in the borderlands and the 'integrity' of the empire.[54] He presented a solution to this problem in terms of his vision of what a future, more heavily Slavicized and

[49] Samarin, *Sochineniia: Okrainy Rossii*.

[50] Karjahärm, 'Ostzeitsy i Baltiitsy': 16.

[51] Brüggemann, 'The Baltic Provinces': 113.

[52] Brüggemann, 'Als Land und Leute "Russisch" werden sollten': 28.

[53] Güttler, 'Scaling the Period Eye': 19.

[54] Sunderland, 'Imperial Space'; Wortman, 'The "Integrity"'.

Orthodox, picture of the Baltic provinces could look like.[55] By creating an image of the potential strength of Russian imperial influence in the Baltic, Rittikh's maps exemplify how, as Thongchai Winichakul writes, maps could be as much 'a model for, rather than model of, what they purported to represent'.[56] Reading Rittikh's maps as aspirations rather than as reflections of a social reality, we can see how colour and toponyms could be used to offer a 'visual normalization' of a certain set of Slavophile beliefs about Russian imperial rule in the Baltic provinces as the natural order.[57]

Not all Rittikh's readers, however, unquestioningly accepted his attempts to use the veneer of objectivity associated with 'map language' to normalize Russian imperial rule in the Baltic. Rittikh's maps struck a chord in the polarizing debates on the Baltic Question in the 1870s and proved highly controversial for the local German-speaking readership in the Baltic provinces. In April 1873, the newspaper *Rigascher Zeitung* ran a lengthy front-page review over three days condemning Rittikh's publication.[58] The defamatory critique by the anonymous reviewer highlighted the many shortcomings of Rittikh's work, including the lack of transparency with the sources of data, the way in which the numbers of different nationalities were based on 'conjecture' by taking old figures and allowing for population growth, and the interpretation of speculative figures as facts. The reviewer pointed to specific examples of Rittikh's careless research practices, such as the failure to record the almost a thousand Estonians living in Riga because he had not turned over a page when consulting a source.

The critical reception of Rittikh's map in the Baltic provinces went beyond discrediting Rittikh for his lax approach to mapmaking and ignorance of local circumstances. The reviewer read Rittikh's maps as a clear reflection of conservative Russian nationalist views towards the Baltic provinces. The reviewer noted with palpable scathing of how '[Rittikh] apparently wrote about the Baltic provinces, because of—or despite—the fact that he did not know anything about them, and he also did not learn much about them while writing. Perhaps he is more at home in Russian history.'[59] By comparing Rittikh's data with the most recent data collected by the provincial statistical committees of Estland, Livland, and Kurland, the reviewer exposed huge discrepancies in Rittikh's figures, which dramatically reduced the number of Germans. Rittikh's maps depicted not what was there but what he wanted to find there. Implicit in the reviewer's criticism was the idea that statistical data collected locally by provincial level institutions was more reliable and trustworthy than ethnographic studies produced in the imperial capital. The

55 Brüggemann, 'Representing Empire, Performing Nation': 233.

56 Winchakul, *Siam Mapped*: 130.

57 On how maps normalize particular ways of seeing and thinking, see Van Houtum & Lacy, 'The Migration Map Trap': 201.

58 'Eine neue Statistik der Ostseeprovinzen I–III', *Rigascher Zeitung*, no. 88–90, 18–20 April 1873.

59 'Eine neue Statistik der Ostseeprovinzen III', *Rigascher Zeitung*, no. 90, 20 April 1873: 1–2.

reviewer questioned the scientific robustness and impartiality Geographical Society's outputs, revealing the extent to which the Geographical Society had come to been seen as an instrument of Russian imperial statecraft by some in the Baltic provinces. This was a dramatic turning of the tables from how the Society was perceived in the 1840s, when Baltic Germans comprised a significant proportion of the organization's founding cohort. By the 1870s, the region's Baltic Germans were now the objects of the Geographical Society's mapmaking projects, rather than the executors, and they did not look so favourably on the organization's work now that its imperial gaze was trained on them. The reviewer concluded that 'Herr Rittich is concerned above all with proving how unfavourably proportionate the Germans are to the other groups of the population', reading Rittikh's map as an explicit political attack on the alleged German colonial influence in the Baltic provinces presented under the guise of a scientific publication.[60]

For the Baltic German readership, the political message underlying Rittikh's representation of the Baltic as a legitimate Russian imperial possession was glaringly obvious. However, several days later an article defending Rittikh appeared in the same newspaper. The anonymous author—we can speculate that it might perhaps have been Rittikh himself or a close acquaintance—lavished praise on Rittikh's work. The author argued that Rittikh's conclusions differed from those of the local provincial statistical committees and previously published works by Baltic German scholars precisely because he was an impartial observer from outside of the region. Rittikh's supposedly neutral position on the Baltic Question meant that he was 'unaffected by the previous results of science' and free of 'the restrictive influences of a particular university education and a fixed moral worldview', a reference to Dorpat academicians such as Iurii Samarin's adversary Carl Schirren. Instead, the author praised Rittikh's methodological innovations of estimating current population statistics based on past data. In the reviewer's opinion, the publication was proof not of Rittikh's 'scattiness', but rather a sign of his brilliance.[61]

The divisive reactions to Rittikh's approach to visualizing ethnic and religious data on the Baltic provinces did not stop with his critics in the Baltic provinces. Shortly after completing his ethnographic study of the Baltic provinces, Aleksandr Rittikh was appointed by the Imperial Russian Geographical Society to lead the team to produce a new *Ethnographic Map of European Russia* (1875) (Plate 9). Since its publication in 1851, Peter Köppen's *Ethnographic Map of European Russia* (Plate 2) had remained the main cartographic reference point for thinking about the distribution of ethnic groups in the western part of Russian Empire. However, commentators had also begun to critically engage with Köppen's

[60] Ibid.

[61] 'A. F. Rittich und seine neueste statistische Erfindung'. *Rigascher Zeitung*, no. 95, 26 April 1873: 1–2.

work and point out ways that his map could be improved and updated, prompting the Geographical Society to finance the production of a new map in the 1870s.[62] As in his earlier mapping projects, Rittikh used ethnographic mapmaking to make a statement about the Russian Empire's Slavic majority. He rendered a large portion of European Russia in shades of red and pink as a Pan-Slavic space. Rittikh went to great lengths to depict tiny islands of Russian inhabitants that were scattered across regions primarily inhabited by non-Russians (Figure 3.1). The overall effect was to create an image of the empire comprising concentric circles of Russianness radiating outwards from Moscow towards the imperial borderlands.[63]

The emergence of the Baltic provinces as a cartographic battleground for different approaches to ethnographic data visualization became apparent in 1877, when the famous Gotha-based cartographer August Petermann published a German-language version of Rittikh's ethnographic map of European Russia (Figure 3.2). Petermann had been closely following the cartographic outputs of the Imperial Russian Geographical Society since the 1850s, eager to make its publications available to German audiences.[64] However, when it came to producing a German edition of Rittikh's ethnographic map, Petermann introduced some key changes. On Rittikh's map, Germans were shaded in a light blue/grey which made them almost invisible to spot amidst the kaleidoscope of other colours on the map. By contrast, Petermann took steps to reverse what he saw as the undermining of the Russian Empire's German population and instead sought to reclaim the Baltic as part of Germandom. Petermann's edition depicted Germans in a conspicuous red, drawing readers' attention to what Petermann considered to be the most important information on the map. The change of colour transformed the Germans 'colonies' in the Baltic provinces, on the Volga, along the Black Sea shore, Volhynia, and other areas of the Russian Empire into the most pronounced feature of the map, subverting Rittikh's use of red to indicate the Slavic dominance in the Russian Empire. Petermann greatly increased the number of Germans in the Baltic in the German edition, provocatively recasting the Baltic provinces as a legitimate object of German interests. Petermann also resolutely stuck to using German-language toponyms. The reworking of Rittikh's *Ethnographic Map of European Russia* by the German map publisher Petermann in Germany, for a German audience, points to the intertextual nature of ethnographic map production and reception during this period. The case underscores how maps emerged as focal points for controversies over different ways of seeing and thinking about the Baltic Question.

[62] 'Zhurnal Zasedaniia Otdeleniia Ethnografii, 8-go ianvaria 1872 g.', *Izvestiia IRGO* 8, sec. 2 (1872): 196–200.

[63] On the idea of Russia as an empire of concentric circles, see Gorizontov, 'The "Great Circle" of Interior Russia'.

[64] SPbF ARAN f. 30, op. 3, d. 213, l.2–4; Tammiksaar & Sukhova, 'Avgust Peterman i Rossiia'.

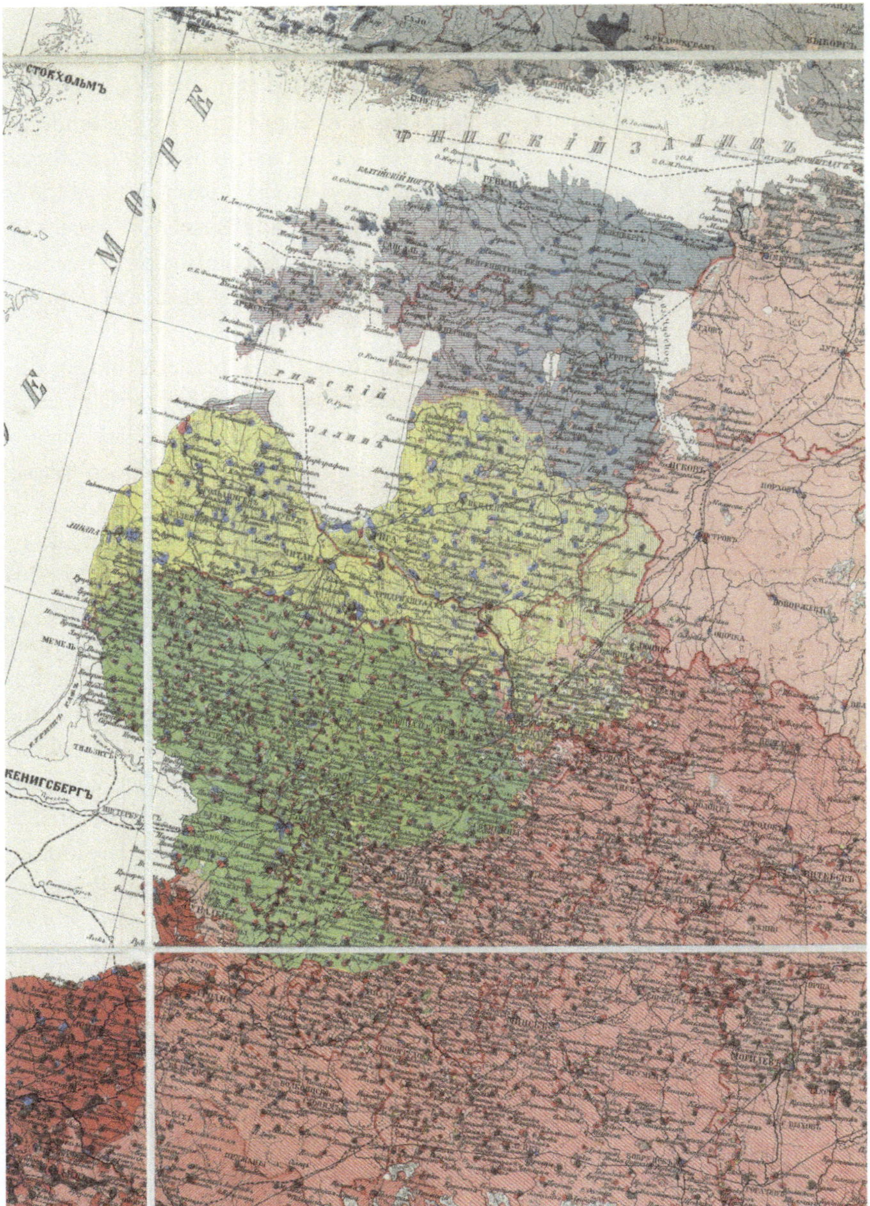

Figure 3.1. Detail of the Baltic provinces and surrounding regions from Aleksandr Rittikh's *Ethnographic Map of European Russia* (1875). Germans are marked in pale blue and are barely visible. See Plate 9 for full map.

Source: Aleksandr Rittikh, *Etnograficheskaia karta Evropeiskoi Rossii* (St. Petersburg: Kartograficheskoe zavedenie A. Il'ina, 1875). Reproduced with permission from the Russian State Library.

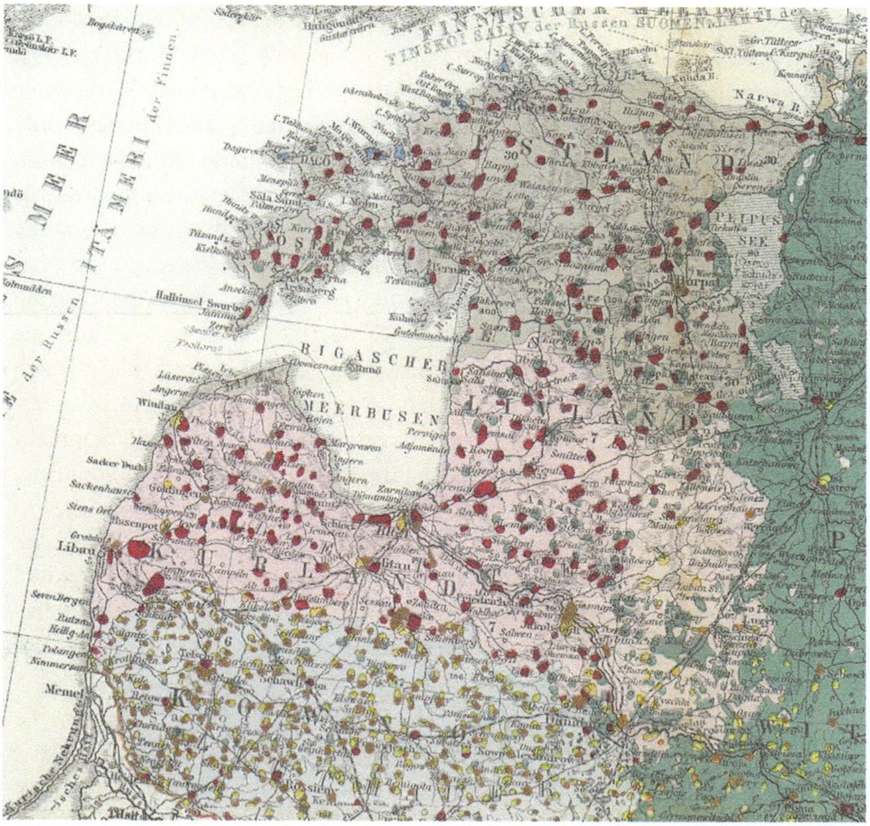

Figure 3.2. Detail of the Baltic provinces from August Petermann's *Ethnographic Map of Russia* (1878), based on Aleksandr Rittikh's *Ethnographic Map of European Russia* (1875; see Figure 3.1 and Plate 9). Notice how Petermann changed the colours from Rittikh's original map, from blue/grey to bright red, so that Germans are made to stand out.

Source: August Petermann & A. F. Rittich, *Ethnographische Karte von Russland* (Gotha: Justus Perthes, 1878). Image courtesy of David Rumsey Map Collection, David Rumsey Map Center, Stanford Libraries.

Choropleth Mapping: Debating National Domains and Attributes

Another technique garnering traction in ethnographic cartography during this period was choropleth mapping. The technique involved shading pre-defined areas, often administrative units, to indicate the proportion of a particular ethnolinguistic group living there.[65] This new, more statistically informed approach to mapping marked a shift away from designating territories as the domain of the majority population to depicting the total extent of a territory inhabited by a particular

[65] Palsky, 'Connections and Exchanges'.

group, to a greater or lesser degree. In the context of debates about the Polish, German, and Pan-Slavic Questions, choropleth mapping gained popularity as a way to make a powerful visual statement about overlapping spheres of influence in the Baltic, even if in numerical terms Poles, Germans, and Russians only amounted to a small percentage of the region's inhabitants. In this section, I examine how choropleth mapping expanded the scope of questioning towards more expansive, transnational scales of thinking about the Baltic provinces at the intersection of various spheres of influence. Mapmaking gave expression to what Ann Laura Stoler and Carole McGranahan call the multiple 'force fields' exerting influence over a particular imperial territory. They remind us how 'at any one time, the designated boundaries were not necessarily the sole force fields in which imperial formations operated or their limits of governance and authorization'.[66] In the course of moving up through ever-increasing scales of thinking, mapmakers moved the conversation away from debating the intricacies of ethnolinguistic taxonomies and defining the precise territorial borders between nationalities to using ethnographic maps to allocate territory according to broader overarching concepts about affinity and belonging to the Polish, German, or Slavic 'worlds'.[67] In the context of these spatial discourses, the Baltic provinces came to be represented on maps as a liminal territory located at the fringes of Polish, German, and Russian space.

The 'Phantom Borders' of the Polish-Lithuanian Commonwealth

In 1830–1 and 1863–4 the western borderlands of the Russian Empire were rocked by violent uprisings against imperial rule which reverberated across the former territories of the Polish-Lithuanian Commonwealth. While the insurrections were concentrated to the south of the Baltic provinces in the predominantly Lithuanian, Belarusian, and Polish regions governed as part of the North-Western Territory, there were several smaller incidents in Kurland in 1831 and western Vitebsk province in 1863–4.[68] Despite this, historians have rarely associated the Baltic Question with the Polish Question. Edward Thaden's influential study of Russian imperial policy towards the Baltic provinces and Grand Duchy of Finland in the second half of the nineteenth century dismissed any connection, claiming that 'in the nineteenth-century Russians made a distinction between Poland-Lithuania and the other Baltic areas within the Russian Empire, and they looked

[66] Stoler & McGranahan, 'Introduction: Refiguring Imperial Terrains': 9.

[67] For a discussion of 'worlds' by historians of the British Empire, see Bridge & Fedorowich, 'Mapping the British World'; Fedorowich & Thompson, eds. *Empire, Migration and Identity*.

[68] Strods, 'Rozruchy chłopskie we wschodniej Łotwie i Kurlandii'; Jēkabsons, 'Stosunki polsko-łotewskie': 31; Samborska-Kukuca, '1863. gada Poļu sacelšanās notikumi Latgalē'; Zajas, *Absent Culture*: 153–61.

at the Polish question as being quite different from that of the Baltic provinces and Finland'.[69]

Cartography by contrast, encouraged such comparative thinking across imperial spaces. To make a map, mapmakers had to define the spatial boundaries of the phenomenon they were mapping. Decisions about what should lie at the map's centre and where to position the map's frame enabled mapmakers to construct narratives about centres of power and influence, on the one hand, and borderlands or peripheries, on the other. Over the course of the second half of the nineteenth century, approaches to framing the Baltic provinces on ethnographic maps underwent notable shifts. Peter Köppen had been influenced by the early links between ethnography and Finno-Ugric studies, which led him to research the Baltic provinces—on the basis of the region's Estonian- and Livonian-speakers—in conjunction with other cartography projects focusing on the northern Finno-Ugric regions of the Russian Empire. Yet, in the context of debates about the Baltic Question, subsequent generations of mapmakers shifted their gaze southward. By the 1870s, the trend was to envelope the Baltic provinces within linguistic, cultural, political, and historical spaces to the south, west, and east.

The question of the relationship of Kurland to the larger territory of Polish and Catholic influence in the Russian Empire's North-Western Territory entered the limelight in the context of discussions about defining the geographical scope of the Polish Question following the uprising of 1863–4. Although the Polish-Lithuanian Commonwealth had been abolished in the second half of the eighteenth century, the polity continued to leave traces in its former territories through the continued presence, for example, of Polish language and high culture, Catholicism, and the memory of Commonwealth's legal, political, and historical traditions among the nobility and intelligentsia.[70] The nineteenth-century discussions about the Commonwealth's legacies in imperial Russia a century later resonates with the historiographical literature on 'phantom borders' (*Phantomgrenzen*) in today's Central and Eastern Europe and the body of scholarship examining 'political borders that once were, are no more, but—nevertheless—somehow still are'.[71] In the context of discussions about imperial Russia's Polish Question, mapmakers debated the extent to which the Commonwealth's 'phantom borders' provided a spatial framework for understanding society that was at odds with present-day administrative boundaries. Mapmakers were forced to consider in which contexts these alternative ethnolinguistic, religious, and political spaces might re-emerge as relevant and meaningful.

[69] Thaden, *Russification in the Baltic Provinces*: 4.

[70] Gibson, 'The Polish Livonian Legacy'.

[71] Zajc, 'Contemporary Borders as "Phantom Borders"': 298. The concept was developed by the German research network 'Phantom Borders in East Central Europe' and is outlined in the first volume of the network's book series on the subject: Von Hirschhausen et al., eds. *Phantomgrenzen*.

The tendency to think about the overlapping territories of the Baltic and Polish Questions was pronounced in Rittikh's geographical imagination. Rittikh perceived the Polish influence in the Russian Empire's western borderlands as seeping over administrative borders into Kurland. Kurland's predecessor, the Duchy of Courland and Semigallia, had remained under the political and cultural influence of the Polish-Lithuanian Commonwealth throughout the seventeenth and eighteenth centuries. Unlike the northern polities of Estland and Livland, which had been ruled by Sweden in the seventeenth century before being formally transferred to the Russian Empire after the Great Northern War in 1721, the Duchy of Courland was only incorporated into the Russian Empire later as part of the lands annexed in 1795 as a result of the third partition of the Commonwealth. From Rittikh's perspective, Kurland's ethnic and religious composition clearly bore the mark of the historical ties with other territories of the former Commonwealth, to the extent that he considered Kurland the 'sister' of Kovna province.[72] Rittikh argued that during the seventeen century 'Polish nobility began to appear, followed by the *szlachta*, and then the Jews, and the whole of Kurland, in ethnographic character and appearance, became like the West-Russian region.'[73] The historical links with the Commonwealth left a visible mark on Kurland's inhabitants. Rittikh's ethnographic and religious maps of the Baltic provinces from 1873 (Plates 7 and 8) clearly portray the Polish and Catholic presence in Kurland, unmissable in bright red. The same red colour used to indicate both the presence of Poles and Catholics on the ethnographic and religious maps reinforced the association between Polishness and Catholicism.[74] In doing so, Catholic Latvians, Lithuanians, and Belarusians living in Kurland's south-easternmost counties of Zel'burg/Selburg (Sėlpilis) and Illustke/Illuxt (Ilūkste) were, by association, portrayed as potential Polish allies and tainted with the same pervasive stereotype of 'the Pole-as-enemy' that was applied by imperial administrators to the North-Western territory.[75]

Rittikh was not alone in advocating for a more expansive definition of the territory associated with the Polish Question that included Kurland. In the spring of 1863 Roderich von Erckert (Erkert in Russian) published an ethnographic atlas of the North-Western Territory (Figure 3.3). Erckert trained as a military cartographer in the Prussian army and then around 1850 he was loaned to the Russian army by the Prussian King, Friedrich Wilhelm IV. Like Rittikh, Erckert pursued an interest in ethnography, linguistics, and statistics alongside his military career and became a member of the Imperial Russian Geographical Society in 1860.[76] Erckert's ethnographic atlas of the North-Western Territory, published in both French and Russian editions,

[72] Rittikh, *Materialy dlia ethnografii Rossii: Pribaltiiskii krai*: 14; Janicki, *Kurlandia*: 680–752.
[73] Rittikh, *Materialy dlia ethnografii Rossii: Pribaltiiskii krai*: 13.
[74] Weeks, 'Religion, Nationality, or Politics'.
[75] Dolbilov, 'The Stereotype of the Pole'.
[76] Petronis, *Constructing Lithuania*: 199.

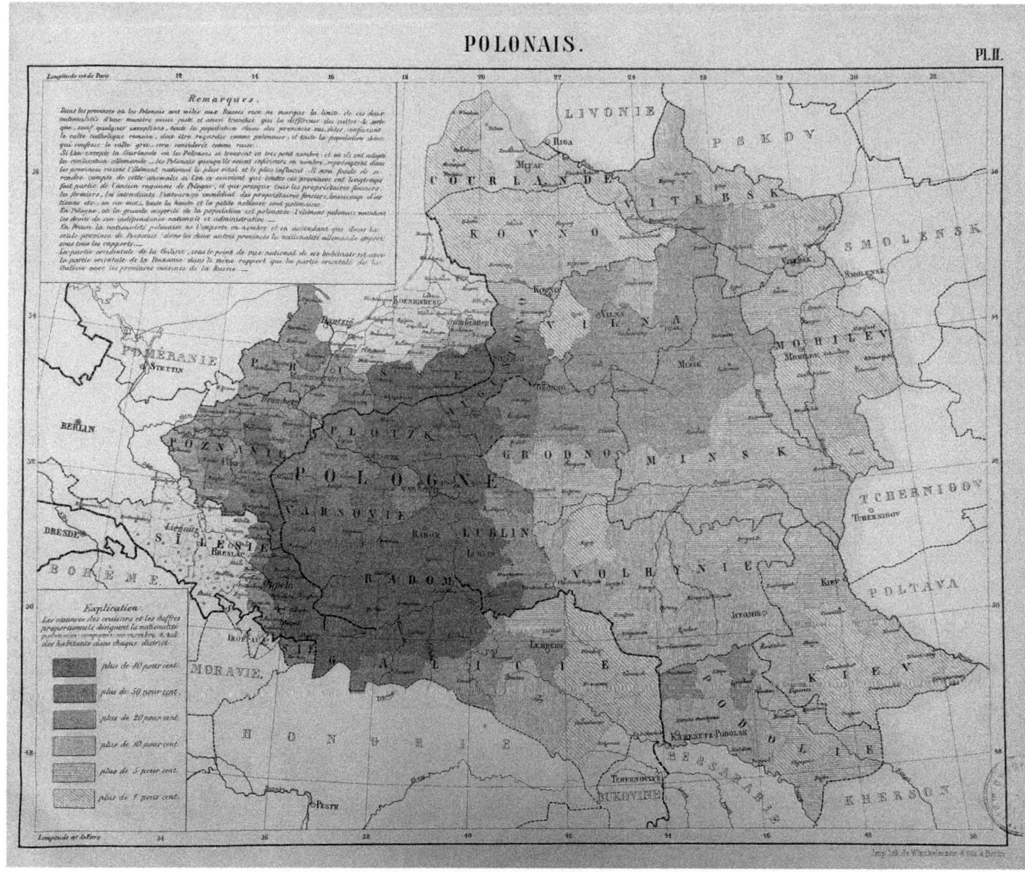

Figure 3.3. Roderich von Erckert's choropleth map of Poles from his *Ethnographic Atlas of the Provinces Inhabited Wholly or Partially by Poles* (1863).

Source: Roderich von Erkert, *Atlas ethnographique des provinces habitées en totalité ou en partie par des Polonais* (St. Petersburg; Berlin: Winckelmann & fils, 1863). Reproduced with permission from the University of Tartu Library.

presented the Russian perspective on the ethnic situation in the Russian Empire's western borderlands to European and Russian readers.[77] Erckert's point of departure for defining the extent of the Polish influence in the Russian Empire was the former territories of the Commonwealth. The atlas included a general ethnographic map of the region as well as a series of choropleth maps showing the presence of Poles, Russians, Germans, Latvians and Lithuanians, and Jews throughout the Commonwealth's former territory. Erckert's conceded that Kurland was the least Polish part of the territories formerly associated with the Commonwealth. Nevertheless, the presence of a Polish-speaking socioeconomic elite, the historical connections with the Commonwealth, and the memory of the Commonwealth's former borders was enough for Kurland to feature on Erckert's mental map of Polish Question. The Polish Question reverberated across administrative regions within the empire around the former territories of the Commonwealth, even if in these peripheral territories the Polish influence existed in a more diluted form.[78]

By drawing Kurland into wider conversations about the Polish Question, mapmakers such as Rittikh and Erkert disrupted the division of administrative space in the Russian Empire's western borderlands—separating the Baltic and North-Western Territory—and encouraged readers to think in transregional ways. They challenged the particularism of the Baltic provinces within the Russian Empire and reminded readers how historical borders differed from those in the present-day. By expanding the scope of the Polish Question to Kurland, mapmakers such as Rittikh 'camouflage[d] ethnicity', weakening the ethnic meaning of the term 'Polish' and instead presenting 'the Pole' as an ideologically subversive and potentially disruptive presence in the empire.[79] In doing so, the maps were designed to cultivate political support for a tougher stance towards the Baltic.

Despite Rittikh's assertion of the enduring Polish and Catholic influence in some regions of Kurland, ascribing ethnocultural labels was not a straightforward process. As Brian Porter-Szűcs notes, the stereotype of the 'Polak-Katolik' suggested 'an inseparable bond between these two communities of belonging, most commonly posited as an ideal (be it utopian or dystopian) rather than as a description'.[80] In reality, many noble families who had migrated to Kurland during the time of the Duchy of Courland and Semigallia had since converted to Protestantism and now used German as their everyday language. Numerous noble and landowning families living in the provinces of Vitebsk, Kovna, and Vil'na had ancestors with historical links to the Baltic provinces but had become Polonized to varying degrees as a result of converting to Catholicism and

[77] Erkert, *Etnograficheskii atlas*. The following year, Erckert published a Russian-language commentary on the atlas. Erkert, *Vzgliad na istoriiu i etnografiiu*.

[78] Le Foll, *La Biélorussie*: 346–51.

[79] Dolbilov, 'Russification and the Bureaucratic Mind': 246.

[80] Porter-Szűcs, 'The Birth of the Polak-Katolik': 2.

their political involvement in the Commonwealth. In the 1860s, many of these Polonized old Baltic German families in the North-Western provinces sought to prove their Baltic ancestry to avoid taxes placed on Polish-Lithuanian nobility after 1863.[81]

The complex relationship of the region's elites to notions of Poland and Polishness was also a subject of concern for intellectuals from these regions, most notably the Vitebskian nobleman Gustaw Manteuffel (Gustav von Manteuffel), who lived in the territory which had formerly been the Polish voivodeship of Inflanty (often translated as Polish Livonia) and which is now today's Latgale region in Latvia.[82] Running through Manteuffel's writings was the idea of Inflanty as part of historical territories of the former Commonwealth, sharing a common history, culture, and political influences. However, for most Poles living far away in the Kingdom of Poland, this voivodeship lying at the remote north-eastern extent of the former Commonwealth's territory had become an 'absence' in collective memory.[83] Manteuffel famously lamented in a letter to his friend Józef Ignacy Kraszewski in 1877 that 'About [Polish] Livonia we [the Poles] know less than of Sumatra and Borneo'.[84] At the same time, Manteuffel also made it clear that the Polish Livonian nobility had a distinct local identity as they were Polonized Teutonic knights who distinguished themselves from both Poles and Baltic Germans. Similar sentiments were echoed in a 1912 collection of essays published in Vitebsk, *From the Vicinity of the [river] Daugava*, written by a group of local elites to reflect on the complex legacies of the Polish-Lithuanian Commonwealth in the region.[85] Writing in Polish, the contributors proudly asserted themselves as part of a common Polish high culture and reached out to their perceived Polonophone Catholic kin to remind them that they too were part of the historical, social, cultural, and political space of the former Commonwealth. At the same time, the contributors expressed a strong sense of self-identification with their homeland on the shores of the river Dźwina/Dvina (Daugava), which they argued had developed its own unique characteristics over time as a result of being on the geographical periphery of Polish cultural and political life within the Russian Empire.

By the early twentieth century, the ambiguous ethnoreligious character and political-economic orientation of parts of Kurland and western Vitebsk province became hotly debated issues during the border negotiations between Latvia, Lithuania, Poland, and Belarus.[86] Eugeniusz Romer, the most renowned Polish

[81] Staliūnas, *Making Russians*: 88.

[82] Manteuffel's publications on Polish Livonia include: Manteuffel, *Polnisch-Livland*; Manteuffel, *Inflanty Polskie*; Manteuffel, *Zarys z dziejów krain dawnych*: viii–xiv.

[83] Zajas, *Absent Culture*. [84] Zajas, *Absent Culture*: 238.

[85] *Z okolic Dźwiny*. On the publication's initiator, see Ajewski, 'Wacław Fedorowicz'.

[86] Seegel, *Mapping Europe's Borderlands*: 89–109; Eberhardt, *Polska i jej granice*; Eberhardt, 'Zasięg terytorialny Polski'.

cartographer of the early twentieth century, famously used ethnographic maps in his *Geographical and Statistical Atlas of Poland* (1916) to popularize the idea of a far-reaching cultural sphere of Polish interests in the east.[87] Using a similar approach to Erkert's atlas from 1863, Romer presented the restoration of the Commonwealth's borders as the recreation of a single political, economic, and cultural unit which had been fragmented by external powers in the eighteenth century.[88] By contouring East Central Europe in bright orange and yellow colour-coded intervals of 25, 50, and 75 per cent "Polish", Romer presented an aggrandized image of the Polish settled territories that included 'all the neighbouring countries who, by the proportion of Poles inhabiting them, also form part of the Polish question' (Figure 3.4).[89] Romer shaded territories inhabited by smaller proportions of Poles (1–5 per cent and less than 1 per cent) in muted blue hues, thereby creating an impression that the whole of the former Commonwealth's lands were Polish to a degree. The western part of Vitebsk province was rendered as part of a corridor of 5–25 per cent "Polishness" stretching northward from Polish Wilno to Polish Dźwińsk, and to the administrative boundary with Estland province. Kurland was shaded blue, associating it with the East Prussian enclave around Königsberg, Silesia, the region south of Kiev, and north-east Vitebsk province as being the least Polish parts of the former Commonwealth.

Romer's conception of what it meant to be Polish rested on the complex intersection of political, social, economic, and cultural legacies of the former Commonwealth. Romer offset the relatively low Polish-speaking population in Kurland by arguing that their socioeconomic influence in the region was more important than their numerical presence. The inhabitants' Catholic religion and lack of strong national consciousness meant that they could be shaped relatively easily into a Polish national mould. Local peasants could become Poles through education and social advancement, just as upwardly mobile Estonians or Latvians in the Baltic provinces had historically often become Germanized.[90] Romer manipulated population data, bending it in time and space, to show how administrative, socioeconomic, and cultural boundaries from the past continued to impact society in Kurland. Being Polish was not an identity that was rooted in a person's native language, something that was absolute and indelibly fixed to a person from the moment of birth, and geographically determined. Instead, Polishness was an attribute or identification that one could assume or lose, in varying degrees. Choropleth mapping gave these ideas a powerful momentum as a strategy for laying claim to territories and peoples.

[87] Seegel, *Map Men*; Labbé, 'Eugene Romer's 1916 Atlas'; Osowska & Przybytek, 'Thematic Maps'.
[88] Eberhardt, 'Zasięg terytorialny Polski', 8.
[89] Romer, *Geograficzno-statystyczny atlas Polski*, plate ix.
[90] Henriksson, *The Tsar's Loyal Germans*: 12.

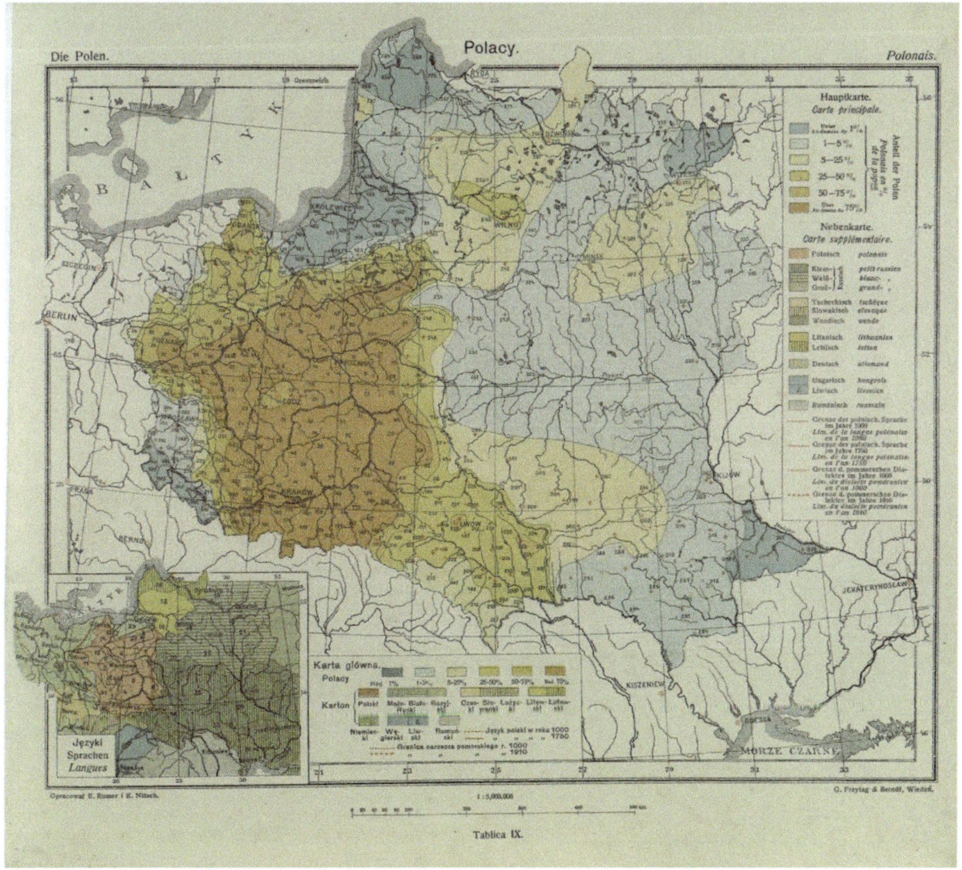

Figure 3.4. Ethnographic plate from Eugeniusz Romer's *Geographical-Statistical Atlas of Poland* (1916). Notice how the map's gradations of Polish space include Kurland and western Vitebsk, both of which were incorporated into the Republic of Latvia after the First World War.

Source: Eugeniusz Romer, *Geograficzno-statystyczny atlas Polski/Geographisch-statistischer Atlas von Polen/Atlas de la Pologne: Géographique et statistique* (Warsaw & Krakow: Gebethner & Wolff, 1916). Public domain.

Overlapping Pan-German and Pan-Slavic Ambitions in the Baltic

In 1871 the German Chancellor Otto von Bismarck succeeded in unifying the loose conglomeration of sovereign states comprising the German Confederation into a new German Empire. Just as the Polish-Lithuanian uprising of 1863–4 sparked cartographic engagement with the Polish Question, the creation of a German Empire on the Russian Empire's doorstep fanned the flames of debates among mapmakers about the status and potential loyalty of German-speakers living within Russia's borders. Mapmakers turned to choropleth maps to debate the Germanic 'force fields' exerting influence over the Baltic provinces.[91]

Terry Martin delineates three distinct but interrelated German Questions in imperial Russia in relation to groups of German-speakers living in different regions of the empire: the Baltic Germans, German colonists in southern Russia, and German migrants in the western provinces.[92] However, similar to the reflections of Polish Livonians who combined a sense of belonging to a larger historical and cultural Polish space with a strong local or regional identification, feelings of Germanness among German-speakers living outside of the German state—the so-called *Auslandsdeutsche*—were more complex, hybrid, and situational than modes of representation on ethnographic maps often gave credit for.[93] Imperial Russia's German-speakers often had very little sense of connection with one another and comprised a mixture of sub-groups.[94] While these people often subscribed to Germanness as a marker of cultural superiority, the way in which they expressed and thought about their Germanness was usually localized and particularist.[95] In the Baltic provinces, concepts of Baltic Germanness were shaped by limited political autonomy, German language and literary culture, Lutheran faith, strong ties to their local province or home region (*Heimat*), and cultural heritage.[96] Germanness was often perceived as a social status, a marker of nobility or bourgeoisie, or of differentiation from Russians. Some Baltic Germans became 'accultured' into high echelons of the imperial administration through military and civil service.[97] Yet even those living in the Baltic provinces who claimed a sense of cultural belonging to the German *Volk* often felt little association with the newly founded German state.[98]

[91] Stoler & McGranahan, 'Introduction: Refiguring Imperial Terrains': 9.

[92] Martin, 'The German Question'.

[93] On the wider concept of *Auslandsdeutsche*, see: Naranch, 'Inventing the Auslandsdeutsche'; Maxwell & Davis, 'Germanness beyond Germany'.

[94] Daija & Kalnačs, '"Small" Germans and "Half"'.

[95] Judson, 'When Is a Diaspora Not a Diaspora?'.

[96] Jõekalda, *German Monuments*: 52; Jansen, 'Baltus': 40–2; Sommerlat-Michas, *Das Baltikum als Konstrukt*; Andresen, 'Formal Stipulation'.

[97] Armstrong, 'Acculturation to the Russian Bureaucratic Elite'; Hennriksson, *The Tsar's Loyal Germans*; Brüggemann, 'Representing Empire, Performing Nation'.

[98] On the complexity of Baltic German identifications, see Plath, Rändrahnud ja majaisandad'; Plath, '*Heimat*'; Wezel, 'Introduction: German Community—German Nationality Belonging'; Brüggemann & Wezel, 'Nationally Indifferent or Ardent Nationalists'.

Nevertheless, after the founding of the German Empire in 1871, mapmakers increasingly came to perceive imperial Russia's German-speakers in terms of their common kinship and potential foreign loyalty. Along with groups such as Armenians and Muslims, the Russian Empire's German-speakers were cast by some conservative thinkers in imperial Russia as a transimperial population who were susceptible to foreign influence from conationals or coreligionists living across imperial borders and thus posed a potential threat to the cohesion and stability of the empire.[99] Mounting anxieties about the Baltic Germans and potential foreign influence coincided with a shift in tsarist policies towards governing the Baltic provinces. Following the assassination of Tsar Alexander II in 1881, his son and successor Alexander III reconsidered his father's liberal policies towards the western borderlands and took the warnings of Samarin and others more seriously. Under Alexander III, new policies were introduced to strengthen imperial rule in the Baltic provinces. This process began in earnest in 1882, following the senatorial inspection of Livland and Kurland conducted by Minister of Justice Nikolai Manasein in response to peasant unrest.[100] The tsarist government placed police and judicial institutions in the region under the Ministry of Interior, Russian language was mandated in schools and courts, and the number of Russian-speaking administrators in the region was substantially increased. Russian Orthodoxy was actively promoted and a large number of new Orthodox churches and schools were built throughout the Baltic littoral.[101] These broad-ranging initiatives aimed to consolidate the 'integrity' (*tselost'*) and cohesiveness of the Russian Empire.[102]

In the context of these policy shifts in the 1880s, Rittikh became obsessed with mapping the existential division of Europe into Slavic and Germanic spheres of influence. He worked on several maps, including an ethnographic map of Slavs produced with fellow military officer Mikhail Fedorovich Mirkovich for a prominent Slavophile society, the St. Petersburg branch of the Slavic Charitable Committee.[103] In 1885 this interest culminated in a book-length publication entitled *The Slavic World* (1885). In line with its avowed Slavophile mission, Rittikh's book provided a detailed overview of the territories inhabited by different Slavic and non-Slavic peoples and the borders between them, accompanied by 'geographical sketches' that gave powerful cartographic expression to ideas about the Baltic as a space of rival Pan-Slavic and

[99] On other transimperial populations living in imperial Russia's border regions, see Riegg, *Russia's Entangled Embrace*; Meyer, *Turks Across Empires*; Kane, *Russian Hajj*.

[100] Thaden, 'N. A. Manaseins Senatorenrevision'.

[101] For an overview of these various policies, see Thaden, *Russification*; Brüggemann, 'Ein Fall von "Verschmelzung" mit Russland'.

[102] Wortman, 'The "Integrity"'; Isakov, *Ostzeiskii vopros*: 28, 36.

[103] Mirkovich & Rittikh, *Ethnograficheskaia karta slavianskikh narodnostei*.

Pan-German ambition. Unlike Rittikh's earlier ethnographic maps which received funding from the Ministry of Interior, Geographical Society, or Slavophile learned societies, *The Slavic World* was published with a private printing house in Warsaw. Rittikh's increasingly zealous views moved his mapmaking away from any pretence of scientific authority and towards peddling mythmaking.

The leitmotif running through *The Slavic World* was the forced Germanization of the Slavs. Rittikh juxtaposed the positive Russian influence in the Baltic provinces with the German 'thrust to the east' (*Drang nach Osten*), characterized by 'violent seizure and war'.[104] Rittikh's maps demonized the German Reich by presenting Germans as a menacing dark presence in the west. He portrayed the Germans as an aggressive intrusion into the naturally 'solid' or 'continuous' distribution of Slavic peoples, thereby creating artificial 'depressions' (*vpadiny*) within the Slavic lands (Figure 3.5).[105] From Rittikh's Venn diagram-like conception of a Europe divided into German and Slavic spheres of influence, the territories inhabited by Romanians and Lithuanians (here signifying Lithuanians and Latvians) fell into the overlap. Although Rittikh regarded Romanians and Lithuanians as distinct ethnic groups, he argued that the centuries of close contact with Slavs and the mixing of Slavic blood had brought them very close to Slavs, especially those who professed the Orthodox faith.[106] However, the continued German influence in these territories also presented a geopolitical threat to Slavs, as they restricted access to the Baltic and Black seas.[107]

The abstract design of Rittikh's 'geographical sketches' in *The Slavic World* presented a dramatic contrast to the data-dense ethnographic maps that he had been accustomed to making for the imperial administration or Geographical Society earlier in his career. Approaching the Baltic from the perspective of different rival pan-movements, Rittikh used visuals to deliver a simple message about the Baltic as the site of a civilizational clash between Germans and Slavs. The maps also departed from Rittikh's earlier practice of creating separate thematic maps of religion and nationality. Instead, Rittikh exemplifies the increasing tendency among radical nationalists to combine these different identities into a 'single, integrated conceptual field' that comprised 'race, ethnicity and nationhood'.[108] The iconographic design of the maps reaffirmed how ideas about the shape of Pan-Slavism were only abstractly defined, meaning that belief in Slavic unity rarely translated into concrete political or territorial ambitions to create a Slavic state.[109]

[104] Rittikh, *Slavianskii mir*: 190.
[105] Rittikh, *Slavianskii mir*: 288; Alexander Rittikh, *Obizhennyi krai*: 399.
[106] Rittikh, *Slavianskii mir*: 18. [107] Rittikh, *Russkaia torgovlia*: 3.
[108] Tolz, 'Constructing Race': 33.
[109] On the failure of Pan-Slavism to cohere into a concrete political movement, see Maxwell, 'Effacing Panslavism': 634.

Figure 3.5. Aleksandr Rittikh's map of 'German Depressions in the Slavic Lands' from *The Slavic World* (1885). Slavs are indicated with horizontal lines and Germans with more closely spaced diagonal lines. According to Rittikh, the Baltic and Bessarabia were two anomalous regions resulting from the spread of German influence into Slavic territories.

Source: Aleksandr Fedorovich Rittikh, *Slavianskii mir: Istoriko-geograficheskoe i etnograficheskoe issledovanie* (Warsaw: Izdanie V. M. Istomina, 1885): 36. Reproduced with permission from the Russian State Library.

Approaching the Baltic Question from a larger overarching framework of the Pan-Slavic Question raised the profile of the Baltic Question within the context of discussions about how empires should deal with ethnic and religious heterogeneity. As Holly Case notes, this was a common strategy among querists who 'tied their questions to larger ones and to ones that had been solved the way querists wanted theirs to be solved. Size mattered for querists, who often declared their questions to be of *Europe-* or *worldwide* significance and therefore "everyone's" problem'.[110] On the one hand, this approach challenged the exceptionalism of the Baltic Question by encouraging readers to think in a broader comparative perspective. On the other hand, by panning out to consider the Baltic as just a small piece in a larger struggle for territorial influence and control, there appeared to be no space or future for the small groups of Estonians and Latvians who were neither Germanic nor Slavic, and religiously divided between Lutheranism and Orthodoxy. Rittikh's maps suggested that they were destined to be assimilated into Slavs or Germans, either for better or worse.

The Baltic Question, which had emerged in the mid-nineteenth century primarily as a domestic issue, was increasingly seen in the light of broader German and Slavic Questions with international dimensions.[111] The German Question gained expediency as mapmakers in the new German Empire experimented with different ways of mapping the Russian Empire's Germans. Nationalist newspapers, travel literature, and images developed a sense of connection between Germany, Austria (that is, Cisleithania within the Dual Habsburg Monarchy), and other more abstractly defined German-speaking spaces, such as the Saxons of Transylvania or the Danube Swabians of south-east Europe.[112] The Baltic provinces featured on several influential maps of Germans in Europe, such as Heinrich Kiepert's *Overview Map of the Spread of Germans in Europe* (1887) and Paul Langhan's *German Colonial Atlas* (1897).[113] Like in the Polish examples discussed above, a distinction emerged in ethnographic mapping practices between mapping German populations, as in the example by August Petermann discussed earlier and mapping Germanness as an attribute. As Kristina Jõekalda argues, Germandom (*Deutschtum*) meaning the territory inhabited by Germans did not necessarily overlap with spatial notions of Germanness (*Deutschheit*), a term used in the late eighteenth and early nineteenth centuries 'in the sense of the quality of being German-like in appearance, or characteristic of German culture'.[114]

The German geographer and cartographer Paul Langhans played an important role in giving cartographic expression to the idea of the Baltic provinces as a space that was fundamentally Germanic in character, if not in numerical terms. In 1897

[110] Case, *The Age of Questions*: 5.
[111] Armstrong, 'Mobilized Diaspora'; Manz, *Constructing a German Diaspora*: 151–3.
[112] Hansen, *Mapping the Germans*: 130–149. See also Liulevicius, *The German Myth of the East*.
[113] Kiepert, *Übersichtskarte der Verbreitung der Deutschen*; Langhans, *Deutscher Kolonial-Atlas*.
[114] Jõekalda, *German Monuments*: 55.

Langhans published an insert map of *Deutschtum* in the Baltic provinces in as part of his *German Colonial Atlas* that shaded the percentage of Germans throughout the region (Plate 11). Symbols indicated the placement of 'pure (or almost pure) German parishes', 'churches in which German preaching is sometimes used', as well as Ordensburgen (fortresses constructed by the Teutonic Order in the Middle Ages) and castles with their founding year, highlighting the cultural influence of Germans even in areas of low German population. Just as Romer had done in his atlas depicting gradations of Polishness, by not depicting other ethnic groups Langhans's maps conveyed the impression that the whole of the Baltic provinces was German to a degree. From 1909–37 Langhans edited the internationally prominent German-language geographical periodical, *Petermanns Geographische Mitteilungen*, which published a series of maps during the war on the ethnographic composition of key areas of German interest, such as northern Italy, Bukovina, and Bessarabia, as well as the Baltic provinces. In a later map of *Languages and Religions in Europe* published in 1917, Langhans pushed this expansive definition of Germanness even further and portrayed the Baltic provinces as situated at the eastern extent of the area where the Protestantism was the dominant religion, Fraktur script was used for writing, German was widely used as a commercial language, German newspapers were published, and even the borders of German town law dating back to the Middle Ages.[115]

Langhans's maps changed the focus of questioning, shifting the conversation from debates about the relative number and distribution of Germans in the Baltic provinces, which had been at the forefront of Rittikh's mapping endeavours in 1873, to the nature and relative strength of the German influence on Baltic society. Langhans's maps of degrees of Germanness in the Baltic suggested that the frontiers of German space in the east gradually petered out, rather than forming a definite border. This approach to ethnographic mapmaking resonates with Stoler's and McGranahan's argument that 'Agents of imperial rule invested in, exploited, and demonstrated strong stakes in the proliferation of geopolitical ambiguities'.[116] By deliberately not providing a definitive solution to the "problem" of where the eastern borders of Germanness lay with respect to the Baltic, Langhans's maps deliberately kept the question unresolved and the subject of continuing debate.

The stream of maps depicting the Baltic provinces as the junction of Pan-Polish, Pan-Germanic, and Pan-Slavic spaces reveal how ideas about ethnographic cartography developed in new directions during these years. Mapmakers used choropleth mapping techniques to render parts of the Baltic littoral as Polish, German, or Slavic using a variety of different parameters: socioeconomic arguments about the distribution of the numerically small, but socially and

[115] Langhans, *Sprachen und Religionen.*
[116] Stoler & McGranahan, 'Introduction: Refiguring Imperial Terrains': 10.

culturally influential landowners, nobility, or imperial administrators; the spread of German Lutheran, Polish Catholic, or Russian Orthodox culture; and German, Polish, and Russian linguistic spheres of influence, where these languages were associated with cultural prestige and upward social mobility.

The turn towards choropleth approaches to ethnographic mapping propelled population density as a crucial variable in ethnographic mapping. Mapmakers applied choropleth mapping techniques to divide populations into shaded intervals and to create images of degrees and gradations of influence over territories in Central and Eastern Europe. While history, religion, and socioeconomic factors had always played a key role in how nineteenth-century mapmakers plotted ethnographic groups on their maps, by the early decades of the twentieth century these aspects were given an elevated importance. Mapmakers used nationality to make bold arguments about the underlying Polishness, Germanness, or Slavicness of the inhabitants living in the historical territories of the Teutonic Order or Polish-Lithuanian Commonwealth. The Baltic littoral, located at the periphery or fringes of the territories of Polish, German, and Russian spheres of interest, functioned as an important testing ground for projecting different visions of Poland, Germany, and Russia, and for negotiating what it meant to be Polish, German, or Russian. Ethnographic mapping underscored the liminality of the Baltic provinces as a cultural and geopolitical borderland. It was presented as neither fully Germanic nor fully Slavic, neither fully western nor eastern. It was precisely this state of 'in-betweenness' that made the Baltic into a focal point for nineteenth-century European querists to project arguments and answers about a wide range of contemporary issues.[117]

Zooming In: Urban Multiculturalism and Social Cartography

Rittikh's approach to mapping questions confronting modern imperial Russian society involved moving up through increasing scales, thinking about the Baltic provinces as an imperial border region or as the meeting point of two opposing Germanic and Slavic worlds. Yet by the late nineteenth century, other mapmakers were coming to the conclusion that insights into nationalities questions and the conditions of modern Russian imperial society could only be found by zooming in on the complexities of multiethnic relations in urban space. In this final section, I examine social cartography as an example of a trend in ethnographic mapmaking in late imperial Baltic society that ran counter to the grand narratives of mapmakers like Rittikh. The question of the relative Polish, Germanic, or Slavic

[117] On the Baltic as a space of 'in-betweenness', see Kaufmann, 'Baltic Reflections': 18, 20.

influence in the region was put to the side by concerned members of civil society who trained their attention on the broader feelings of social crisis and widespread concern about the impact of industrialization, urbanization, and migration on urban life. The previous examples of ethnographic maps discussed in this volume demonstrated how ethnicity was usually examined through the prism of other ideologically or politically charged topics, from matters of language, religion, and social class, to questions of imperial loyalty or geopolitical threats from abroad. Towards the end of the nineteenth century, physicians and public health officials increasingly used urban statistics to make maps that linked ethnicity to social issues such as housing, population density, and public health. In doing so, ethnographic cartography also became a form of social cartography and was used as a tool for tracking 'indexes of progress on a spatial basis'.[118]

Approaches to mapmaking shifted to keep up with broader changes underway in Baltic society. From the 1880s, the pace of industrialization and urbanization in the Baltic provinces picked up and led to steady population growth and dramatic changes in the demographic composition of multiethnic cities. Across the region, the proportion of Germans and Russians in urban spaces decreased as the numbers of Estonians and Latvians increased.[119] The gradual erosion of the traditional, Baltic German, estate-dominated society paved the way for the development of new social and political orientations. The 1877 Municipal Statute, which introduced urban self-government in the form of a city council elected by property-owners, empowered the Estonian and Latvian electorate and led to the emergence of a nascent civil society.[120] A developing strong culture of voluntary associations transformed metropoles into spaces of civic activism and self-organization.

In large European and North American cities during this period, the changing composition of urban space inspired a wave of experimental initiatives to map the spatial syntax of ethnic and national groups, with a particular focus on immigrant communities. Charles Booth's poverty maps of late nineteenth-century London, W. E. B. Du Bois's map of African-Americans living in Philadelphia, maps of ethnic enclaves and ghettos in San Francisco's Chinatown, and Jane Addams's Hull-House maps of Chicago all exemplify how mapping came to be used as a tool of social enquiry to examine urban multiculturalism.[121] Similar efforts to use charts, graphs, cartograms, and maps to visualize population data were underway in the Russian Empire. From the late 1860s, the St. Petersburg Statistical Committee produced

[118] Maier, *Once within Borders*: 191.

[119] On urban changes in late imperial Baltic provinces, see Kruus, *Linn ja küla*; Oberländer & Wohlfart, *Riga: Porträt einer Vielvölkerstadt*; Woodworth, *Civil Society and Nationality*.

[120] Woodworth, 'Patterns of Civil Society': 150.

[121] Bulmer et al., eds. *The Social Survey*; Vaughan, *Mapping Society*: 24–60, 129–44; Battle-Baptiste & Rusert, eds. *W. E. B. Du Bois's Data Portraits*.

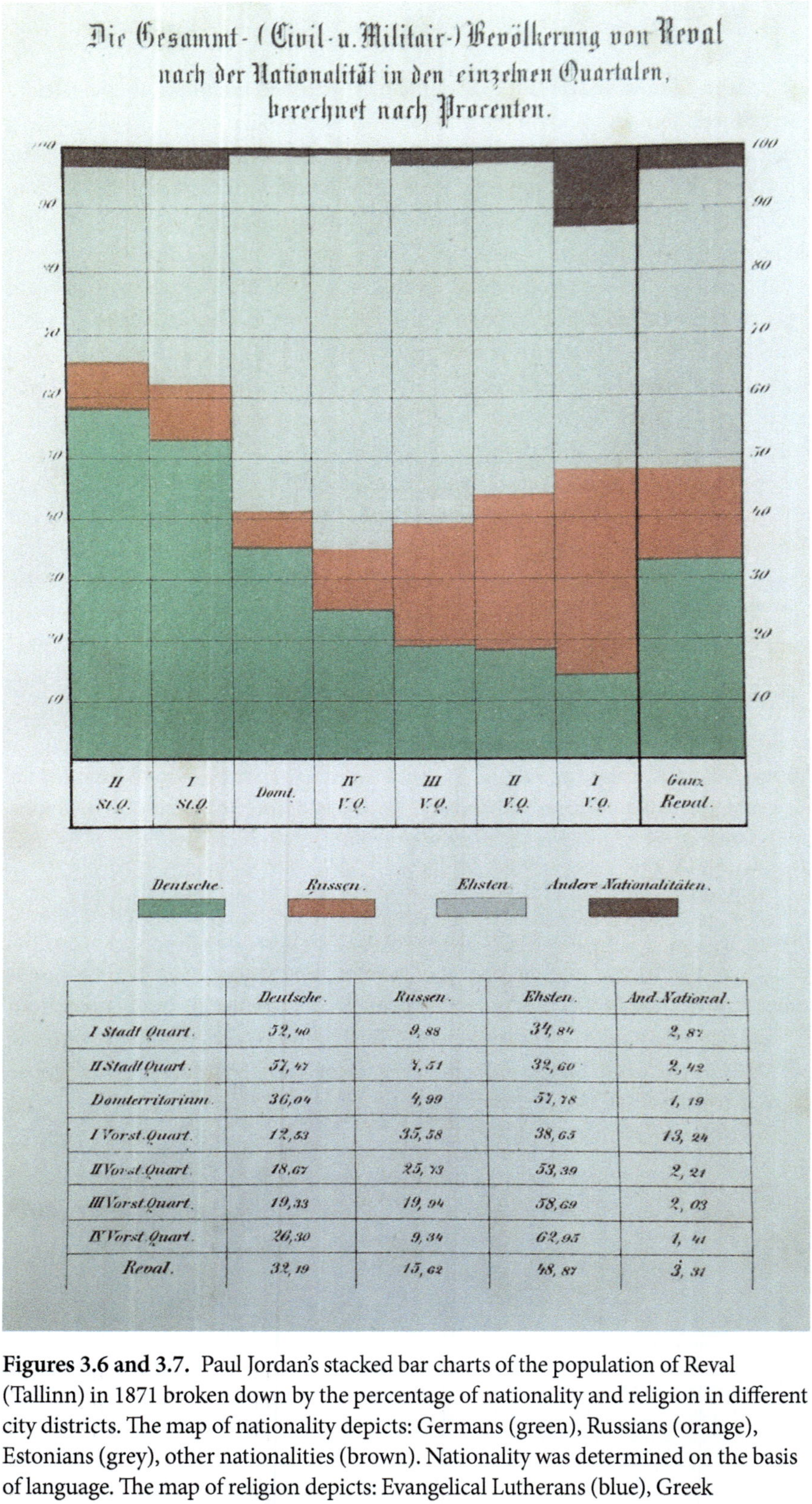

	Deutsche.	Russen.	Ehsten.	And. National.
I Stadt Quart.	52,40	9,88	34,84	2,87
II Stadt Quart.	57,47	7,51	32,60	2,42
Domterritorium.	36,04	4,99	57,78	1,19
I Vorst. Quart.	12,53	35,58	38,65	13,24
II Vorst. Quart.	18,67	25,73	53,39	2,21
III Vorst. Quart.	19,33	19,94	58,69	2,03
IV Vorst. Quart.	26,30	9,34	62,95	1,41
Reval.	32,19	15,62	48,87	3,31

Figures 3.6 and 3.7. Paul Jordan's stacked bar charts of the population of Reval (Tallinn) in 1871 broken down by the percentage of nationality and religion in different city districts. The map of nationality depicts: Germans (green), Russians (orange), Estonians (grey), other nationalities (brown). Nationality was determined on the basis of language. The map of religion depicts: Evangelical Lutherans (blue), Greek Catholics-i.e. Orthodox (orange), Catholics (grey), Jews (brown), Muslims (yellow).

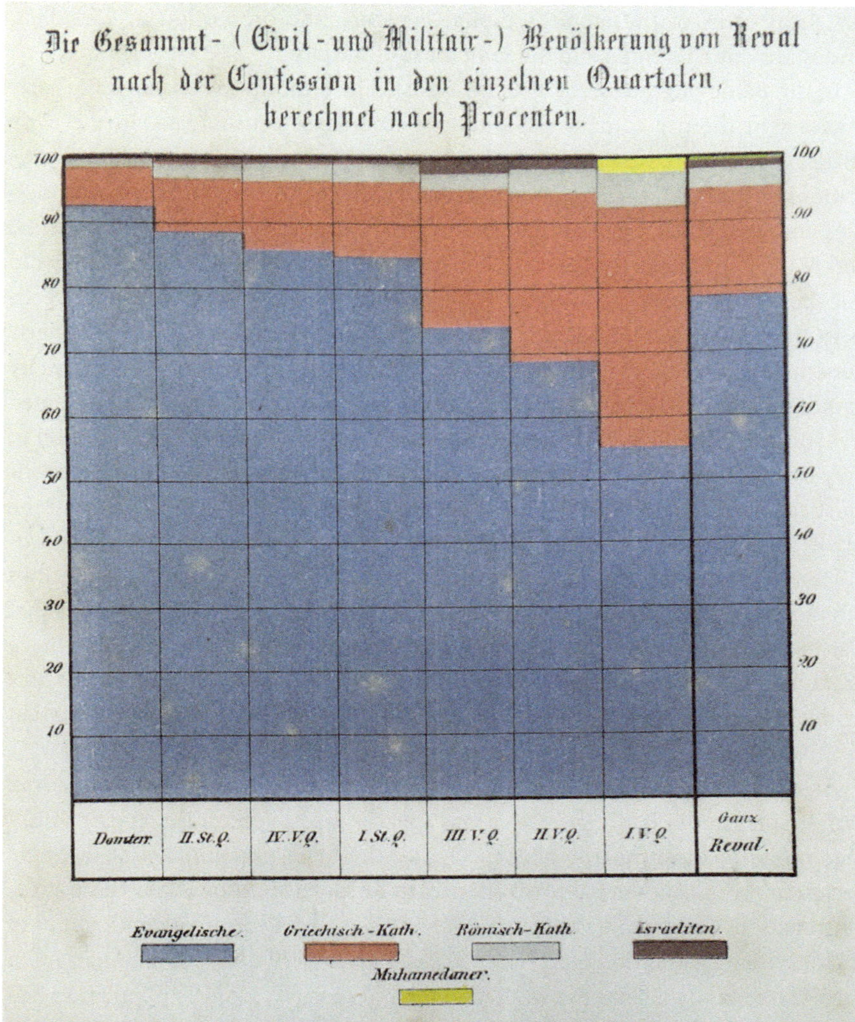

	Evangel.	Griech.	Kathol.	Gregor. Armen.	Israel.	Muham.
I Stadt Quart.	85, 64	11, 35	1, 84	——	1, 05	0, 11
II Stadt Quart.	88, 63	8, 70	1, 74	0, 02	0, 90	——
Domterritorium	91, 80	6, 67	1, 00	——	0, 53	——
I Vorst. Quart.	55, 94	36, 11	5, 80	——	0, 20	1, 94
II Vorst. Quart.	68, 89	26, 23	3, 27	——	1, 55	0, 05
III Vorst. Quart.	73, 83	20, 52	2, 27	——	3, 22	0, 15
IV Vorst. Quart.	86, 82	10, 43	1, 64	——	1 08	0, 01
Reval.	79, 35	16, 62	2, 41	——	1, 34	0, 27

Figure 3.6 and 3.7. Continued

Source: Paul Eduard Jordan, *Die Resultate der Volkszählung der Stadt Reval am 16. November 1871* (Reval: Kluge & Ströhm, 1874). Reproduced with permission from the University of Tartu Library.

statistical plans of the empire's capital depicting the proportion of Catholics, Protestants, and foreigners in different areas of the city.[122]

In the Baltic provinces, the Estland Statistical Committee was one of the early pioneers in the use of data visualizations to analyse census data. In the early 1870s, Paul Jordan, head of the Estland Statistical Committee, published visualizations of the Reval (Tallinn) census data to compare the proportion of nationalities and confessional groups living in different districts of the city (Figures 3.6 and 3.7).[123] Although the Estland Statistical Committee's data visualizations did not show spatial patterns of nationality at the level of detail of individual streets, as in the British and US examples mentioned above, Jordan's stacked bar charts nonetheless communicated ideas about the multiethnic composition of the city and the spatial clustering of groups across the city's quarters at an unprecedented level of detail. For example, Jordan's data visualizations drew readers' attention to how Reval's I quarter (Fischermay, or today's Kalamaja) was the most ethnolinguistically and religiously diverse. Using language as a indicator of nationality, the report highlighted how the population in this quarter comprised the largest proportion of Russian-speakers and other nationalities, as well as non-Lutherans (Orthodox, Catholics, and Muslims). Fischermay was an industrial district that was located next to the port and Patarei Sea Fortress, used at the time as military barracks. Many of the workers, sailors, merchants, and soldiers living in this part of the city had moved to the Baltic provinces from other parts of the Russian Empire, thereby contributing to the quarter's multicultural character.[124]

As in other large European and North American cities, innovation in social cartography in the Baltic provinces was spearheaded by interdisciplinary thinking among medical practitioners.[125] One of the main practitioners of mapping social survey data in the Baltic was Bernhard Körber, Professor of Medicine at the University of Iur'ev (today's Tartu). In 1902 he published an in-depth study of socioeconomic conditions and the spatial distribution of different nationalities in Iur'ev based on data from the 1897 All-Imperial census.[126] Iur'ev's population had doubled in the second half of the nineteenth century due to migration from the surrounding area. By 1897 Iur'ev had over 40,000 inhabitants, making it the second largest urban settlement in Livland after Riga.[127] Whereas in the mid-nineteenth century the city was inhabited by a German-speaking majority, the influx of Estonian-speakers over the second half of the nineteenth century had

[122] *Plan Sankt-Petersburga.* [123] Jordan, *Die Resultate der Volkszählung der Stadt Reval.*

[124] On the multiethnic population of late imperial Reval, see Woodworth, *Civil Society and Nationality*: 14–40.

[125] Morrissey, 'Mapping Civilization'; Henze, *Disease, Healthcare, and Government*, esp. pp.53–79 on cholera mapping in Saratov. On public health and sanitary reforms in late imperial Russia, see Mazanik, *Sanitation, Urban Environment.*

[126] Körber, *Die Stadt Dorpat.*

[127] Berendsen & Maiste, *Esimene ülevenemaaline rahvaloendus.*

dramatically transformed the city's population. Unlike Rittikh's maps, which presented static snapshots of populations, Körber used mapping to analyse societies in flux. He was interested in assessing the impact of migration on demographic changes and ethnic configurations in urban space. Körber was part of group of civically engaged professionals who emerged in the expanding civil society.[128] He directed his maps at an audience of medical practitioners and municipal authorities who, in theory at least, had the ability to take steps to alleviate the negative effects of poverty and overcrowding.

Driven by his interest in forensic medicine, hygiene, and sanitary reform, Körber prepared twenty-eight statistical plans of Iur'ev's medical geography across a range of different themes, ranging from sewerage and water supply to the number of people living in a household, the location of gardens, and access to fresh air and light. Körber cross-referenced sanitary conditions with the population density of the city's four largest nationalities—Estonians, Germans, Russians, and Jews—to analyse the relationships between socioeconomic factors, spatial conditions, and public health (Plate 10). As no separate question was asked about nationality in the 1897 census, data on nationality were produced by aggregating census data on religion and language. Körber's analysis of spatial configurations in Iur'ev revealed that Estonian-speaking labourers in the northern and outer neighbourhoods of Annenhof (Annemõisa), Malzmühle (Meltsiveski), Rosen-Langstrasse (Pikk), Techelfer (Tahtvere), and Bohnenstrasse (Supilinn) lived in low-quality buildings with poor sanitary conditions. By contrast, the districts to the south of the river in the town centre, Dom (Toome) hill and Karlowa (Karlova), had better sanitary conditions and were inhabited by German-speaking social elites, intellectuals, business owners, and merchants, and displayed the lowest rates of disease. Russian-speaking officials lived around the cathedral and Marienkirch (Maarja kirik), while the city's 'indigenous Russians' lived around Fischer and Fortuna streets (today's Kaluri and Fortuuna). The city's Jewish inhabitants mainly lived around Fischer, Alexander (Aleksandri), Karlowa, and Steinstrasse (Kivi).[129] As such, Körber's city plans closely resembled a realization of ideas about mapping nationalities in urban space which Aleksandr Sementovskii had embarked for Vitebsk province's Jewish population thirty years earlier (see Chapter 2).

Körber's statistical city plans highlighted patterns of residential clustering among different ethnolinguistic groups and socioeconomic disparities present even in a relatively compact urban space. Some ethnolinguistic groups were more vulnerable to infection as they had poor diets and lived in densely populated areas with

[128] The growing authority of medical experts in imperial Russia is examined in Becker, *Medicine, Law, and the State*.

[129] Körber, *Die Stadt Dorpat*: 279–80.

limited access to sanitation. Körber approached disease not only as an epidemiological phenomenon, but understood it to be deeply intertwined with social, cultural, and political factors. Körber's snapshot of the city demonstrated that while there were increasing opportunities for Estonians to participate in public life by the end of the nineteenth century, opportunities for socioeconomic advancement in the Baltic provinces still remained quite limited.[130] Telescoping in on Iur'ev, Körber presented the nationalities question in the Baltic as an issue that intersected with socioeconomic inequalities and public health. Körber recognized that questions of place and space were of fundamental importance for understanding variance in public health.

Körber did not conceive of his maps of Iur'ev's principal nationalities as ethnographic maps. However, Körber's work demonstrates how by the early twentieth century the approaches to mapping ethnic groups had broadened out and mapmakers were thinking about the intersections between nationality, socioeconomic inequalities, and pressing social questions such as the relationship between water supply and sanitation on public health.[131] By representing the proportions of nationalities in different neighbourhoods, Körber communicated a more nuanced and sophisticated picture of the complex socio-spatial organization of a multiethnic urban space. Classifying ethnolinguistic groups at the level of streets shifted the emphasis to where people lived in the now (or rather the date of the census), rather than ideas about historical belonging and primordial ethnolinguistic regions that appeared to exist outside of time. Moreover, the maps drew attention to how rural to urban migration intensified ethnic mixing. Körber's penetrating examination of the relationship between nationality and public health was only possible thanks to the availability of statistical data at the scale of individual households. As recounted in Chapter 2, the development of bureaucratic methods for counting and recording imperial subjects in the second half of the nineteenth century through statistics and census-taking ramped up social surveillance on an unprecedented scale and enabled this shift in approaches to ethnographic mapping from the macro to the micro. The maps highlighted the tensions between empirical rootedness of the microscale and the explanatory power of generalization when viewing an issue from a broader perspective. These maps serve as a powerful reminder of how contemporaries examined the relationship between processes taking place at different scales and the need to understand multiple scales simultaneously.

[130] Woodworth, 'Multi-ethnicity and Estonian Tsarist State Officials'; Leppik, 'Social Mobility and Career Patterns'.

[131] Jaadla & Puur, 'The Impact of Water Supply'.

Conclusion

The ethnographic maps discussed in this chapter allow us to reinstate mapmakers as lively and earnest protagonists in histories of questioning. Their maps are revelatory of the active making and remaking of the Baltic provinces as an imperial region across the nineteenth century in response to unfolding political, social, and economic developments in the region. Ethnographic maps were never just straightforward representations of the distribution of ethnolinguistic groups. Mapmakers used ethnicity as a prism through which to explore a wide range of interconnected questions in Baltic society. In asking what such cartographic imaginaries afford us, we can see how maps came to be used as ways of establishing and changing values about the place of the Baltic in Europe and as part of the wider Russian imperial territory.

Examining maps as sources for the intellectual history of nineteenth-century questioning also prompts us to explore the plurality of spatial approaches to the Baltic past. Concepts such as naming, framing, and scale provide us with analytic purchase for thinking about the nineteenth-century history of the Baltic provinces more generally in a way that foregrounds the imperial context. In the process of making their maps, mapmakers considered how widely to compare, what scale allowed or prevented their readers from seeing, and at which scale questions and their answers could be comprehended. These maps invite us to think in different ways about the Baltic: as a borderland on the western edge of the Russian Empire or the eastern extent of German-speaking Europe; as a multiethnic and multiconfessional region at the centre of cross-cutting transimperial influences between Slavs and Germans, and between east and west; as a region defined by the particularism of its legal and political status until the 1880s; as part of a continuum of borderland territories which were governed by a spectrum of different imperial policies; or a place of growing modern cosmopolitan cities inhabited by multiethnic and religiously diverse populations.

By the end of the nineteenth century there was growing evidence that these grand projections about the seemingly inevitable Germanization or Russification of the Estonian and Latvian populations had failed to materialize.[132] The emerging public literary sphere and increasingly self-conscious Estonian and Latvian civic activism harnessed ethnographic mapmaking to give expression to the region's Estonian and Latvian inhabitants and their cultural particularism. Challenging the macro-perspective of the Baltic provinces as lying at the fringes of the Polish, Germanic, and Slavic worlds, pawns in wider imperial and geopolitical games, ethnographers and mapmakers living in the Baltic provinces instead turned their gaze to researching and mapping Estonians and Latvians, peoples

[132] Brüggemann, 'Representing Empire': 239.

who had literally retreated into the background of many ethnographic maps. As will be examined in the next chapter, this inspired a wave of ethnographic map production and popular cartographic consumption in the Baltic provinces in the late imperial period. By the turn of the twentieth century, the 'Baltic Question' began to take on new meanings and was increasingly seen through the lens of the rights and autonomy of the region's Estonian- and Latvian-speakers.[133]

[133] Karjahärm, 'Balti küsimus'; Ījabs, 'Als "Latwija"'.

4

Mapping Latvians in Local and Global Perspectives

In August and September 1896, the landmark Latvian Ethnographic Exhibition was held in Riga. The event was organized by the Riga Latvian Society, a learned and charitable society founded by Latvian intelligentsia in 1868 to further Latvian culture and education. During the exhibition's run more than 45,000 visitors passed through its doors, equivalent to one-sixth of Riga's population at the time.[1] Meandering through the pavilions, visitors marvelled at mannequins dressed in folk costumes, material objects of traditional folk culture, full-scale replicas of traditional peasant dwellings, and theatrical and musical performances. The Riga Latvian Society invited artists to participate in the exhibition, as well as two prominent mapmakers from Kurland province, August Bielenstein and Matīss Siliņš.[2] Bielenstein's and Siliņš's maps encouraged visitors to situate the diverse items on display into an overarching narrative of the long history of a collective Latvian people. Maps imposed a unifying logic on the ethnographic artefacts collected from across Kurland, Livland, and Vitebsk province and consolidated the concept of a Latvian territory that transcended imperial administrative borders.

By the late nineteenth century, society in the Baltic provinces was undergoing rapid changes. Industrialization, the building of the railways, expansion of trade, and rapid urban population growth brought dramatic social and economic transformations in the Baltic provinces. The demographic composition of urban centres shifted as growing numbers of Estonians and Latvians migrated to cities to work in factories. Access to free professions and property ownership afforded opportunities for social mobility and career advancement.[3] These socioeconomic developments gradually eroded the traditional hegemony of the Baltic German elites and brought agitation for political change.[4] The Latvian Ethnographic Exhibition revealed how Latvian intellectuals, scholars, and artists were attempting to overturn the traditional distinction in the Baltic between 'high' German

[1] On the history of the Latvian Ethnographic Exhibition, see Stinkule et al., *Latviešu etnogrāfiskā izstāde*; Stinkule, *Indivīds. Vēsture. Nācija*. On the Riga Latvian Society, see Wohlfart, *Der Rigaer Letten*.

[2] *Katalog der lettischen ethnographischen Ausstellung*: 13, 24, 64.

[3] On the transformation of urban life in the Baltic in the late imperial period, see Woodworth, *Civil Society and Nationality*; Oberländer & Wohlfart, *Riga: Portrait einer Vielvölkerstadt*; Von Hirschhausen, *Die Grenzen der Gemeinsamkeit*; Raun, 'Riga and Tallinn in the Late Tsarist Era'.

[4] Henriksson, *The Tsar's Loyal Germans*: 83–105.

Geographies of Nationhood: Cartography, Science, and Society in the Russian Imperial Baltic. Catherine Gibson, Oxford University Press. © Catherine Gibson 2022. DOI: 10.1093/oso/9780192844323.003.0005

and 'low' Latvian 'peasant' culture by showcasing Latvian visual and material culture as a source of national pride and cultural distinctiveness.[5] After the exhibition closed many of the items were transferred to the collections of the Riga Latvian Society museum, the precursor of today's National History Museum of Latvia.

While the Latvian Ethnographic Exhibition promoted ideas about Latvians as a historically and culturally distinctive people with a definable territory, late imperial Riga hosted multiple other ethnographic exhibitions and shows which provided opportunities for the city's inhabitants to confront ideas about their sense of selfhood in relation to the other inhabitants of the Russian Empire and the wider world. Running in parallel to the Latvian Ethnographic Exhibition in Riga in the summer of 1896 was the Tenth All-Russian Archaeological Congress organized by the Society for Antiquity and History of the Baltic Provinces of Russia.[6] The Archaeological Congress rotated around different cities in the Empire. These orchestrated celebrations of the Russian Empire occurred as a result of collaboration between state initiatives and civil society to realize common aims: popularizing science, fostering civic pride, and showcasing the vast expanses and diversity of the Empire.[7] The organizers of these exhibitions sought to present a coherent and overarching image of the Russian Empire, while retaining a sense of local specificities in a way that was compatible with political conservatism and Pan-Russian nationalism.[8] As Deborah Coen argues, the 'technical challenge of visualizing "unity in diversity"' was a defining characteristic of nineteenth-century empires.[9]

Riga's inhabitants increasingly encountered people from further afield as well. Live ethnographic shows featuring non-Europeans often performed at the port city as part of their European tours. In 1889 Samoyeds from Archangelsk and Ashanti people from modern-day Ghana performed in Riga. In 1895 folk dancers from western Guinea visited the city. In 1901 a troupe of Dahomey Amazons from today's Republic of Benin participated in the programme of events celebrating Riga's 700th anniversary.[10] Against the backdrop of colonial expansion and global migration, these ethnographic shows expanded the geographical horizons of the Baltic provinces' inhabitants and forced them to confront ideas about race and the different colonized peoples of the world.[11] This chapter uses maps to explore these interweaving narratives about the Latvian ethnographic territory and its relationship to wider imperial and global space.

[5] Jõekalda, *German Monuments*: 57; Pourchier-Plesseraud, *Arts and a Nation*: 32. On the links between science and nation-building, see Ash & Surman, eds. *The Nationalization of Scientific Knowledge*.

[6] Pourchier-Plasseraud, *Arts and a Nation:* 53.

[7] Bradley, 'Pictures at an Exhibition'.

[8] Tolz, 'Orientalism, Nationalism, and Ethnic Diversity': 138.

[9] Coen, *Climate in Motion*: 123.

[10] Boldāne-Zeļenkova, 'Others Among Others': 274–5; Novikova, 'Imagining Africa'.

[11] Boldāne-Zeļenkova, 'Others Among Others: Latvians': 280.

Bielenstein and Siliņš were contemporaries, yet their maps have rarely been discussed together. Siliņš's mapmaking activities have been examined in the context of Latvian national historiography, which presents Siliņš as the 'first Latvian cartographer'.[12] Bielenstein, by contrast, has been excluded from the nationalist canon of Latvian history of cartography. Instead, he is presented in the literature as a Baltic German linguist. In addition, Bielenstein is also far more well-known internationally than Siliņš as he is credited with coining the term 'isogloss' to refer to lines drawn on maps to represent the geographical boundaries of different linguistic features.[13] By bringing the work of these two mapmakers into conversation with one another, this chapter broadens our understanding of the multiple spatial concepts of Latvian territory developed by mapmakers living and working in the Baltic provinces of the Russian Empire through the medium of both Latvian and German. As Susan Gal argues in the case of Habsburg Hungary, many '*multilingualist* nationalisms' were proposed and enacted in the nineteenth century. Yet, thanks to narratives which became entrenched during the interwar period, this multilingualism has often been overlooked in the historiography in favour of narrower monolingualist understandings of nationalism movements.[14] This chapter builds on Gal's work by considering the role of multilingual German and Latvian cartographic practices in constructing a Latvian national space. It looks at the role of maps in shaping various spatial perceptions of the Latvian ethnolinguistic territory, as well as the interactions that emerged between map producers and consumers.

Learned Societies and Visions of the Baltic Heimat

By the turn of the nineteenth century, the French Revolution and the ideas of thinkers such as Johann Gottfried Herder had begun to suffuse language with symbolic importance as an essential component of selfhood. Language came to be widely regarded as a powerful way of proving the authentic existence and cultural worth of the common people.[15] Herder worked as an adjunct pastor and schoolmaster at the Cathedral school in Riga between 1764 and 1769 and his work collecting folksongs and other cultural artefacts inspired a generation of churchmen in the Baltic provinces.[16] In the late eighteenth and first half of the nineteenth century, the region's German-speaking Lutheran pastors conducted ethnographic

[12] Štrauhmanis, *Matīss Siliņš*; Štrauhmanis, 'Matīsa Siliņa ietekme'; Soboļeva, ed. *Sava ceļa gājējs*: 33–45.
[13] Kļaviņas, 'Die Werke von August Bielenstein'; Pi, 'Beyond the Isogloss': 177.
[14] Gal, 'Polyglot Nationalism': 31.
[15] McCain, *The Language Question*: esp. pp.6–15.; Kamusella, *Politics of Language and Nationalism*: esp. pp. 29–37.
[16] Jaremko-Porter, 'The Latvian Era of Folk Awakening'; Baycroft & Hopkin, eds., *Folklore and Nationalism*.

observations of their parishioners and published studies about the language and folk culture of the Estonian- and Latvian-speaking peasantry, who in the German Enlightenment imagination were perceived both as peoples in need civilizing and as noble savages.[17] For some, interest in the Estonian- and Latvian-speaking inhabitants of the Baltic provinces was tied to debates about the abolition of serfdom. Baltic German intellectual Garlieb Merkel described the historical contributions and ethnographical distinctiveness of the Latvian-speaking inhabitants of the Baltic littoral as an argument for social and agrarian reform.[18] Meanwhile, others contended that Estonians and Latvians were physically and mentally inferior, using physical anthropology and ethnography to entrench racial hierarchies.[19]

Following the abolition of serfdom in the Baltic provinces between 1816–19, the study of the language, folklore, and culture of the Estonian- and Latvian-speaking populations became more organized with the founding of numerous learned societies.[20] The most active were the Kurland Society for Literature and Art (est. 1817), the Latvian Literary Society (est. 1824), and the Estland Literary Society (est. 1842). The German-speaking pastors, educated elites, and nobles who were members of these societies in the first half of the nineteenth century fashioned themselves as Estophiles and Lettophiles. Motivated by Enlightenment ideas of progress, these German-speaking Estophiles and Lettophiles understood their activities as combining personal interest with public service.[21] As Kristina Jõekalda argues, interest in cultural heritage preservation provided an 'opportunity to solidify the cultural identity of the Baltic Germans in their search for an "anchor" of stability'.[22] In addition to developing written Estonian and Latvian by publishing grammars, dictionaries, and collections of songs and poetry, these learned societies were hubs for a wide range of other activities. From the 1850s members of these societies organized choral singing societies and festivals, a tradition originating in the Romantic nationalism of the German-speaking lands.[23]

Learned societies in the Baltic provinces also channelled funds into map production. In 1839 the Livland Charitable and Economic Society published a detailed topographic map of Livland province. The map's focus on agriculture and trade reflected the Society's mission to use science to bring economic prosperity to the Baltic provinces.[24] Other mapping projects appeared over the subsequent

[17] Plath, 'Euroopa viimased metslased': 41.

[18] Merkel, *Die Letten*. See also Blumbergs, *The Nationalization of Latvians*: 167–96.

[19] Kalling & Heapost, 'Racial Identity and Physical Anthropology'; Pushaw, 'The Visual Whitening of Estonians'.

[20] Hackmann, ed. *Vereinskultur*; Hackmann, *Geselligkeit in Nordosteuropa*.

[21] Applegate, *A Nation of Provincials*: 3.

[22] Jõekalda, *German Monuments*: 24.

[23] Lajosi & Stynen, eds. *Choral Societies*.

[24] Jeske, 'The Map Specialcharte von Livland'; Tammiksaar, 'Liivimaa Üldkasulik ja Ökonoomiline Sotsieteet'; Finlay, 'International Science'.

decades. In 1861 the Naturalist Society in Dorpat covered the publication costs of Constantin Grewingk's geological map of the Baltic provinces. The Livland Economic and Charitable Society published a hypsometric map of relief in Estland and northern Livland in 1876 by Seidlitz Meiershof. The Learned Estonian Society supported the publication of Jaan Sitska's archaeological map of the Baltic provinces in 1896.[25] In the context of these various projects to map various aspects of physical and human geography, interest in ethnographic map-making also garnered interest in the region.

August Bielenstein: Pastor, Lettophile, Mapmaker

August Bielenstein played a key role in the development of Latvian linguistic research and mapping in the second half of the nineteenth century. Bielenstein was born in Mitau (Jelgava) in Kurland province in 1826. He spent his childhood in Neu-Autz (Jaunauce), near the border with Kovna province, where his father was a pastor. Although Bielenstein was sent abroad to Saxony for schooling, he returned to the Baltic provinces in 1846 to study theology at the University of Dorpat (Tartu). Following his father's death in 1851, Bielenstein succeeded him as pastor of Neu-Autz. In 1867 he moved to the slightly larger town of Doblen (Dobele). During the revolutionary violence that erupted in the Baltic provinces in 1905, the parsonage in Doblen was looted and much of the Bielenstein family archive and library was destroyed.[26] After these events, the family moved to an apartment in Mitau, where August lived until his death in 1907.[27]

Bielenstein served as president of the Latvian Literary Society for thirty years, from 1864 until ill health forced him to step down in 1895. Between 1867 and 1903 he edited the most widely circulated Latvian-language newspaper, *Latviešu Avīzes*, which had been founded in Mitau in 1822. He was also involved in organizing many different cultural events to promote Latvian language and culture, including the first Latvian song festival in Kurland at Doblen castle in 1870 and the 1896 Latvian Ethnographic Exhibition. Bielenstein was, however, staunchly opposed to the politicization of Latvian culture and regarded nationalism as a dangerous and destabilizing current in the region. Politically conservative, he fiercely defended Baltic German traditions. Instead, Bielenstein identified himself as a 'Lettophile' (*Lettenfreund*), a devotee and lover of the Latvian people.[28]

Bielenstein's activities as a Lettophile developed at the intersection of professional and personal interests. Bielenstein felt that it was his duty to learn the local

[25] Grewingk, *Geognostische Karte der Ostseeprovinzen*; Meiershof, *Hypsometrische Karte Ehstland's and Nord Livland's*; Sitska, *Archäologische Karte von Liv, Est- und Kurland*.

[26] Štrāle, 'Dobeles apkārtnes mācītājmuižu bibliotēku un arhīvu liktenis'.

[27] Details of Bielenstein's life are taken from his autobiography: Bielenstein, *Ein glückliches Leben*.

[28] Bielenstein, *Ein glückliches Leben*: 287.

vernacular as part of his work as a pastor, revealing how linguistic and ethnographic research was supported and enforced by the work of Lutheran churchmen.[29] As Bielenstein noted:

> We pastors of German nationality were the only men in the country who had the opportunity and the obligation to get acquainted with the Latvian people in their entire external and spiritual life and to work with them daily, especially through the Word. The sanctity of the matter that we pastors have to represent to the peasantry requires that we know the language of the people in the most accurate way and that we use it in the most correct manner.[30]

Bielenstein worked hard to improve his knowledge of Latvian during his first years as a pastor in Neu-Autz to deliver sermons in this language to his parishioners. However, due to the scarcity of books for learning Latvian at the time, acquiring fluency in Latvian was only possible 'by ear'.[31] Bielenstein established a library with Latvian books at the parsonage and worked on revised Latvian translations of the Bible, catechisms, and hymnbooks. His pastoral duties also provided him with ample opportunities to study Latvian language and folk traditions during his visits to households to perform baptisms, marriage ceremonies, and funerals, during which he collected thousands of folksongs and ethnographic observations. Over the course of his lifetime, Bielenstein published numerous works on Latvian language and culture. His earliest publications concentrated on Latvian linguistics and grammars for use in schools.[32] At the request of the Latvian Literary Society, he compiled a study of the Latvian language, for which he awarded the prestigious Demidov Prize by the Imperial Academy of Sciences.[33] In later decades, Bielenstein moved away from a strict focus on language and wrote more broadly on archaeology, ethnography, folklore, and cultural history.[34] Despite his prodigious list of publications, Bielenstein wrote humbly in his autobiography about his scholarly contributions, presenting himself as 'only a dilettante' who dabbled in various areas of research alongside his main day job as a pastor.[35]

Bielenstein's passion for studying the language and culture of the local Latvian-speaking population was also strongly linked to his self-identification as a

[29] On the participation of churchmen in scientific knowledge production, see Brenna, 'Clergymen Abiding in the Fields'.

[30] Bielenstein, *Ein glückliches Leben*: 204.

[31] Bielenstein, *Ein glückliches Leben*: 205.

[32] Bielenstein, *Handbuch der lettischen Sprache*; Bielenstein, *Die Elemente der lettischen Sprache.*

[33] Bielenstein, *Die lettische Sprache.*

[34] Bielenstein, *Die Grenzen des lettischen Volkstammes*; Bielenstein et al., *Studien aus dem Gebiete der lettischen Archäologie*; Bielenstein, *Die Holzbauten und Holzgeräte der Letten.*

[35] Bielenstein, *Die Grenzen des lettischen Volksstammes*: ix.

'*Kurländer*'.[36] Bielenstein felt a strong sense of attachment to Kurland province as his 'fatherland' and 'native homeland' (*Heimat*), presenting his research on Latvian language and folklore as 'a testimony to my warm love for the *Heimatland*'.[37] Bielenstein was like many Baltic Germans who 'clung on to their geographical identity—to their Baltic *Heimat*, a phenomenon that they explicitly distinguished from both the German "motherland" and the Russian "fatherland"'.[38] Bielenstein's strong identification with his home province indicates how the administrative division of the Baltic littoral, based the borders between the fiefdoms of the noble corporations (*Ritterschaften*) since medieval times, continued to exert a powerful influence over geographical imagination and shape individual feelings of selfhood and belonging.

From Ethnographic to Linguistic Mapping

From the 1870s, linguists across Central and Eastern Europe began experimenting with using cartography as a tool for mapping single language areas. In contrast to the ethnographic maps examined in earlier chapters, these new maps concentrated on marking dialectal variations within a single language area. The trend towards linguistic mapmaking was exemplified in the region by the map of the Lithuanian language area published in 1876 by Friedrich Kurschat, a Prussian-Lithuanian theologian and director of the Lithuanian Seminary at the University of Königsberg (Kaliningrad).[39] Another well-known example of the genre was Slavicist Efim Fedorovich Karskii's (Iauchim Fedaravich Karski) map of the Belarusian ethnic group published in 1903, which subdivided the territory inhabited by Belarusian-speakers into six dialectal areas.[40]

In the early 1880s Bielenstein collaborated with the artist Julius Döring to produce two maps of the Latvian-speaking territory. Döring was originally from Dresden and had moved to Kurland in 1845 to work as a drawing teacher at the Mitau gymnasium. Döring is best known in Latvia today as a portrait painter, but he also worked as librarian and secretary of the Kurland Literary and Artistic Society, editor of its journal, and director of the Kurland Provincial Museum from

[36] Bielenstein, *Ein glückliches Leben*: 75, 128. On provincial-scale identifications in the Baltic, see Sommerlat, 'L'élément national': 110. On the concept of *Heimat*, see Blackbourn & Retallack, eds. *Localism, Landscape*.

[37] Bielenstein, *Die Grenzen des lettischen Volkstammes*: vii.

[38] Jõekalda, *German Monuments*: 24; Plath, '*Heimat*'.

[39] Kurschat, *Grammatik der littauischen Sprache*; Safronovas, *The Creation of National Spaces*: 84–5. On the role of German linguists in making Lithuanian 'visible', see Leopold, 'Lithuanian Made "Visible"'.

[40] E. F. Karskii, *K voprosu ob etnograficheskoi karte*; Petronis, *Constructing Lithuania*: 167–70.

1865 to 1893.[41] In February 1881, Döring gave a lecture to the Latvian Literary Society on the 'Origins of the Kurland Latvians', where he discussed the link between the 'indigenous' inhabitants of Kurland in the twelfth and thirteenth centuries and present-day Latvians.[42] Two maps were published to accompany the published version of his lecture. The first was a map of the 'Latvian language area' around 1860, which was 'mostly based on' Bielenstein's research (Figure 4.1). The second map depicted the proto-ethnic groups or 'tribes' inhabiting the East Baltic lands between 1170 and 1300 based on archaeological findings. Döring extensively referenced Bielenstein's work and claimed that the language map was based on a sketch Bielenstein had given him and permitted him to use.[43] Matīss Siliņš referred to the maps' authors as 'Bielenstein and Döring', highlighting the collaborative nature of map production (cf. Figure 4.7). Situated within the broader context of developments in cartographic thinking, the maps marked a crucial shift in approaches to mapping Latvians.[44]

The sequence of the two maps constructed a narrative of the history of the Latvian language, compressing seven hundred years of language change into a series of still images.[45] They portrayed continuities in the ethnographic composition of the territory's inhabitants over time and provided a historical explanation for dialectal differences occurring across the contemporary Latvian language area. The addition of a historical dimension to ethnographic mapping made a strong statement about the autochthonous nature of the Latvian-speaking population and their historical rights to the territory dating back to the twelfth century. At the same time, in keeping with Bielenstein's conservative political views, Latvians were not depicted as a single unified nation in the modern sense, but rather as an object of ethnographic study. Place names are given in their German forms, underscoring the strong German influence on the social and economic structure of Baltic society. Similar to practices in American mapping at the time which depicted Native Americans as part of the natural landscape rather than as actors in political history, by omitting references to administrative and political borders, the Bielenstein-Döring map conveyed the impression that Latvians lived in passive detachment from the world of politics.[46]

The Bielenstein-Döring map of the Latvian language area in 1860 (Figure 4.1) subdivided the Latvian language into three 'main dialects'. The use of colour suggested the relative proximity and distance between different dialects and languages. The orange-coloured Upper or High Latvian in the east was presented as being closest to the red of the Russian-speaking territory. The turquoise Livonian-Latvian area appeared to bridge the yellow Middle or Written Latvian dialects and

[41] On Döring's life, see his published autobiography: Döring, *Was ich nicht gern vergessen möchte*.
[42] Döring, 'Über die Herkunft der kurländischen Letten'.
[43] Döring, 'Über die Herkunft der kurländischen Letten': 48.
[44] Trumpa, 'Karte des lettischen Sprachgebiets'.
[45] Goffart, *Historical Atlases*: 8.
[46] Schulten, *Mapping the Nation*: 24.

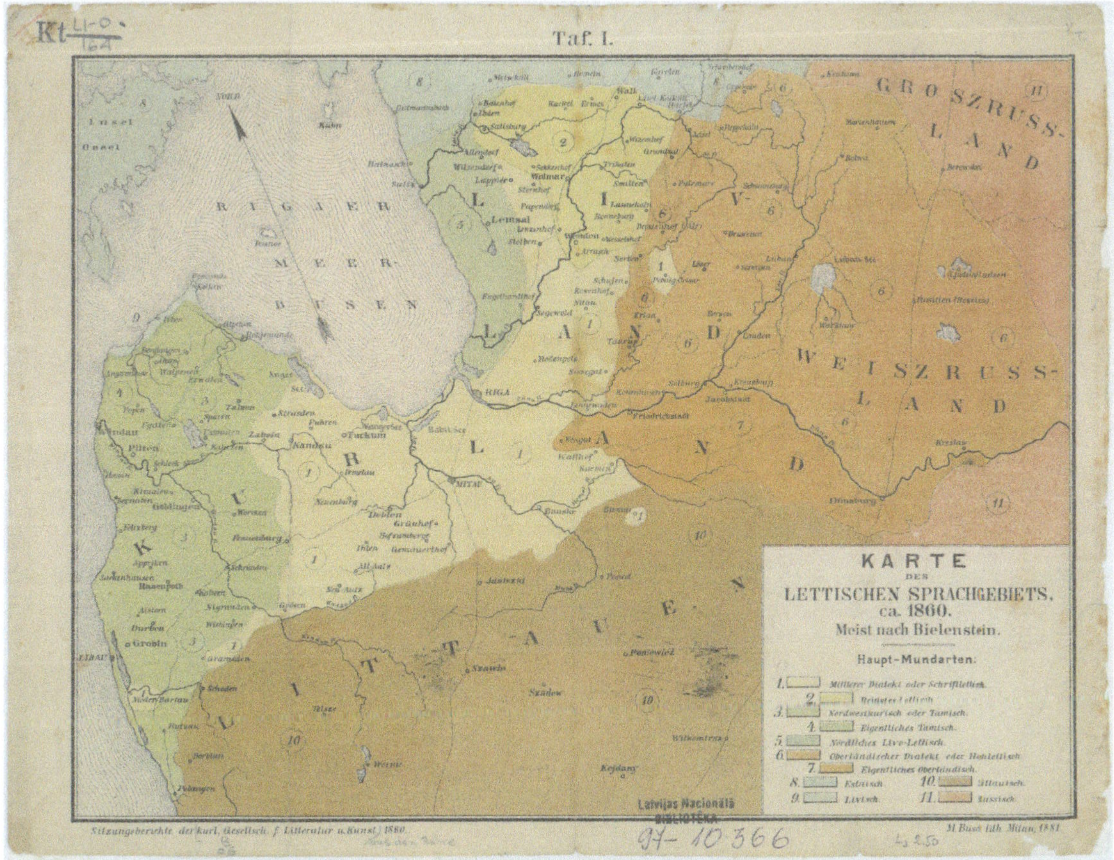

Figure 4.1. *Map of the Latvian Language Area ca. 1860* by Julius Döring and based on the work of August Bielenstein. The legend describes the following languages and dialects using a combination of numbers and colours: (1: yellow) Middle dialect or written Latvian; (2: dark yellow) Pure Latvian; (3: green) North-West Courish or Tamisch; (4: green) Only Tamisch; (5: dark green) Northern Livonian-Latvian; (6: orange) Highland dialect or High Latvian; (7: dark orange) Upland dialect; (8: blue) Estonian; (9: light blue) Livonian; (10: brown) Lithuanian; (11: pink) Russian.

Source: *Karte des Lettischen Sprachgebiets, ca. 1860, nach Bielenstein* (Mitau: M. Busch lith., 1881). Reproduced with permission from the National Library of Latvia.

the blue Estonian-speaking area to the north. The map also subdivided the dialects further, indicating areas where 'pure Latvian' and 'only' North-West Kurlandish and Upper Latvian were spoken. These areas were juxtaposed with regions with an 'impure dialect', such as the Latvian-Livonian area on the Livland coast.[47] Numbers drew readers' attention to 'language islands' among other dialectal areas and indicated regions of 'transitions from one dialect to another'.[48] These distinctions between pure and impure dialects reflect the emergence of value judgements about a hierarchy of the dialects. Although all the various dialects were shaded on the map, the map and accompanying article communicated ideas about Middle Latvian as the gravitational centre of the standard Latvian language (and the basis of the literary standard).[49] Döring's text gave no indication as to the linguistic reasoning behind how the boundaries between the dialects were determined and instead the map delineated dialects based on the mapmakers' perceptions of difference.[50]

The map of the Latvian area language bears further traces of its makers. In addition to marking major provincial and district cities, the map included the names of places where Bielenstein conducted his fieldwork, including the small village of Neu-Autz where he held his first appointment as pastor. The density of place names noticeably decreased the further eastward one moved across the Latvian language area, with only the major towns and cities marked in western Vitebsk province. In 1882 Bielenstein travelled to western Vitebsk to study the province's Latvian-speakers and described the region as a terra incognita: 'a departure into regions, some of them completely unknown, into an area that was not visited by tourists, who have endowed history and legend, even the youngest, with all the horrors of insecurity and ineffability.'[51] The eastern border of the Upper or High Latvian dialectal area is drawn as an abrupt line at the map's right-hand edge, based on the former borders of Polish Livonia. This cartographic approach resembles the colonial practice of drawing straight-line borders on maps of the Middle East and Africa where imperial administrators could not obviously identify geographic and ethnographic borders.[52]

The Bielenstein-Döring map of the Latvian language area signalled a key change in approaches to ethnographic mapmaking in the Baltic provinces. The map challenged the idea of imperial administrative borders as the main frame of reference for thinking about the Latvian-inhabited territory. Instead, the map's frame was guided by the limits of the Latvian language area. The top half of Livland province was cut off to the north and the map extended eastward to encompass a large part of western

[47] Döring, 'Über die Herkunft der kurländischen Letten': 49.
[48] Ibid: 48–9.
[49] Marten, 'Latgalian Is Not a Language'.
[50] On the concept of perceptual dialectology, see Iannàccaro & Dell'Aquila, 'Mapping Languages from Inside'.
[51] Bielenstein, 'Reiseskizzen aus dem Oberlande': 570.
[52] Green, 'Lines, Traces, and Tidemarks': 68.

Plate 1. Peter Köppen's *Ethnographic Map of Finland* (1846). The map depicts the geographical distribution the following ethnolinguistic and religious groups: Finns (yellow), Karelians (orange, subdivided into Lutherans, mixed Lutheran and Orthodox, and Orthodox), Russians (red), Finns and Swedes (green), Swedes (blue), and Sámi, listed as Lapps (grey). The pigment used to colour Sámi (grey) and Lutheran Karelians (light orange) has faded over time. Notice how the map also indicates the Russian presence in nearby St. Petersburg, which was not part of the Grand Duchy of Finland.

Source: Peter von Köppen, 'Finnland in ethnographischer Beziehung (avec une carte coloriée)', *Mémoires de l'Académie Impériale des Sciences de St. Pétersbourg. Sixième Série. Sciences politiques, histoire, philologie* 7 (1848): 452–64. Reproduced with permission from the National Library of Finland.

Plate 2. Peter Köppen's *Ethnographic Map of European Russia* (1851).

Source: Petr Keppen, *Etnograficheskaia karta Evropeiskoi Rossii* (St. Petersburg: Russkoe geograficheskoe obshchestvo, 1851). Reproduced with permission from the Russian Geographical Society.

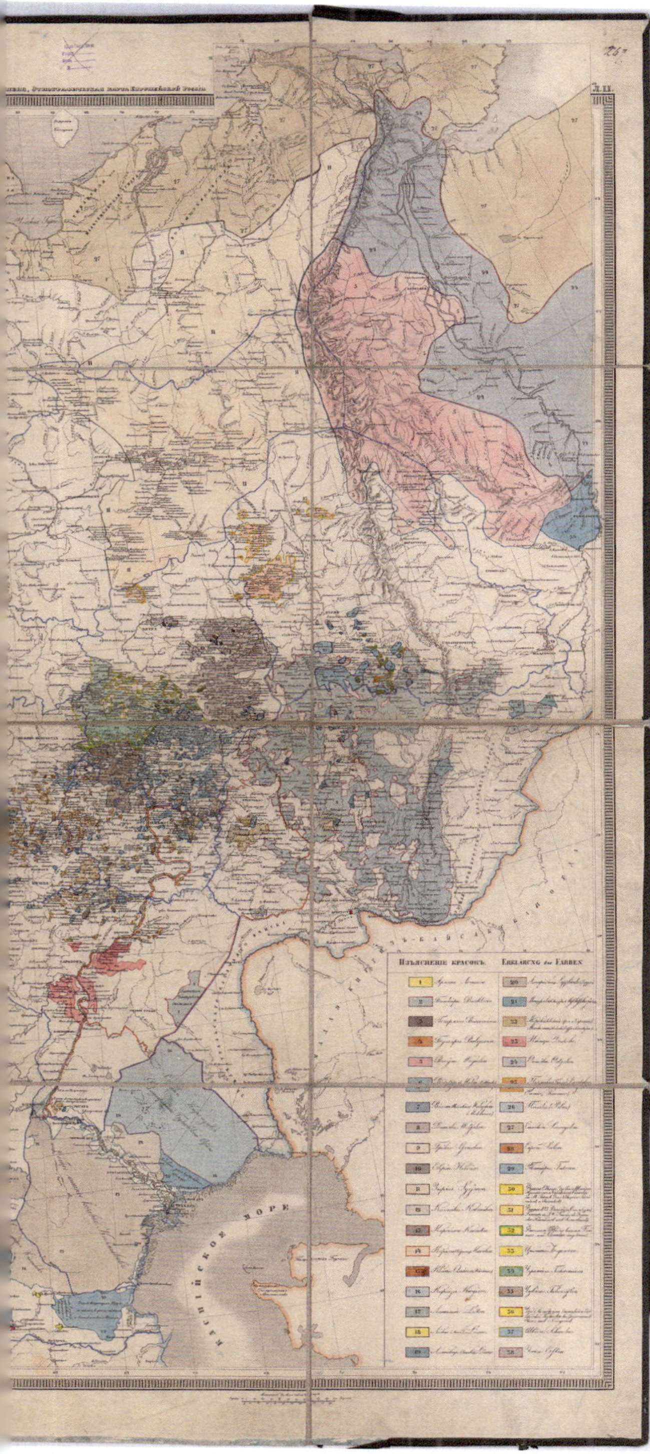
Этнографическая карта Европейской Россiи
Изъяснение красокъ
Erklärung der Farben
Каспийское море

Plate 3. Hand-coloured map of Livland province from Peter Köppen's *Ethnographic Atlas of European Russia* (1848). Latvians are depicted in green and Estonians in orange. The island of Runö (Ruhnu) in the Gulf of Riga is coloured blue to denote its Swedish inhabitants. Curiously, only one German enclave is shaded in red in the south: the Ogershof manor (Ogresmuiža) estate owned by the Von Pander family, perhaps a place that Köppen was familiar with. Notice how the framing of the topographic base map is at odds with distribution of ethnolinguistic groups.

Source: List XVII. Guberniia Lifliandskaia. Petr Keppen, *Ethnograficheskii atlas Evropeiskoi Rossii* (St. Petersburg: 1848). Reproduced with permission from the Russian Geographical Society.

Чудское
Озеро
Озеро
Псковское

Plate 4. Draft of Peter Köppen's *Ethnographic Map of St. Petersburg Province.*

Source: © SPbB ARAS—St. Petersburg Branch of the Archive of the Russian Academy of Sciences. SPbF ARAN f. 30, op. 1, d. 102, l. 2.

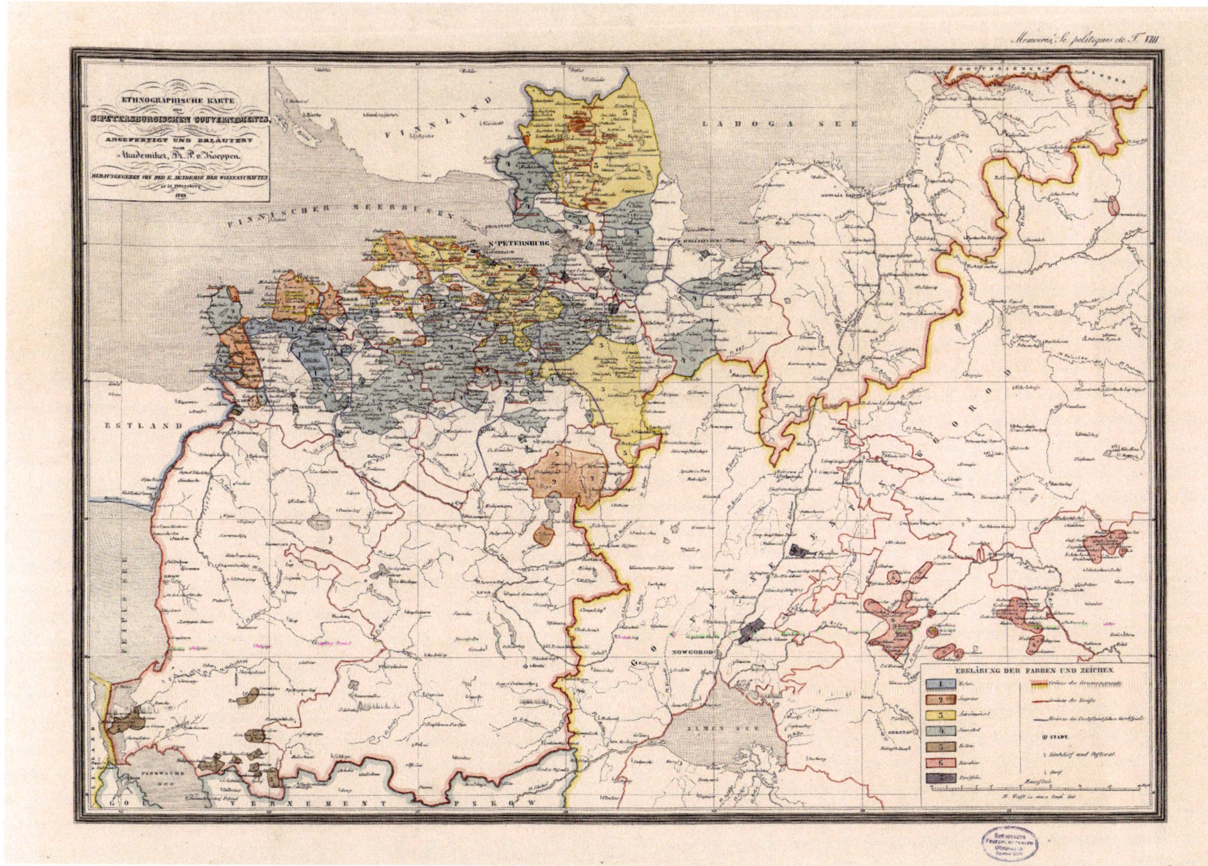

Plate 5. Peter Köppen's *Ethnographic Map of St. Petersburg Province* (1849).

Source: Peter von Köppen, *Ethnographische Karte des St. Petersburgischen Gouvernements* (St. Petersburg: Kaiserliche Akademie der Wissenschaften, 1849). Reproduced with permission from the Russian Geographical Society.

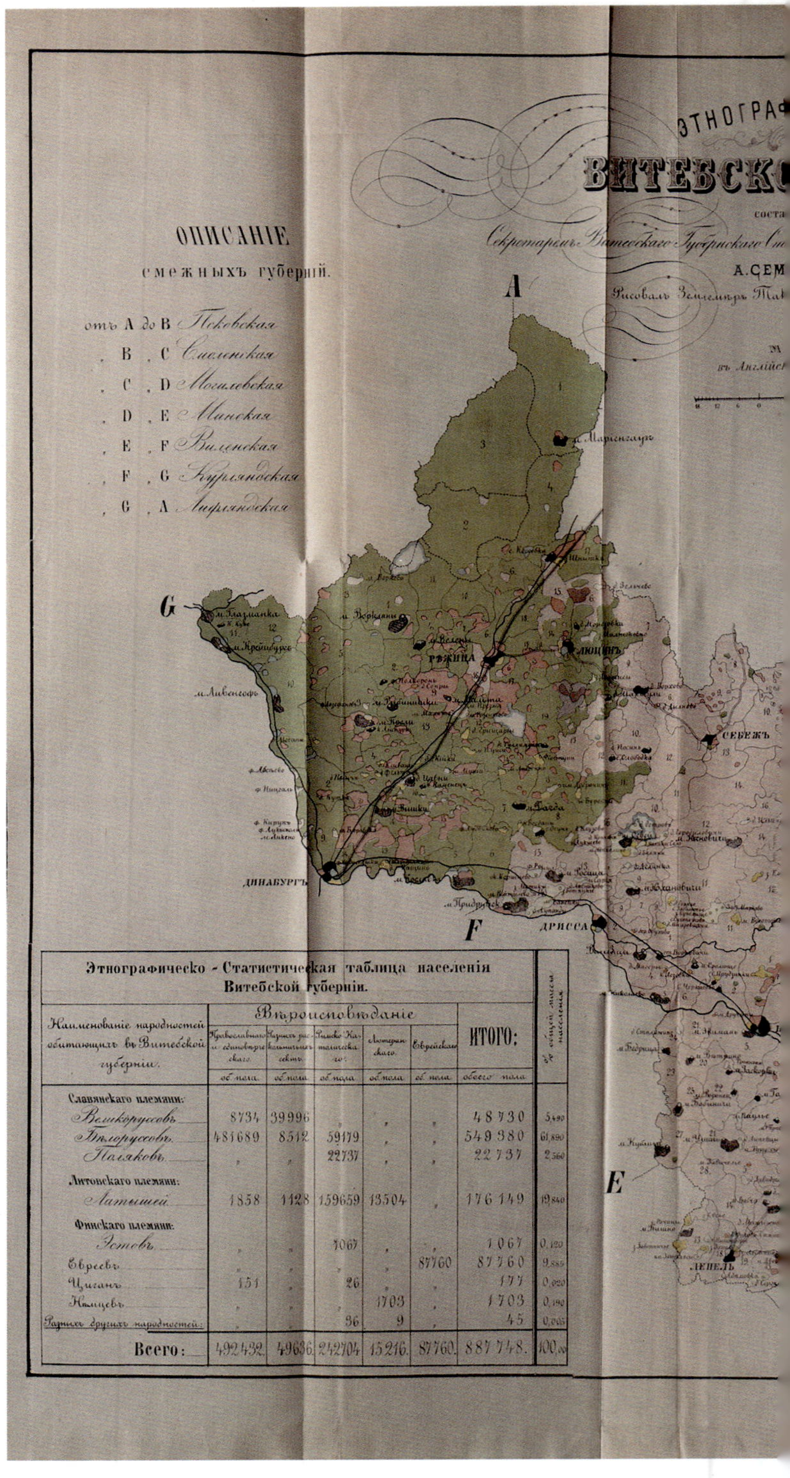

Plate 6. Aleksandr Sementovskii's *Ethnographic Map of Vitebsk Province* (1872). The map depicts Vitebsk province as starkly divided in ethnographic terms between its western Latvian (or, more precisely, Latgalian) third depicted in green and the Belarusian majority, indicated in light pink, inhabiting the eastern two-thirds. Jews (black), Russians (dark pink), and Poles (yellow) are also shown scattered throughout the province in smaller numbers.

Source: A. Sementovskii, *Etnograficheskii obzor Vitebskoi gubernii* (St. Petersburg: Tipografiia M. Khana, 1872). Reproduced with permission from the Russian State Library.

…КАЯ КАРТА
ГУБЕРНІИ
…1 году.
…скаго Комитета Статскимъ Совѣтникомъ
КИМЪ.
…нстантинъ Кондратовичъ
…ЗЪ,
… 18 верстъ.
УСЛОВНЫЕ ЗНАКИ:
МѢСТА НАСЕЛЕННЫЯ НАРОДАМИ.
Славянскаго племяни
Великорусcами
Бѣлоруссами
Поляками исключительно
вмѣстѣ съ другими народами
Литовскаго
Латышами
Эстами (Чухнами) исключительно
вмѣстѣ съ другими народами
Нѣмцами
Циганами
Евреями исключительно
вмѣстѣ съ другими народами
Дороги
Желѣзные
Шоссейные
Большие
Озера
НЕВЕЛЬ
ГОРОДОКЪ
ВИТЕБСКЪ
СУРАЖЪ
ВЕЛИЖЪ
B
C

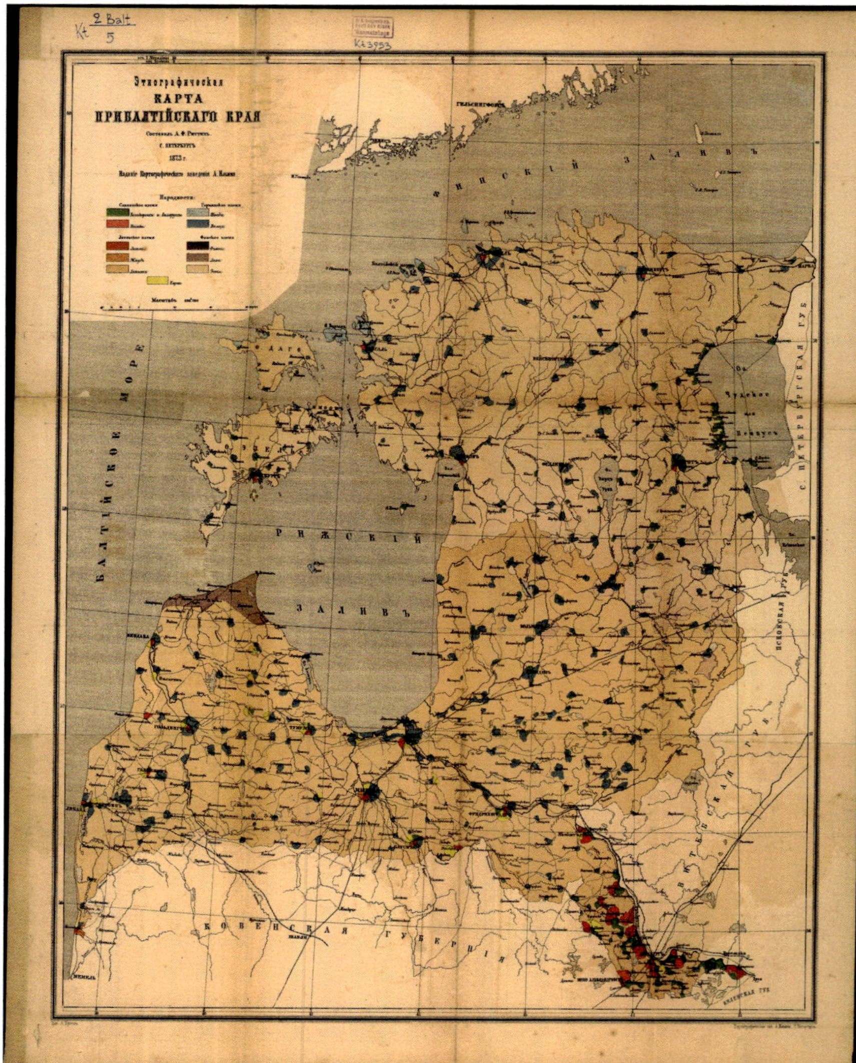

Plate 7. Aleksandr Rittikh's *Ethnographic Map of the Baltic Territory* (1873). The colours correspond to the following ethnolinguistic groups: Russians & Belarusians (together in green), Poles (pink), Swedes (blue hatched), Germans (blue), Lithuanians (orange), Samogitians (orange hatched), Latvians (dark beige), Finns (Finns), Livonians (cross-hatched brown), Estonians (light beige), and Jews (yellow).

Source: Aleksandr Rittikh, *Etnograficheskaia karta Pribaltiiskago kraia* (St. Petersburg: Kartograficheskoe zavedenie A. Il'ina,1873). Reproduced with permission from the National Library of Estonia.

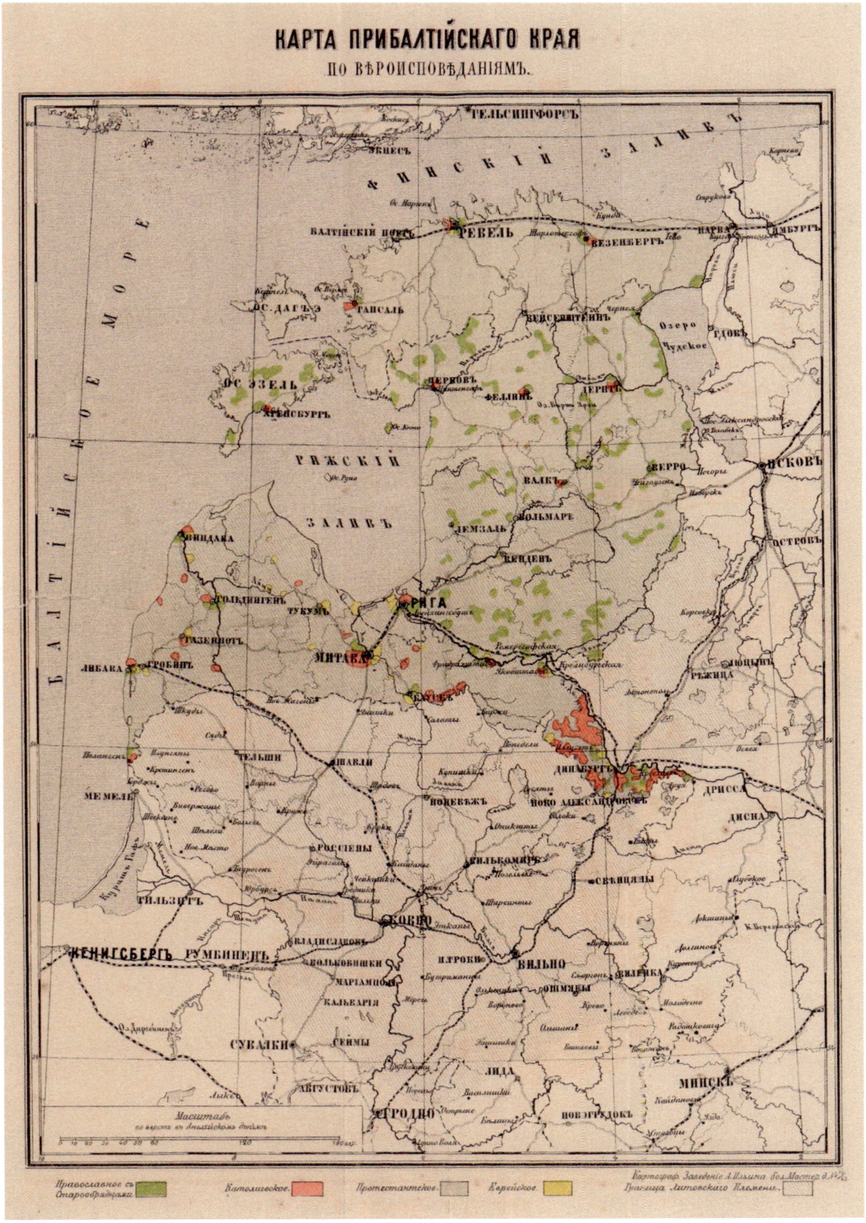

Plate 8. Aleksandr Rittikh's *Map of the Baltic Territory* according to Confession (1873). The colours correspond to the following religious groups: Orthodox and Old Ritualists (dark green); Catholics (red); Protestants (pale green); Jews (yellow). The cross-hatched area in the southern half of the map indicates the 'borders of the Lithuanian ethnic group'.

Source: Aleksandr Rittikh, *Karta Pribaltiiskago kraia po veroispvedaniiam* (St. Petersburg: Kartograficheskoe zavedenie A. Il'ina, 1873). Reproduced with permission from the National Library of Estonia.

Plate 9. Aleksandr Rittikh's *Ethnographic Map of European Russia* (1875).

Source: Aleksandr Rittikh, *Etnograficheskaia karta Evropeiskoi Rossii* (St. Petersburg: Kartograficheskoe zavedenie A. Il'ina, 1875). Reproduced with permission from the Russian State Library.

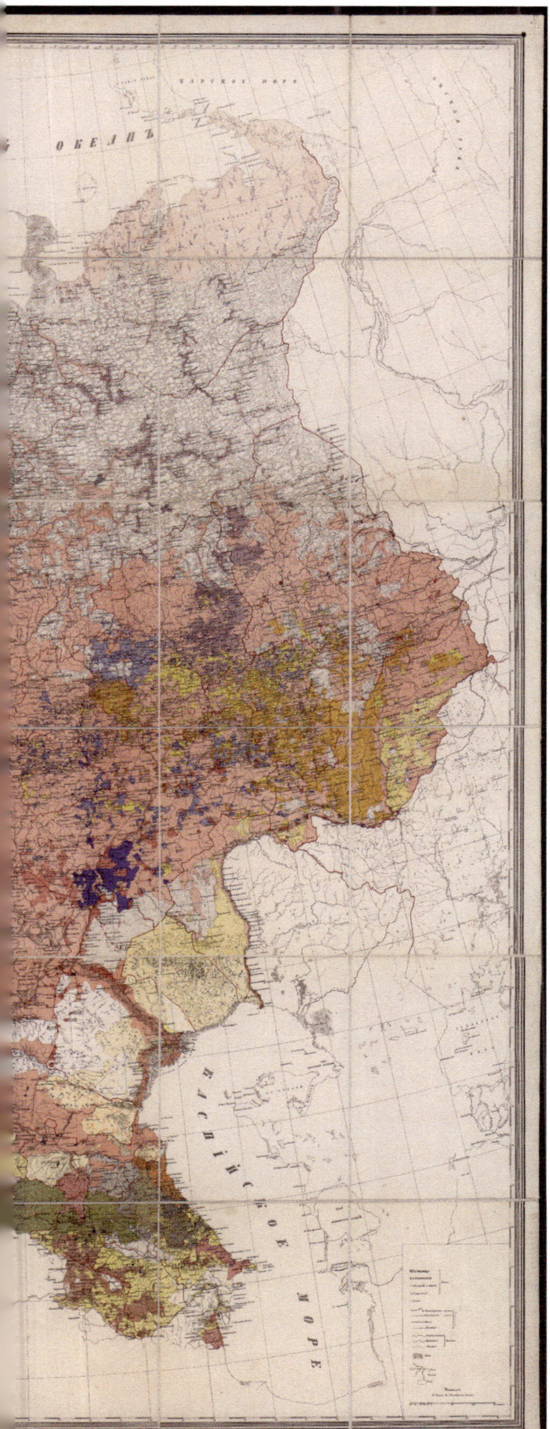

КАРСКОЕ МОРЕ
ОКЕАНЪ
КАСПІЙСКОЕ МОРЕ

Kartogramm 23. Bezirke geordnet nach der Zahl der estnischen Einwohner.

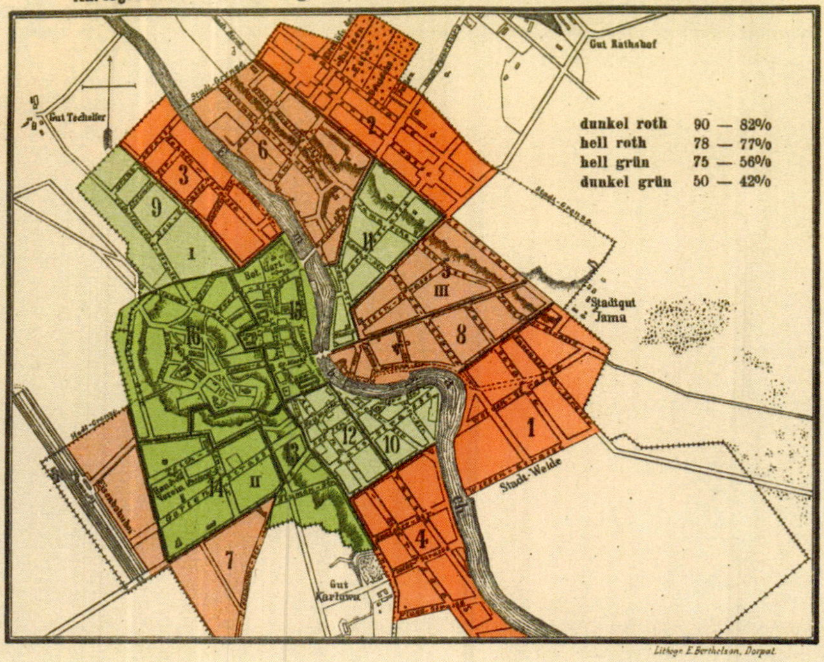

Kartogramm 25. Bezirke geordnet nach der Zahl der deutschen Einwohner.

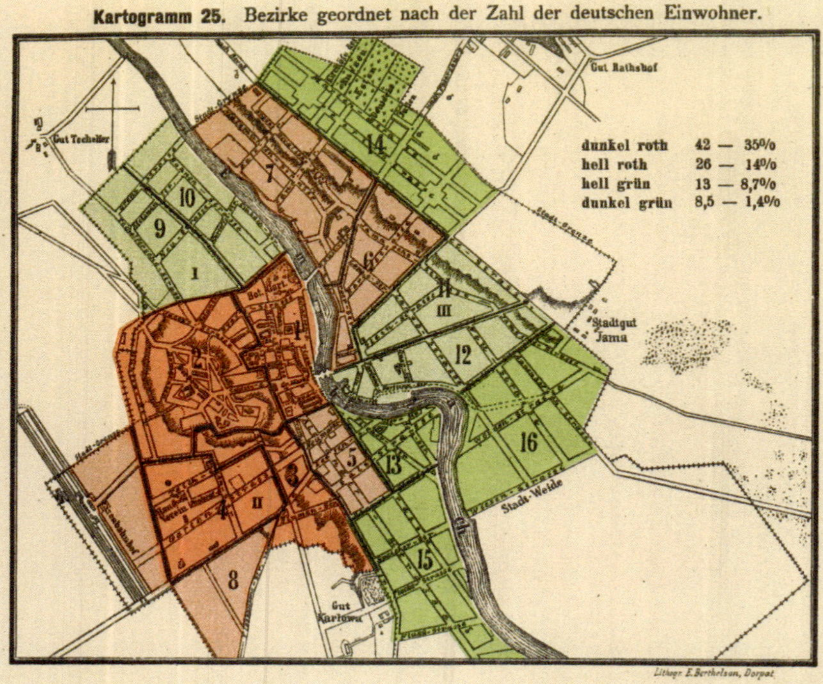

Plate 10. Bernhard Körber's city plans depicting the population density of Estonians, Germans, Russians, and Jews in Iur'ev (Tartu) in 1897. Dark red represents a high concentration of a particular ethnolinguistic group, while dark green represents a low proportion. Note how the percentages represented by the colours are not consistent so the plans are not directly comparable.

Source: Bernhard Körber, *Die Stadt Dorpat (Jurjew) in statistischer und hygienischer Beziehung* (Jurjew: C. Mattiesen's Buckdruckerei, 1902), Kartogramm 23, 25, 27, 28. Reproduced with permission from University of Tartu Library.

Kartogramm 27. Bezirke geordnet nach der Zahl der russischen Einwohner.

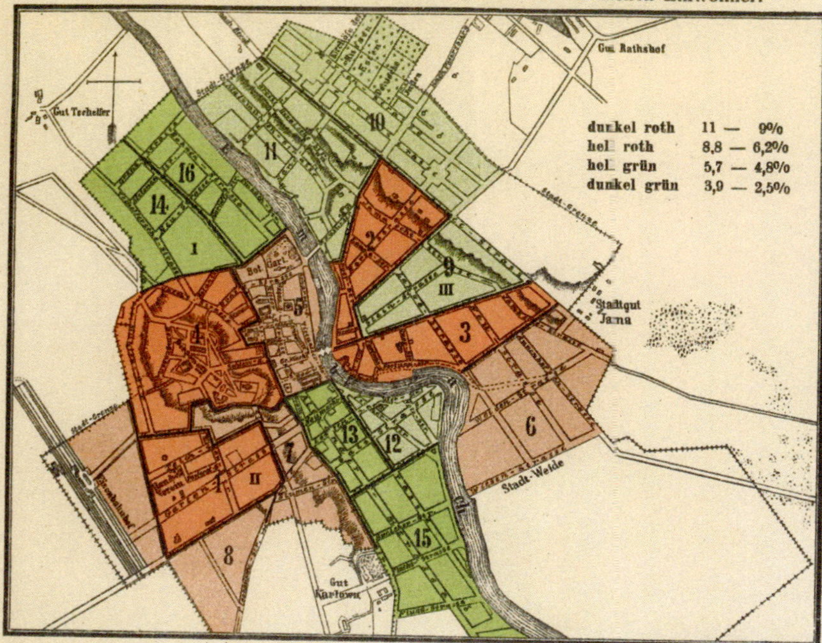

Lithogr. E. Berthelson, Dorpat

Kartogramm 28. Bezirke geordnet nach der Zahl der jüdischen Einwohner.

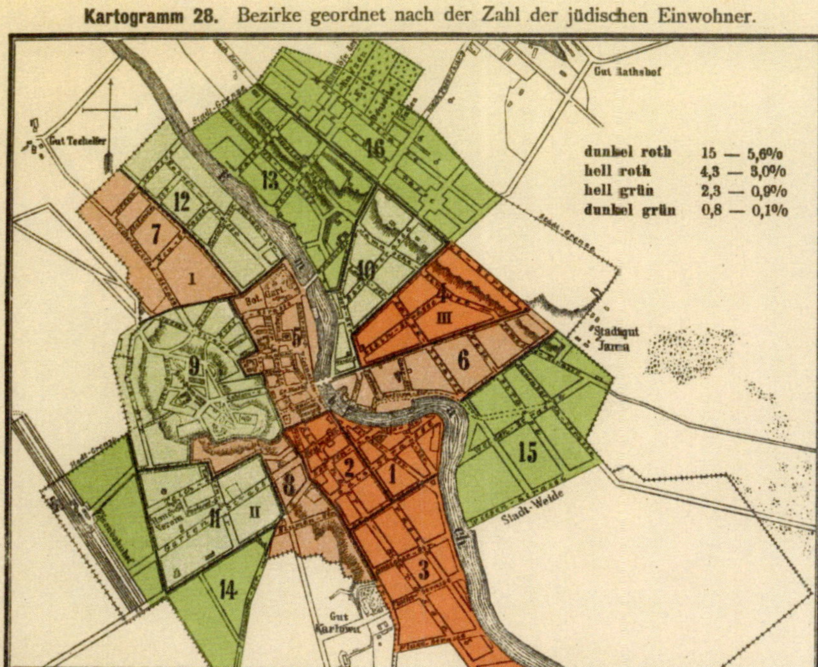

Lithogr. E. Berthelson, Dorpat

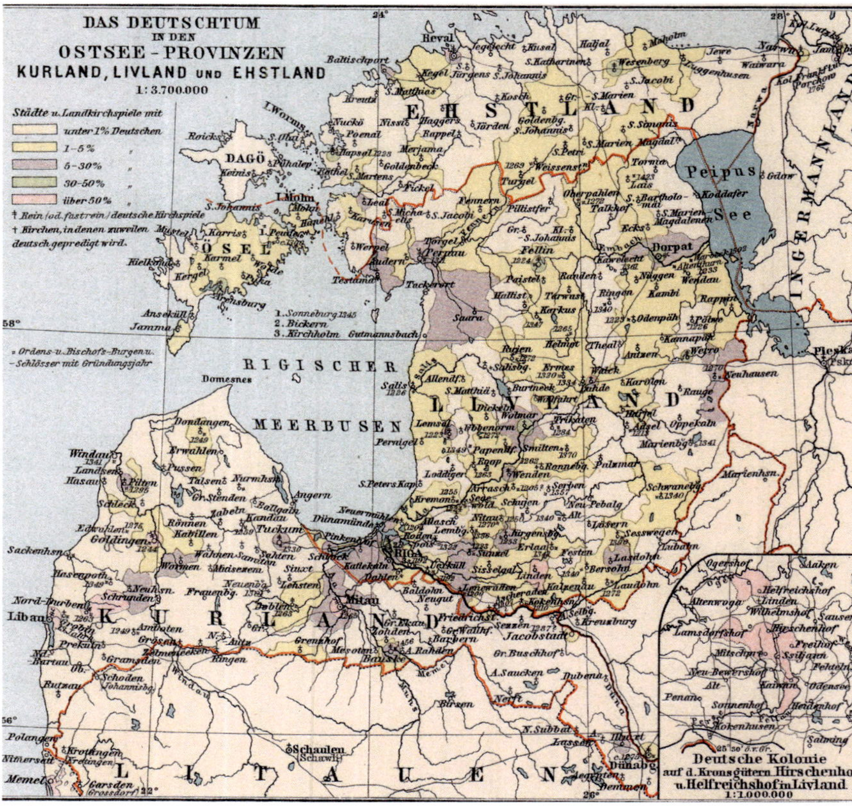

Plate 11. Excerpt of the Baltic provinces from Paul Langhans's map of *German Colonisation in the Slavic East* (1897). Beige and yellow areas of shading indicate that Germans comprise less than 1 percent and less than 5 percent of the population. Purple, green, and pink represent areas of higher concentration. Symbols indicate the placement of German parishes, parishes where preaching is sometimes in German, and the location of Ordensburgen (fortresses constructed by the Teutonic Order in the Middle Ages) and castles with their founding year.

Source: Paul Langhans, Nr. 7. Deutsche Kolonisation im Osten. II. Auf Slavischem Boden, In *Deutscher Kolonial-Atlas* (Gotha: Justus Perthes, 1897). Image courtesy of David Rumsey Map Collection, David Rumsey Map Center, Stanford Libraries. https://purl.stanford.edu/rk587fn9541.

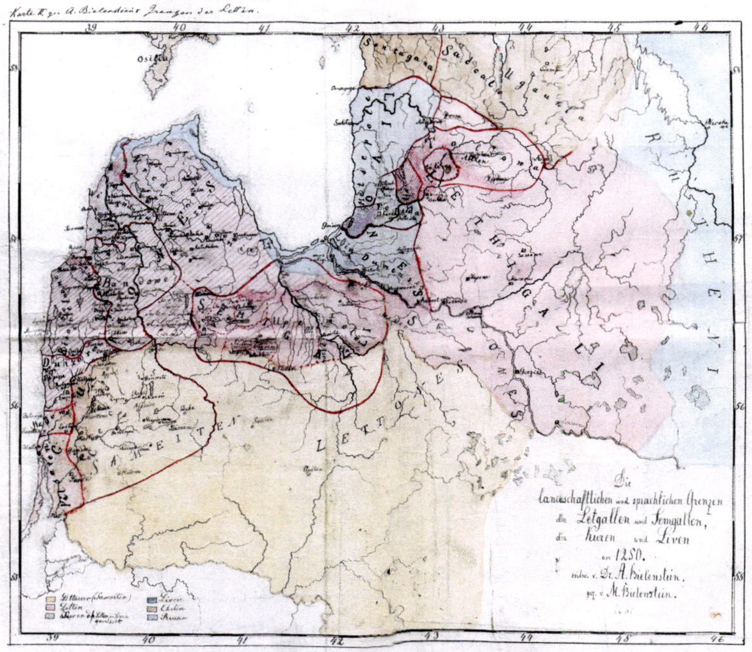

Plate 12. Martha Bielenstein's manuscript of the *Map of the Landscape and Geographical Borders of the Latvian Language Area c.1250.*

Source: LVVA f. 3946, arp. 1, l. 2, lp. 1. Reproduced with permission from the Latvian State Historical Archives.

Plate 13. August and Martha Bielenstein's map of *The Latvian Language Area in 1884.*

Source: August Bielenstein, *Atlas der ethnologischen Geographie des heutigen und des praehistorischen Lettenlandes: Beilage zu den Grenzen des lettischen Volkstammes und der lettischen Sprache in der Gegenwart und im 13. Jahrhundert* (St. Petersburg: Eggers & Co.; J. Glasunof, 1892). Reproduced with permission from the National Library of Latvia.

Plate 14. Matīss Siliņš's *Map of Latvia* (1890).

Source: Matīss Siliņš, *Latvijas karte (Kurzeme līdz ar Vidzemes un Vitebskas guberņu latviešu daļu)* (Riga: M. Siliņš, A. V. Grotusa litogrāfija, 1890). Reproduced with permission from the National Library of Latvia.

ns 1890. g.

Дозволено цензурою.–Рига 8. Сентября 1889. г.

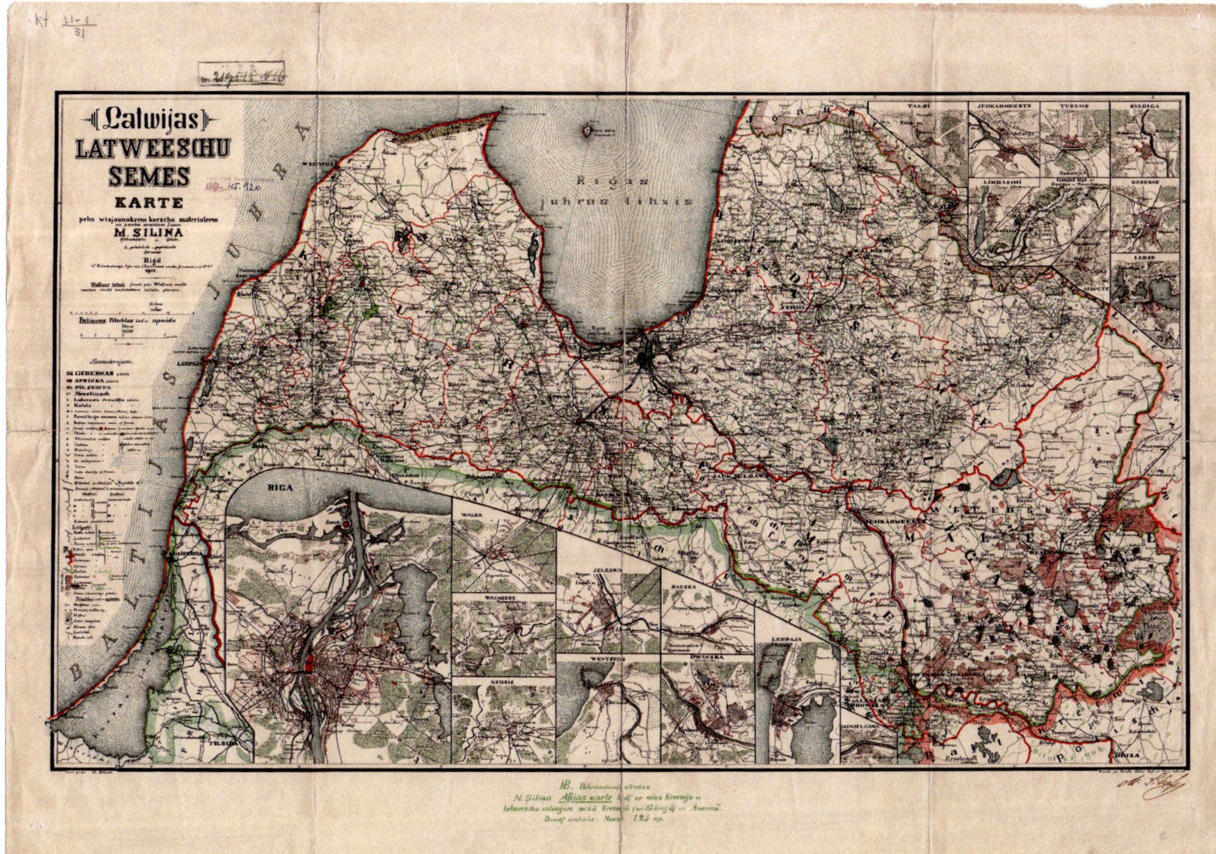

Plate 15a. Matīss Siliņš's *Map of the Latvian Land* (1911 edn).

Source: Matīss Siliņš, *Latvijas latviešu zemes karte pēc visjaunākiem karšu materiāliem un paša krātām ziņām M. Siliņa izstrādāta un izdota* (Riga: M. Siliņš, 1911). Reproduced with permission from the National Library of Latvia.

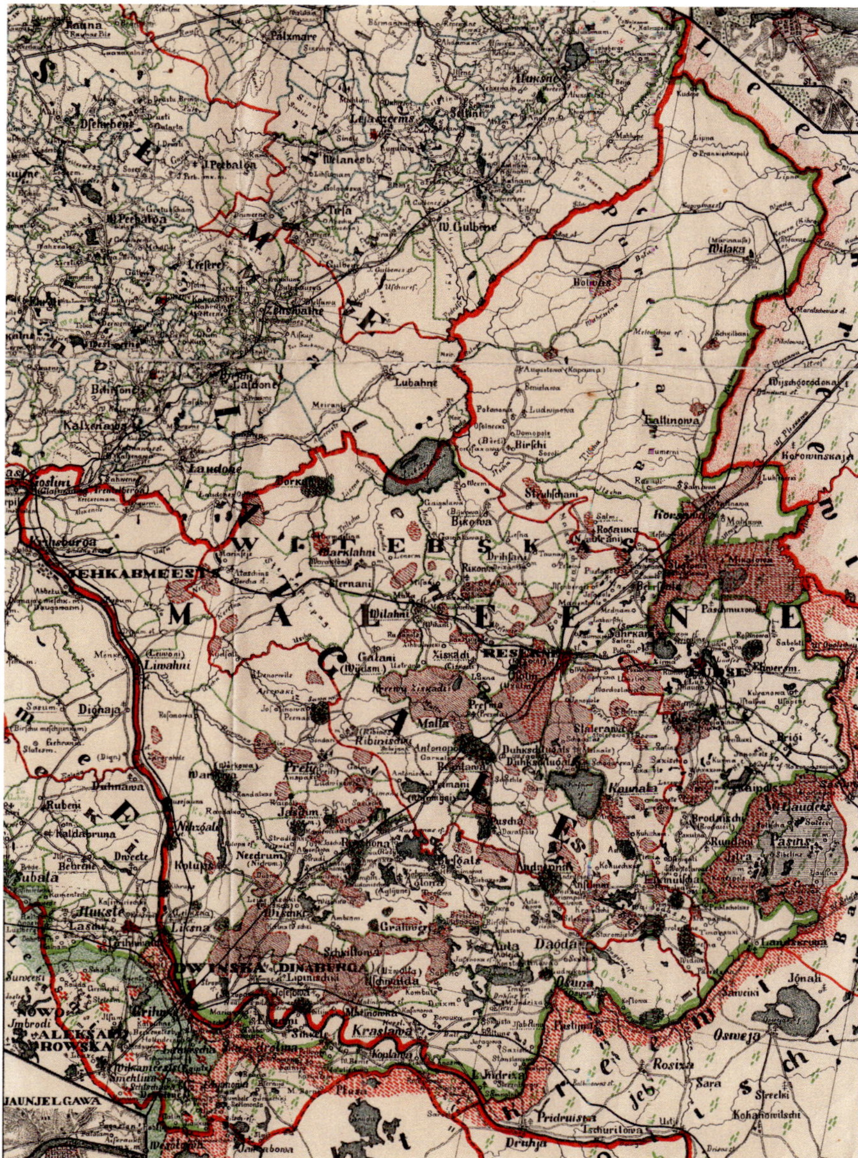

Plate 15b. Excerpt from Matīss Siliņš's *Map of the Latvian Land* (1911 edn) depicting the eastern region of Latgale. The thick green line delimits the external border of the Latvian territory. Areas inhabited by various minorities are indicated using a complex combination of colours and patterns: Belarusians (black horizontal lines overlayed with thick red diagonal dashes), Russians (black diagonal lines overlayed with red dots), Poles (black crosshatched lines covered in red, blue, and green), Estonians (blue background overlayed with small black dots and larger red dots), and Lithuanians (blue background overlayed with green dots).

Source: Matīss Siliņš, *Latvijas latviešu zemes karte pēc visjaunākiem karšu materiāliem un paša krātām ziņām M. Siliņa izstrādāta un izdota* (Riga: M. Siliņš, 1911). Reproduced with permission from the National Library of Latvia.

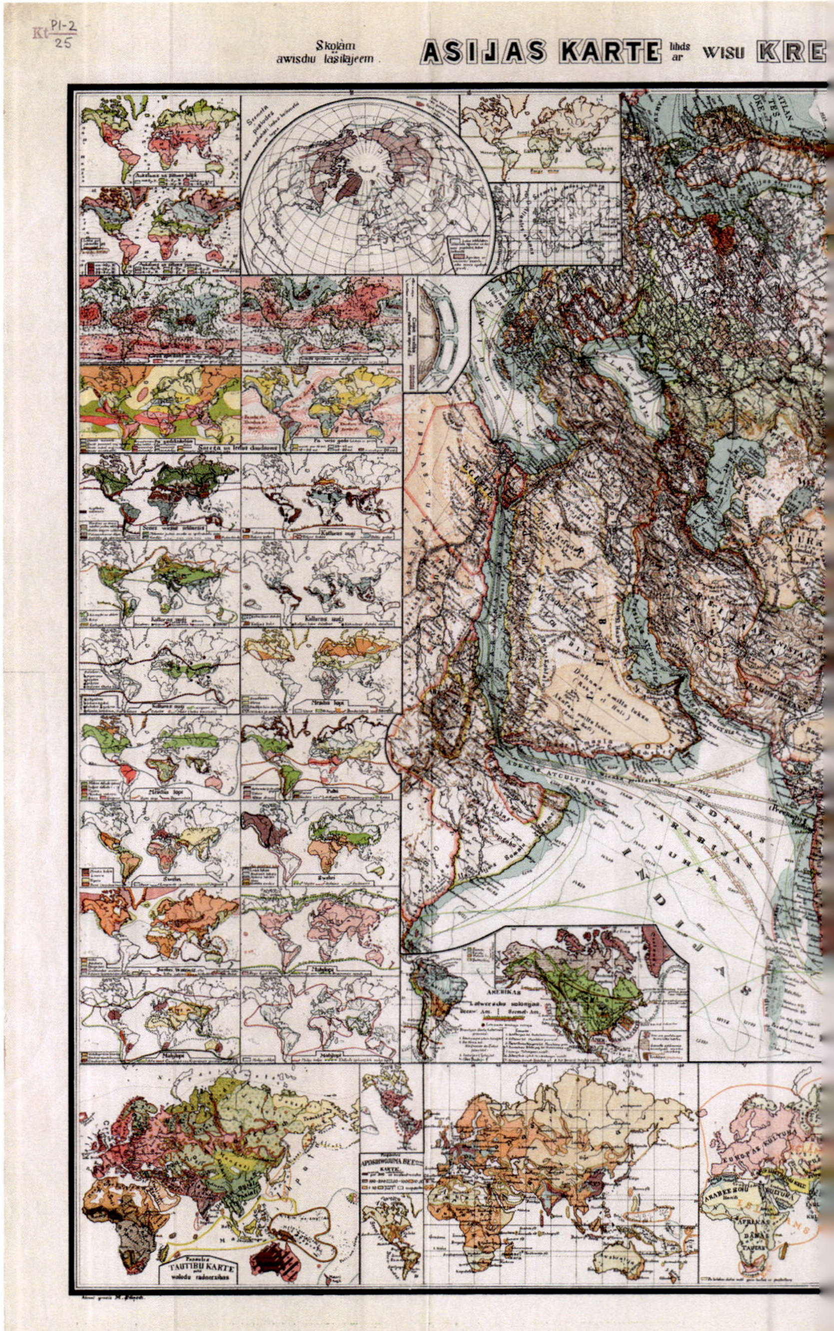

Plate 16. Matīss Siliņš's *Map of Asia* (1904).

Source: Matīss Siliņš, *Āzijas karte līdz ar visu Krieviju un latviešu kolonijām: skolām, avīžu lasītājiem* (Riga: Šnākenburga drukātava, 1904). Reproduced with permission from the National Library of Latvia.

JU un LATWEESCHU KOLONIJÀM, M. Siliņa isstrahdata un isdota.

Plate 17. Excerpt from the *National Language Border* map produced by the Estonian representatives to the Estonian-Latvian Boundary Commission in 1919. The northern line represents the 1917 administrative border, while the southern line indicates the Estonian language border.

Source: RA, ERA.T-6.1.173 sheet 1. Reproduced with permission from the National Archives of Estonia.

Листъ 4 Рядъ VI
Polli vald
Taagepere vald
Loodi vald
ПЕЧАТАНО ВЪ VIII—1915 г.

Plate 18. Map of Valga/Valka produced by the Estonian-Latvian Boundary Commission showing the distribution of property by nationality. Estonian-owned property is indicated in blue and property owned by Latvians is shaded in red.
Source: RA, ERA.1604.1.90 sheet 25. Reproduced with permission from the National Archives of Estonia.

Bahn

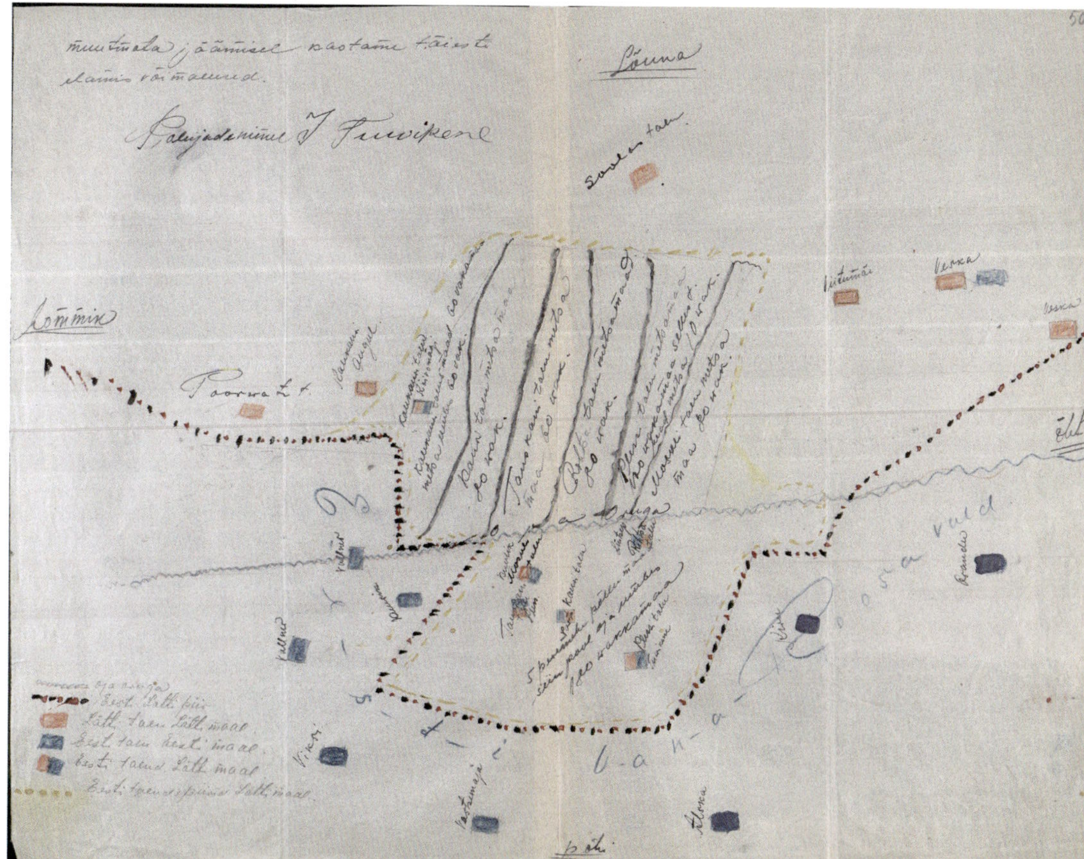

Plate 19. Petition map of the border region in Vastse-Roosa. The map depicts the Estonian-Latvian border (black and red dotted line) and contested area (yellow dotted line). The rectangles indicate Latvian farms in Latvia (red), Estonian farms in Estonia (blue), and Estonian farms in Latvia (red-blue). The map is undated, but it was sent to the Estonian-Latvian Boundary Commission sometime between August 1920 and February 1922.

Source: RA, ERA.1604.1.81 sheet 50. Reproduced with permission from the National Archives of Estonia.

Vitebsk province. Moreover, the Latvian language area and its borders were drawn based on the furthest extremes where dialects of Latvian were spoken. As a result, the Bielenstein-Döring map presented an aggrandized image of the Latvian language area that extended past Ljudsin (Ludza) and Kreslaw (Krāslava), much further eastward than on previous maps. The Bielenstein-Döring map left no doubt in viewers' minds that the western region of the map labelled Belarus (Weissrussland) was inhabited by Latvian-speakers.

Map Production in the Home and Family

The 1881 Bielenstein-Döring map of the Latvian language area has been overshadowed by Bielenstein's later, more famous cartographic work, the *Atlas of the Ethnological Geography of Present-Day and Prehistoric Latviann Lands* (1892). The atlas was published alongside Bielenstein's magnus opus on Latvian ethnography and language, *The Borders of the Latvian Ethnic Group and the Latvian Language in the Present and 13th Century* (1892).[53] Bielenstein embarked on the project to make a Latvian linguistic atlas following his research expedition through eastern Kurland and western Vitebsk province in 1882.[54] He gathered data to make the maps using a questionnaire distributed to Lutheran pastors 'south and north of the Düna' which asked the respondents to provide information about the dialectal characteristics of their locality.[55]

The process of preparing the atlas took place in the Doblen parsonage in the second half of the 1880s. The domestic setting of the atlas's production is important for rethinking places of map production in the Russian Empire.[56] There were no formal higher educational establishments in Kurland province at this time. The closest institutions of higher education were located in Livland province to the north: the Riga Polytechnic Institute (est. 1862) and the prestigious University of Dorpat (officially renamed Iur'ev between 1893–1918). From his vantage point in the parsonage on the outskirts of the small town of Doblen, Bielenstein felt that he worked 'in the rural seclusion' far away from important libraries and the 'scientific centre', which he variously identified as the Kurland's provincial capital of Mitau, Riga, or further afield in Königsberg and St. Petersburg.[57] The scientific activities of the Bielenstein household in rural Kurland affirm the continued importance of domestic settings for the production of knowledge in an era that

[53] Bielenstein, *Atlas der ethnologischen Geographie*; Bielenstein, *Die Grenzen des lettischen Volksstammes*.

[54] Bielenstein, 'Reizeskizzen aus dem Oberlande'.

[55] Bielenstein, *Ein glückliches Leben*: 293.

[56] On places of knowledge production in history of science, see Ophir & Shapin, 'The Place of Knowledge'; Livingstone, *Putting Science in Its Place*; Withers, 'Place and the "Spatial Turn"'.

[57] Bielenstein, *Die Grenzen des lettischen Volksstammes*: vi.

has traditionally been characterized by the gradual professionalization and institutionalization of scientific disciplines.[58]

The site of the atlas's production in the home provided opportunities for family members who would not otherwise have had access to academic, military, or bureaucratic places of cartographic production to participate in mapmaking processes.[59] The names listed on the maps' cartouches indicate that the process of drawing the maps was a family endeavour.[60] Six maps were drawn by Martha Luise Sophie (Figure 4.2 and Plate 13), the third eldest of the Bielenstein family's nine children, and the isogloss map was drawn by her older brother, Emil Louis Johann, a pastor like his father (Figure 4.3). Only one complete manuscript map of the atlas survives, the map depicting the language areas of different ethnic groups living in the region in the twelfth and thirteenth centuries prepared by Martha Bielenstein in 1886 (Plate 12).

Martha Bielenstein's role in drawing the maps was not unusual, but the inclusion of her name on the maps' cartouches was.[61] In the nineteenth century it was common for women to produce technical scientific drawings, especially in the field of botany, as it allowed them to apply skills they were traditionally taught as part of their education to scientific labour.[62] However, the skills typically associated with female labour, such as drawing or colouring, were often perceived as artisanal labour or handicrafts and thus not recorded or deemed worthy of attribution.[63] By contrast, we are able to learn about Martha Bielentein's role in making the maps as her father credited her contributions in his acknowledgements and on the cartouches of the printed maps.[64]

Examining Martha Bielenstein's hand-drawn manuscript map enables us to write her back into the history of the atlas's production (Plate 12). Traces of a pencil grid on the manuscript reveal how she drew the map in square sections to maintain accurate scale and proportions, thereby demonstrating her technical knowledge of cartographic draughting methods. The engravers who later printed the atlas at A. Il'in's Cartographic Establishment in St. Petersburg, discussed in Chapter 2, almost exactly reproduced her original design, from the colours and hatching used to denote the different ethnic groups to the style and positioning of the text. Some slight aesthetic modifications were made to tidy up the published map,

[58] On the home in history of science, see Opitz et al., eds. *Domesticity in the Making*.

[59] On approaches to rewriting women back into the history of geography, see Rose, *Feminism and Geography*; Maddrell, *Complex Locations*; Dando, *Women and Cartography*; Lindsay, 'Intimate Inmates'; Von Oertzen et al., 'Finding Science in Surprising Places'.

[60] On the involvement of different family members in knowledge production, see Pycior et al., *Creative Couples in the Sciences*.

[61] Güttler, 'Unsichtbare Hände'; DenBeste, 'NOT Finding Women in the Archives'.

[62] Güttler, 'Scaling the Period Eye': 14. On the increasing participation of women in science in imperial Russia from the 1860s and 1870s, see Koblitz, *Science, Women and Revolution*; Blakesley, 'Women and the Visual Arts'; Valkova, 'The Conquest of Science'.

[63] On the lack of credit given to women's craftwork and handicrafts, see Salmond, *Arts and Crafts*. On the gendered practices of working with paper in the history of science, see Bittel et al., eds. *Working with Paper*.

[64] Bielenstein, *Die Grenzen des lettischen Volksstammes*: vi.

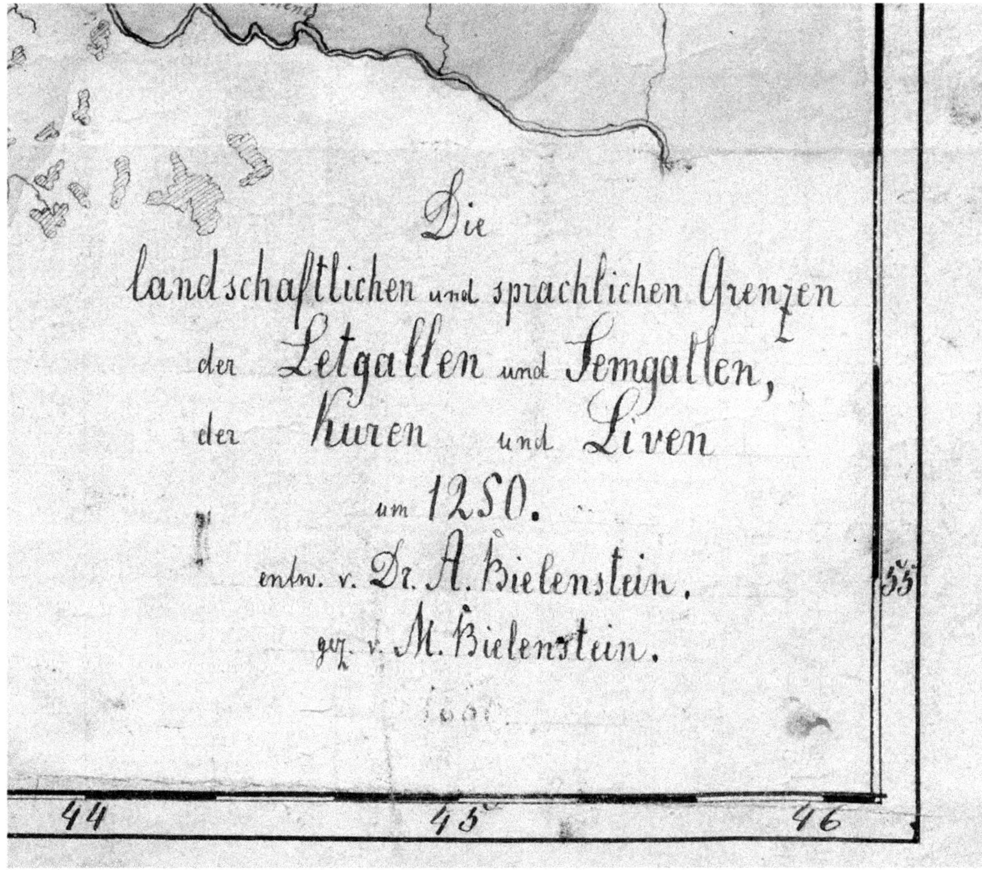

Figure 4.2. Cartouche from Martha Bielenstein's manuscript of the 'Map of the Landscape and Geographical Borders of the Latvian Language Area c.1250'. See Plate 12 for full map.

Source: LVVA f. 3946, arp. 1, l. 2, lp. 1. Reproduced with permission from the Latvian State Historical Archives.

such as in cases of placement where she had run out of space and a place name ran onto two lines or was very cramped. Nevertheless, the close resemblance between the manuscript and published version of the map indicates her key role in determining the final visual appearance of the maps in the atlas. For Martha Bielenstein—a pastor's daughter from a small provincial town—the home provided her with informal ways to be involved in mapmaking at a time when opportunities for women to pursue scientific knowledge in formal educational settings were limited.[65] Her father was blind for the last fifteen years of his life and came to completely depend on his daughter to continue researching and writing. Martha Bielenstein's responsibilities as a caregiver and her involvement in scientific work were thus deeply intertwined.[66]

The atlas manuscript was completed in the late 1880s. However, August Bielenstein struggled to find a publisher in the Baltic provinces or Germany willing to take on the project as it was unlikely to be commercially profitable. Bielenstein conceded that, unlike popular novels, the atlas was not the sort of publication that would be widely bought as a Christmas gift.[67] Instead, in April 1890 Bielenstein decided to send the atlas to the Imperial Academy of Sciences, hoping that the academicians there would recognize the scientific value of his work and provide financial support for publication. As with Aleksandr Sementovskii's *Ethnographic Map of Vitebsk Province* (1872) examined in Chapter 2, Bielenstein's atlas manuscript had to be transported by rail to St. Petersburg for printing and publication. Bielenstein recounted the thrilling tale of how his precious manuscript travelled from Kurland to St. Petersburg:

> In the middle of April 1890, I gave my manuscript to the post office and insured it for 1,000 roubles, because I did not even possess a rough draft of the manuscript. The postmaster was startled; he had never had such a precious package pass through his hands. On the platform of our station that day it was said to have been a humorous sight, as the postman of the post waggon stood with his drawn weapon in one hand and my parcel in the other.[68]

Bielenstein's investment paid off and the Academy of Sciences funded the publication of five hundred copies of the atlas.

The Bielenstein family's linguistic atlas not only contributed to the definition of the Latvian language area (Plate 13) and its dialects, but also made important theoretical contributions to the wider genre of linguistic cartography. The atlas dealt

[65] Whelan, 'The Debate on Women's Education'; Wilhelmi, 'Hauswirtschaft fur Frauen?'; Wilhelmi, 'Mädchenbildung'.

[66] Korsaka, 'Martha Bielenstein'. For a more in-depth discussion of Martha Bielenstein and the role of woman in map production in imperial Russia, see Gibson, 'Mapmaking in the Home': 71–78.

[67] Bielenstein, *Ein glückliches Leben*: 325.

[68] Bielenstein, *Ein glückliches Leben*: 325.

with the internal subdivision of the Latvian language area into dialects separately on the 'Isogloss-Map of Latvian Dialects in the Present-Day' (Figure 4.3). Bielenstein coined the term isogloss to refer to the technique of using lines to mark the geographical boundaries of certain linguistic features, such as grammatical case endings and vowel and consonant changes. The technique continues to be used on linguistic maps today. Bielenstein based this technique on the isotherms on maps of climatic conditions, another example of how cartographic techniques used on thematic maps in the natural sciences were appropriated for ethnolinguistic mapping. However, the process of adapting isotherms to languages also changed the meaning of the lines: whereas isotherms denote gradations of a continuum on weather maps, isoglosses describe the limit of a certain linguistic feature. The isoglosses were drawn from data Bielenstein gathered through correspondence with local pastors, published monographs, and oral reports of travellers, rather than pacing the border along these lines. Bielenstein thus emphasized that these lines were approximations that created the illusion of a seemingly continuous and standardized line of observations. The accumulation of lines illuminated bigger patterns and attempted to explain complex ideas to a non-specialized audience through data visualization.

The timing of the atlas's publication in 1892 coincided with the period when the imperial government implemented policies of so-called 'Russification' to erode the traditional autonomy of the Baltic provinces' German-speaking elites and to politically and culturally integrate the three provinces more closely into the empire.[69] Yet the circumstances of the atlas's publication reinforce more recent historiographical reassessments which see these policies as less one-dimensional than historians previously suggested.[70] Bielenstein's maps were published and financially supported by the Academy of Sciences, yet he continued to publish in German and use German-language versions of place names. The atlas conveyed a strong sense of the influence of the Baltic province's German-speaking elite on the social, economic, and cultural life in the region, an image which, as we saw in Chapter 3, was increasingly viewed by conservative Russian nationalists and members of the imperial administration as highly problematic.

The transnational sphere of German-language map publishing and the strong connections between the Baltic provinces and German-speaking Central Europe had proved to be an advantage for aspiring mapmakers such as Peter Köppen in the 1830s and 1840s. However, by the 1890s the changing political circumstances meant that mapmakers such as Bielenstein working and publishing in a minority language in the Russian Empire struggled to reach wider audiences. Bielenstein conceived of the atlas as having a dual aim: to contribute to knowledge of his local home region (*Heimat*) and to make a broader theoretical contribution to the

69 Hatzel, 'The Baltic Germans': 153–160.
70 Brüggemann, *Licht und Luft*: 209–10.

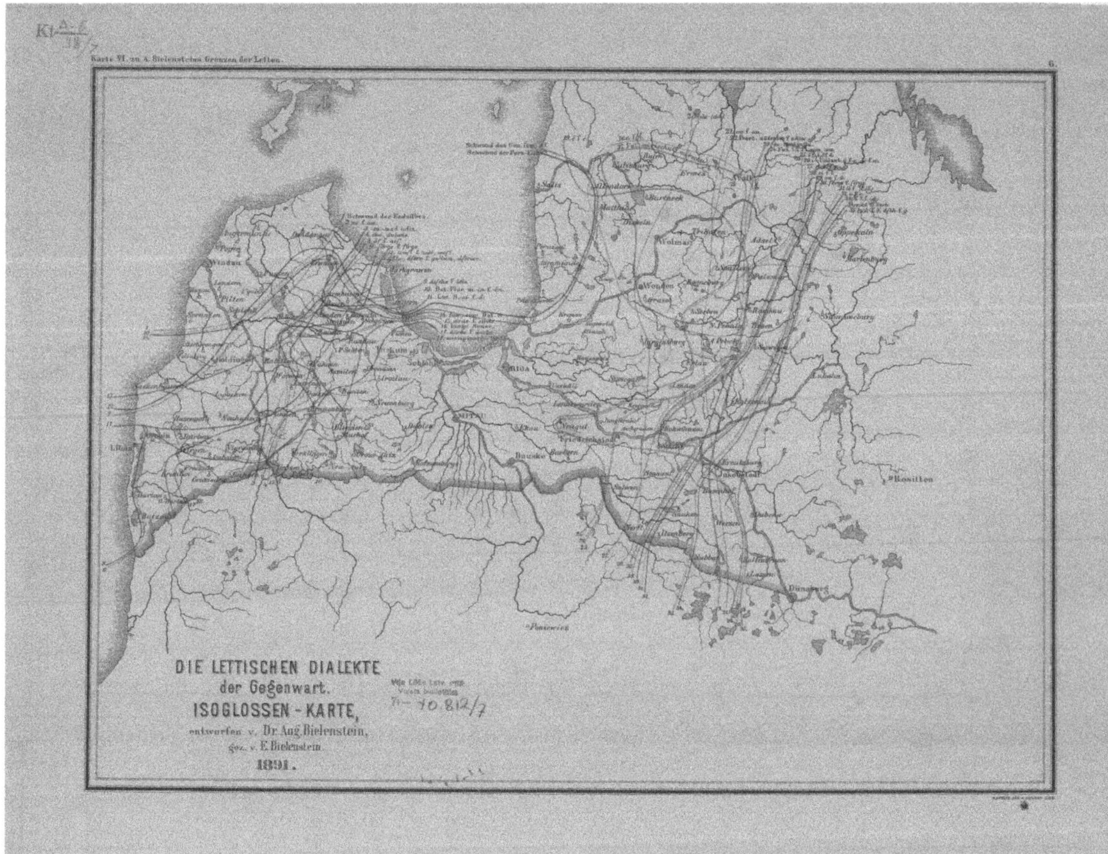

Figure 4.3. August and Emil Bielenstein's isogloss map of 'The Latvian Dialects in the Present'.

Source: August Bielenstein, *Atlas der ethnologischen Geographie des heutigen und des praehistorischen Lettenlandes: Beilage zu den Grenzen des lettischen Volkstammes und der lettischen Sprache in der Gegenwart und im 13. Jahrhundert* (St. Petersburg: Eggers & Co.; J. Glasunof, 1892). Reproduced with permission from the National Library of Latvia.

emerging field of linguistic mapping as 'a small building block to the comprehensive science of ethnological geography and history'.[71] For Bielenstein, the local focus of his maps did not mean that the subject matter was peripheral and he sought to position his work within wider European debates. However, the atlas missed the mark on both accounts. On the one hand, Bielenstein's maps were met with a mixed reception in the Baltic by Latvian-speaking intellectuals in Kurland and Livland, for whom Bielenstein represented a key pillar of a conservative, socially elite, German-speaking community in the region. Members of the Riga Latvian Society were outspoken in their criticisms of Bielenstein's publications, which they perceived as a continuation of the early nineteenth-century tradition of German-speaking elites researching the Latvian peasantry rather than working towards promoting Latvian national consciousness.[72] On the other hand, while the atlas was met with positive acclaim in St. Petersburg by the narrow circle of experts in the Academy of Sciences and Geographical Society interested in the Latvian language, they cited it as a source of empirical information.[73] The atlas did not gain widespread recognition for its theoretical and methodological contributions to linguistic mapping during Bielenstein's lifetime.

Popular Cartographic Print Culture and Map Literacy

At the turn of the twentieth century the Estonian writer Oskar Luts included a memorable scene in his popular novel, *Spring* (1912–13), about the use of maps in daily school life in the Baltic provinces. The young Estonian protagonist Toots is transfixed by a large map of Europe hanging in the classroom: 'the map held a particular fascination for him, the myriad of lines, and names strewn over it like sand on the shore! What patience it must have cost its makers!'[74] Later, Toots decides to create his own globe by carving out a block of wood, saying to his friend Imelik:

> 'I'll make one. Mark my words, and I'll paint a large red cross on it where our school stands, so we can see where we are. And I'll draw the river and....'
>
> 'Oh, don't forget the churchwarden's potato patch', mocked Imelik. 'And once you draw the river, don't forget the raft in the bottom of the river, to make the globe complete.'
>
> 'You fool, a globe is not meant for such things.'[75]

[71] Bielenstein, *Die Grenzen des lettischen Volksstammes*: vii.
[72] Ķikuts, '1896. gada Latviešu etnogrāfiskā': 47.
[73] Karskii, *K voprosu ob etnografiheskoi karte*: 2.
[74] Luts, *Spring*: 230. [75] Luts, *Spring*: 235.

The discussion between Toots and his friends reveals how maps had become familiar objects of visual and material culture by the early twentieth century, along with normative ideas about what should and should not appear on a map. Even more significantly, Toots and his friends are convinced that anyone could make a map or globe. Maps could be used to gain knowledge about faraway places, as well as being made meaningful by adding local points of reference. The scene illustrates, as Martin Brückner argues in his work on the spread of geographical literacy in early America, how education internalized mass geographical literacy in the nineteenth century and mapping came to be used by non-elites 'to define their own voice and sense of place'.[76]

Pastor August Bielenstein and his Kurlandish family of mapmakers produced maps for an audience of fellow Baltic German Lettophiles and an international audience of scholarly experts. However, in the second half of the nineteenth century the consumption of cartographic print culture was rapidly rippling through a much broader spectrum of Baltic society. The high literacy levels among the inhabitants of the Baltic provinces—which according to the 1897 census were 95 per cent for Estland, 92 per cent for Livland, and 85 per cent for Kurland—led to a growing demand for printed press.[77] The drop in prices thanks to developments in the paper industry and printing technology led to a booming market for cheap paper products.[78] By the middle of the nineteenth century commercial lithographic firms in Reval, Dorpat, Riga, and Mitau began publishing maps in Estonian and Latvian, in addition to German, to cater for local demand.[79] Newspapers, illustrated journals, and missionary publications increasingly featured maps as part of their visual content.[80] The rise of Estonian and Latvian cartographic publishing allowed mapmakers to challenge the Baltic German and Russian monopoly over map production and present alternative ways of seeing and thinking about the world.[81]

The beginnings of Estonian and Latvian vernacular cartography are usually traced back to the printing of place names in these languages on seventeenth-century maps of the Sweden Empire.[82] At the end of the eighteenth century, Ludwig August von Mellin published a fourteen-page atlas of Livland province between 1781–1810, which marked place names in Estonian and Latvian

[76] Brückner, *The Geographic Revolution*: 238.

[77] Raun, 'Literacy in the Russian Empire': 66. On the growth of popular print culture in the Baltic provinces and wider Russian Empire, see Zelče, *Latviesu avizniecība*; Brooks, *When Russia Learned to Read*.

[78] Mashkina, 'The Pulp and Paper Industry'; Rowley, *Open Letters*.

[79] *Ausgewählte Werke aus dem Verlage*: 3; *Werden und Gedeihen der graphischen Kunstanstalt*; Jantson, 'Booksellers, Publishers and Press Workers'.

[80] Chuchvaha, *Art Periodical Culture*; Kosonen, 'Making Maps and Mental Images'. For a broader overview of the emergence of the illustrated press during this period, see Mainardi, *Another World*.

[81] For a discussion of the 'print revolution' in the Russian Empire's border regions, see Khalid, 'Printing, Publishing, and Reform'.

[82] Mardiste et al., 'The Development of Estonian National Cartography': 108.

vernaculars.[83] From the mid-nineteenth century the first maps solely with Estonian and Latvian text began to be published, usually as components of geography textbooks or as school atlases. The first map published fully in Estonian, *Europe and Palestine*, appeared as an appendix to a geography textbook in 1854.[84] The map was published in Dorpat (Tartu) and was paid for by the nobleman Karl Eduard Liphart, an avid art collector and founding president of the Estonian Naturalists' Society. The subject of the map, which positioned Europe in relation to Palestine and the Holy Land, reflected the close relationship between religious pedagogy and the emphasis in geographical education on learning the location of places mentioned in the Bible.[85] Several Estonian-language geographical school atlases were published in the following years.[86]

The earliest Latvian-language maps developed closely alongside their Estonian counterparts.[87] In 1859, the first Latvian-language geography textbook, *Our Self-Description of the Land*, was published by writer and collector of Latvian folk-songs (*dainas*) Krišjānis Barons, who was a student at the University of Dorpat at the time.[88] The map included in the textbook was the first to explicitly depict a territory defined as the 'Latvian lands', comprising the administrative districts of Kurland and southern Livland. Notably, western Vitebsk province was not included within Baron's conceptualization of Latvian territory. The first full Latvian-language geographic atlas published in Mitau in 1861 by teacher and publicist Ernests Dinsbergs was almost identical to the Estonian-language atlas published two years prior as the two maps were both printed in Dorpat by the same firm and used the same lithographic stones.[89] Unlike Lithuanian-language geography textbooks and maps, which were forced to circulate illegally due to the ban on printing in Latin letters in the North-Western Territory between 1864–1904, Estonian- and Latvian-language maps were not subject to the same restrictions.[90] Until the 1860s map production in the Baltic provinces was subject to the centralized system of censorship, meaning that the publication of *Geographic Map Book* (1859) had to be approved by the Military Topographic Corps in St. Petersburg.[91] From the 1860s, the censorship of school atlases and wall maps was transferred to the local general censor in Riga.

[83] Mellin, *Atlas von Liefland*. On Mellin as an Estophile, see Jürjo, 'Ludwig August Krahv Mellin'.

[84] *Keograhwi*; Mardiste et al., 'The Development of Estonian National Cartography': 109.

[85] Lembit, *Estonskie narodnye shkoly*: 162.

[86] Other notable examples of Estonian-language geographical atlases for schools include: Jakobson, *Kooli atlas*; Kentmann, *Geograahwia Kaardid*.

[87] Apīnis, 'Latviešu valodā sarakstīto ģeogrāfijas mācības grāmatu'.

[88] Barons, *Mūsu tēvzemes aprakstīšana*; Dzlyuma, 'Vklad absolventov'.

[89] Krūmiņa, 'Zinātniskas'; Karma & Pātsi, 'Kā tapis pirmais ģeogrāfiskais atlants'.

[90] Petronis, *Constructing Lithuania*: 229; Dolbilov, 'Vicissitudes of Cyrillization'.

[91] Schultz, *Maa kaardi-ramat*. On the granting of approval to publish this geography textbook, see RA, EAA.321.1.228.75; *Allgemeines Kurländisches Amts- und Intelligenzblatt* 52, 24 June 1852: 1.

School atlases and maps proliferated with growing efforts to teach geography as part of an enlightened educational curriculum and emerged as powerful shapers of geographic imagination.[92] The focus on producing geographical materials for schoolchildren reflects the importance attached by contemporaries to the role of 'local studies' (*kraevedenie* in Russian, *Heimatkunde* in German) as a means of heightening 'regional consciousness'.[93] In the Baltic provinces, geography was gradually introduced as a compulsory part of the school curriculum from the 1870s.[94] At first, teaching materials were often imported from abroad. In October 1872 the Estland Provincial Consistory placed an order to Berlin for 221 maps of the globe and 243 maps of Europe to supply Lutheran schools in Estland province.[95] Thanks to changing language policies in the 1880s—whereby Russian replaced German in universities, secondary schools, and some primary schools in the Baltic—by the 1890s Russian-language maps published by A. Il'in's Cartographic Establishment could also be found in schools in the Baltic provinces (Figure 4.4).[96] However, as Matīss Siliņš noted in 1890: 'So far we have a complete German-language map about our homeland, or Russian... but it is, of course, not useful for Latvians: the names of places are in a foreign language, but they sound very different in the Latvian language.'[97]

The maps and textbooks produced for these schools were priced cheaply to achieve maximum dissemination. The publisher of the first Estonian-language geographical atlas advertised that the price of the map had been set 'as cheap as possible' (1 rouble and 20 kopeks) and free copies were distributed to schools.[98] Yet, the initial selling price was still prohibitive and several months later the publisher halved the price to 60 kopeks.[99] Similarly, Carl Robert Jakobson's sixteen-sheet Estonian-language *School Atlas* (1872) was also published in a cheaper four-sheet version for students to purchase.[100] Nevertheless, the prevalence of German- and Russian-language maps and atlases suggests that the distribution of Estonian- and Latvian-language cartographic print culture remained relatively limited.

Teachers also frequently complained that these wall maps, atlases, and geography textbooks were not very effective pedagogical tools. In 1880, Hugo Lieven, headteacher of the gymnasium in Pernau (Pärnu), complained that geography textbooks presented students with lists of facts and statistics. Pupils passively memorized this information and then quickly forgot everything, 'without bringing the student any benefit'.[101] Instead, many teachers advocated a more active

[92] Baranskii, *Istoricheskii obzor uchebnikov geografii*; Baár, *Historians and Nationalism*: 96.
[93] Loskutova, 'S chego nachinaetsia rodina?'; Applegate, *A Nation of Provincials*.
[94] Lembit, *Estonskie narodnye shkoly*: 162.
[95] RA, EAA.321.1.228.75.
[96] On the spread of Russian-language textbooks, see Paert, 'The Russian Imperial Practices'.
[97] Siliņš, 'Pie kartes'. *Altbass kalendārs*, 1890: 89–90.
[98] *Pernausches Wochenblatt*, no. 21, 23 May 1859: 9.
[99] *Perno Postimees ehk Näddalileht*, no. 46, 17 November 1865: 374.
[100] Lembit, *Estonskie narodnye shkoly*: 191.
[101] Lieven, *Die Revision des Lehrplanes*: 30.

Figure 4.4. Pupils at Peningi village parish school in Estland province. The map hanging on the wall is the *Map of European Russia*, published in Russian in multiple editions in the second half of the nineteenth century by A. Il'in's Cartographic Establishment. The photo was taken by teacher and photographer Heinrich Tiidermann c. 1896. Along with the blackboard, abacus, desks, and schoolbooks, the map is presented in the staging of the photo as a key classroom prop which, as Sumathi Ramaswamy argues, 'materializes the aspirational intentions' of using maps to teach young children about their place in the world.[102]

Source: Vallakool Peningil, Harju-Jaanis, AM _ 13741:342 F 11684:168. Eesti Ajaloomuuseum SA, http://www.muis.ee/en_GB/museaalview/2037590. Public domain.

approach to using cartography in the classroom. In his instruction manual for teaching 'local studies' (*Heimatkunde*), Gustav Blumberg, director of the German-language gymnasium in Dorpat (Tartu), encouraged teachers not only to display maps on the classroom wall, but to assign students the task of producing their own maps. Blumberg specified that pupils should be taught aspects of technical drawing, such as how to work with quadrants and scales. Cartographic literacy and mapmaking were incorporated into the school curriculum. Blumberg instructed teachers to include mapmaking in their lesson plans to develop pupils' geographical outlook on the world, as 'especially through their drawing and the related discussions, the youths learn a lot'.[103]

[102] Ramaswamy, *Terrestrial Lessons*: 209.

[103] Blumberg, *Heimathskunde*: 21.

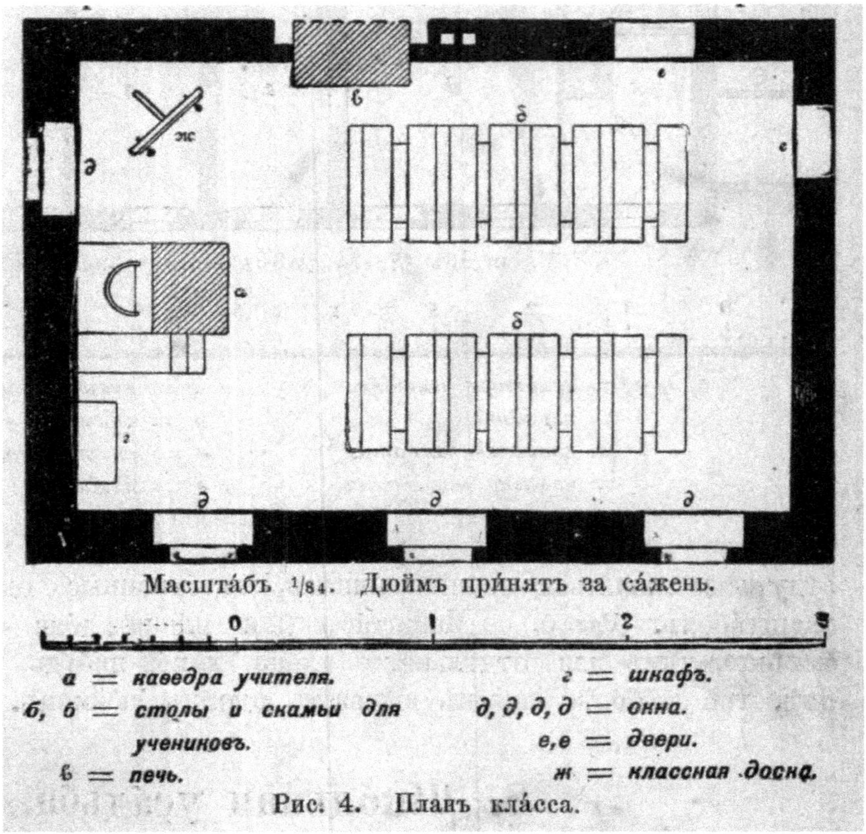

Figure 4.5. Instructional plan from a Russian-language geography textbook published in Riga in 1914 teaching primary school pupils how to draw a plan of their classroom.

Source: E. Andreev & K. Stepanov, *Kratkii uchebnik geografii Rossii dlia nachal'nykh uchilishch: primenitel'no k programme dlia volostnykh i gorodskikh nachal'nykh uchilishch* (Riga: K. H. Zikhman, 1914): 5. Reproduced with permission from the National Library of Estonia.

Map literacy was not limited to secondary level educational establishments. Martha von Grot, a pedagogical reformer and director of the Dorpat teacher seminary, argued in her lecture on teaching *Heimatkunde* at the meeting of Baltic German teachers in Riga in 1907 that schoolchildren as young as 9 or 10 should be able to 'understand maps of unseen landscapes'. According to von Grot, teachers could measure a pupil's understanding of cartography by whether the child was able, when 'seeing the map of their hometown, they imagine the hometown itself, just as they saw it in nature', but also 'vice versa, when the child is able, when looking at the hometown out there, to think of how to transfer it themselves to map symbols'.[104] Geography textbooks for primary schools introduced pupils

[104] *Arbeiten und Ergebnisse*: 79.

to the concept of mapping by first getting the children to draw plans of everyday objects, such as a book, and then apply the same principles to draw maps of their classroom, schoolhouse, and school grounds (Figure 4.5).[105] Teachers aspired to impart a deeper understanding of cartography, whereby children had the knowledge and skills to both read maps made by others and to create maps themselves of the world around them. Map literacy was perceived by teachers to be a key component of geography teaching over the course of the second half of the nineteenth century. Maps gradually entered into the familiar material and visual culture of everyday life in the Baltic provinces and 'map-mindedness' began to spread among the general population.[106]

The Commodification of Cartography: Selling Maps to Latvian Readers

The cartographic activities of Matīss Siliņš in the 1890s and 1900s played a key role in the development of Latvian-language cartographic print culture and map consumption. Siliņš was a generation younger than Bielenstein. He was born in Strutele parish in western Kurland province in 1861 into a Latvian-speaking family.[107] He trained at the teaching seminary in Irlava and at a pedagogical institute in Halberstadt in Saxony, before working as a teacher in Kurland and Moscow. Upon his return to the Baltic provinces in 1888, he became an active member of the Riga Latvian Society. From 1903–21 he was the sole employee of the Riga Latvian Museum, organized by the Riga Latvian Society, acting as the museum's director, curator, and tour guide. He was renowned for guiding groups of visitors around the museum every Sunday and sharing his knowledge through public education. Siliņš conducted numerous ethnographic expeditions to gather artefacts for the museum, for which he gained a reputation as a magpie-like scavenger of Latvian material and oral culture.[108] Alongside his work at the museum, Siliņš regularly contributed to the Latvian-language newspaper *Rīgas Avīze*, wrote poetry, and produced numerous translations of classical literature into Latvian.

Over the course of his lifetime Siliņš published nineteen large and medium-scale maps, and over fifty inset maps, earning him the accolade of the 'first Latvian cartographer'.[109] Siliņš's main years of cartographic activity in the 1890s and 1900s coincided with a period of expanding audiences for maps through the rise of popular press cartography in newspapers, journals, and magazines.[110] To the

[105] Andreev & Stepanov, *Kratkii uchebnik geografii*: 5.

[106] Kivelson, *Cartographies of Tsardom*: 27.

[107] Details of Siliņš's biography are taken from Karnups, 'Matīss Siliņš'; Karulis, 'Matīss Siliņš'; Štrauhmanis, *Matīss Siliņš*; Štrauhmanis, 'Matīsa Siliņa ietekme'; Stinkule, 'Skolotājam, literātam un tulkotājam'; Soboļeva, ed. *Sava ceļa gājējs*.

[108] A caricature of Siliņš depicted as a scavenger appeared in: *Lietuvēns*, no. 9, 27 February 1915.

[109] Štrauhmanis, *Matīss Siliņš*.

[110] Kosonen, 'Making Maps and Mental Images': 27.

north in the Grand Duchy of Finland, the Geographical Society of Finland famously published the *Atlas of Finland* (1899), considered the first national atlas in the world and which played an important role in emerging ideas about Finnish nationhood in the Russian Empire.[111] Siliņš's maps exemplified the movement in the late nineteenth century towards the vernacularization of cartography, the use of maps as pedagogical tools, and as accessible form of geographic knowledge for the wider Latvian reading public.

Siliņš started making maps following his move from Moscow to Riga in 1888, when he began publishing a Latvian-language popular science calendar almanac, *Atbalss kalendārs* (Echo Calendar). Calendars were one of the most popular literary genres aimed at lower-class readers and could be found in households throughout the Russian Empire.[112] Like many other nineteenth-century calendars, *Atbalss kalendārs* not only included calendar tables and lists of important dates, but also fiction and popular science articles about a wide range of different practical topics, such as how to make butter, trim hedges, grow tobacco, and preserve fruit for winter.[113] From 1893, Siliņš incorporated a series of articles entitled 'Collection of patriotic descriptions', spanning history, natural history, ethnography, and geography. Unusually for the time, Siliņš published his calendar with supplementary maps. By doing so, Siliņš was seeking to fill an important gap in the literary marketplace. While Latvian-language newspapers and book publishing proliferated in the 1880s, Siliņš felt that cartographic publishing still lagged behind. Siliņš wanted to encourage Latvian-speakers to explore their homeland in all its human and physical geographical dimensions. By incorporating information about hydrology, geography, geology, and ethnography, he conceived of the calendar and maps as popular science literature and as instruments of public enlightenment.

Like many of those involved in the book trade in late imperial Russia, Siliņš was both the publisher and primary author of the calendar. Reviewers frequently characterized Siliņš as 'a local cartographer' and praised the 'self-made' nature of his maps.[114] Siliņš contracted the printing of the calendar and maps to the A. Stahl Printing House, one of several printing houses located on the Domplatz in the Riga's Old Town.[115] As Siliņš was a private entrepreneur, his maps had to be commercially viable. He had a strong business acumen and sought to combine high-quality and innovative content with low prices. Siliņš regarded the high quality of his maps as a symbolic resource for nationalistic discourse, noting that 'the

[111] Monmonier, 'The Rise of the National Atlas'.

[112] For a discussion of Estonian calendars, see Annus, *Eesti kalendrikirjandus*; Talts, 'The Role of Popular Science Literature': 64–5. On popular print culture in imperial Russia, see Brooks, *When Russian Learned to Read*.

[113] Stinkule, 'Skolotājam, literātam un tulkotājam': 39.

[114] 'Neue Karte von Libau un Umgebend in lettischer Sprache,' *Libauische Zeitung*, no. 288, 18 December 1893: 3.

[115] Buchholtz, *Geschichte der Buchdruckerkunst in Riga*: 249.

publisher of the map has spared neither time, nor effort, nor success to achieve something worthwhile.'[116]

Comparing Siliņš's maps with August Bielenstein's atlas published in 1892 illustrates the different ways in which these mapmakers conceived of their wares within the literary marketplace. Whereas Bielenstein's atlas was priced at 2 roubles, Siliņš's 1890 issue of *Atbalss kalendārs*, including its supplementary *Map of Latvia*, was sold for a quarter of the price at just 40 kopeks. Just 500 copies of Bielenstein's *Atlas* were printed, while the 1890 edition of *Atbalss kalendārs* sold 12,000 copies in 1891 alone.[117] Siliņš explained that while the map alone would normally be sold for 2 roubles, the 'unusually large circulation, by which *Atbalss kalendārs* is distributed among the people, allows for the price to be so low'.[118] Moreover, Siliņš responded to the empire's developing consumer culture by producing different versions of his maps, ranging from a simple two-colour print of the map published as a supplement to the calendar, to a separate wall map sold for 75 kopeks, and a multicolour version for 1 rouble.[119] Selling maps across a range of different price points allowed Siliņš to generate the greatest amount of income from customers. Nevertheless, maps were still expensive purchases for working-class Latvian-speakers. For reference, the above-mentioned Latvian-language atlas by Ernests Dinsbergs sold for 1 rouble and 20 kopeks in 1861, which was equivalent at the time to either one- or two-week's wages for a servant, two pairs of boots, or three pairs of trousers.[120] Yet, as one Latvian-language newspaper proclaimed in 1895, Siliņš's calendar and maps were excellent value for money: 'It is really a small price! If you notice the price of the publisher's tireless work and diligence, with which he tries to introduce Latvians to Latvia, then we can wish for the calendar a lot of buyers and a lot of admirers.'[121]

As a small-scale publishing operation that had no guarantee of being lucrative, Siliņš heavily relied on advertising to persuade customers to buy his calendar and maps. In the 1890s Siliņš placed advertisements in all the major Latvian-language newspapers in Livland and Kurland. The advertisements featured alongside announcements for the meetings of the Riga Latvian Society, theatre performances, and society balls, as well as adverts for consumer goods such as cigars and chocolate. The juxtaposition of Siliņš's maps alongside luxury commodities suggests that he aimed to reach people who had some disposable income and leisure time, as well as more educated readers who were concerned with political and social issues. The advertisements promoted the map supplements as one of the calendars' unique selling-points, which made Siliņš's product superior to his

116 Stinkule, 'Skolotājam, literātam un tulkotājam': 39.
117 Bielenstein, *Ein glückliches Leben*: 325; *Balss*, no. 43, 21 October 1892: 10 (advertisement).
118 *Altbass* (1894): back cover.
119 On the development of consumerism in late imperial Russia, see Hilton, *Selling to the Masses*.
120 Karma & Pātsi, 'Kā tapis pirmais ģeogrāfiskais atlants': 35.
121 'Grahmatu galds', *Balss*, no. 6, 8 February 1895: 4.

competitors.[122] In December 1896, the German-language paper *Düna Zeitung* included the calendar and map supplement on its list of recommended Latvian-language books to buy as Christmas gifts that year. If we remember, this stood in direct contrast to Bielenstein's understanding of the market for his atlas when he explicitly stated that he could not imagine it being bought as a Christmas present.[123] In 1914, an article in a Latvian-language newspaper for farmers about a shortage of agricultural workers in Vitebsk province (Latgale) noted that workers from Livland and Kurland travelling to seek out employment are 'definitely advised to bring M. Siliņš's map of the Latvian land, where all parishes and roads are finely marked. The card can be obtained at bookstores in Riga.'[124]

Siliņš attempted to seduce potential buyers by providing detailed descriptions of the maps, seeking to cultivate a cartographic sensibility in his readers. The maps condensed vast amounts of knowledge into a synoptic view and instilled a sense of national consciousness by presenting geography and history in an accessible visual format where the details could be situated and remembered. The advertisements educated customers about how to judge the quality of a map by emphasizing factors such as the newness of the data and the credentials of the mapmakers. Siliņš drew readers' attention to both the educational and aesthetic value of his maps, noting how 'the map, richly presenting different geographical information, at the same time remains beautiful, clear, easy to read and will give... an extensive and accurate picture of all the regions of our homeland'.[125]

The rhetorical devices used by Siliņš to market his maps reveal how he aimed both to satisfy an alleged demand for vernacular cartographic materials among Latvian-readers and inspire further interest and engagement with geography and cartography. As Siliņš wrote in an advertisement in 1892: 'Loyalty creates loyalty. Following this conviction, the publisher presents the result of his difficult and expensive undertaking in the hope that the Latvian people will support it now and always by buying a calendar and promoting its dissemination.'[126] By attempting to persuade his Latvian readers of the importance of cartography for the development of Latvian culture and national consciousness, he politicized personal consumption by connecting it to the wider needs of an imagined Latvian community. He hoped that loyal customers would become self-conscious and patriotic Latvians. When in 1894 Siliņš was forced to increase his prices, he appealed to his readers to continue support the venture:

> 'Dear readers of *Atbalss kalendārs* will find that, starting this year, the price of the calendar has increased from 25 kopecks up to 35 kopecks. There is no need

[122] *Baltijas Vēstnesis*, no. 261, 17 November 1889: 4 (advertisement).

[123] 'Zu Weihnachtsgeschenken geeignete lettische Bücher diese Jahres,' *Düna Zeitung*, no. 281, 13 December 1896: 2.

[124] 'Ausim kājas uz Latgali!', *Zemkopis*, no. 3, 15 January 1914: 53.

[125] *Balss*, no. 43, 21 October 1892: 10 (advertisement).

[126] Ibid.

> for a long explanation of the reasons. Beautiful maps and other supplements require not only long, hard work, but above all—a lot of money. In order to continue publishing maps as usual, the publisher is forced to turn to the Latvian people for support: it is good to buy a calendar, despite the increased price.'[127]

We must be wary, however, of taking Siliņš's claims at face value. These appeals to readers tell us more about the commercial interests of Siliņš as a publisher and bookseller than the views of the consumers themselves. Moreover, Siliņš's cartographic endeavours were often hampered by his struggles to gather data and finance his projects. Unlike many other mapmakers encountered in this book thus far, Siliņš's mapmaking activities were not supported by a subvention from a learned society. Some years Siliņš was overly ambitions and he was unable to deliver the maps he promised to his readers. In the 1894 issue of *Atbalss kalendārs* he apologized to readers that he had not finished a map of the Cēsis-Limbaži area, explaining that: 'This is a big job that takes time. For now, dear readers, please be patient.'[128] At the same time, this public apology to his readers may also have simply been a marketing strategy to emphasize the popularity of his work and connect with his customers. Despite Siliņš's considerable efforts to publicize his wares, he was forced to stop publishing *Atbalss kalendārs* in 1898 due to economic difficulties.[129]

Siliņš's maps were regularly reviewed in the periodical press, giving us some sense of what readers thought about his maps. In contrast to the reviews that appeared in scientific periodicals of ethnographic maps discussed in earlier chapters, which usually concentrated their comments on questions of data and methodology, Siliņš's customers were more concerned with the maps' aesthetic qualities and value for money. Having little knowledge on which to judge the scientific accuracy of the maps themselves, they instead valued maps that were legible, well-engraved, beautifully coloured, and reasonably priced. An anonymous review in the German-language *Libauische Zeitung* praised the aesthetic appearance and the 'great care' with which one of Siliņš's maps had been made, but noted how the inclusion of so much information about political, human, and physical geography left the map too cramped. The use of hatching was felt to have compromised the 'readability' of place names.[130] Siliņš was attuned to his customer's tastes and in subsequent advertisements for new publications he made sure to emphasize the aesthetic qualities, including the 'beautiful engraving' and 'clear print on thin paper'.[131]

It is also important to add a caveat about the limitations to the circulation of cartographic knowledge about the territory of Latvia and the geographical spread of Siliņš's customers. Siliņš's calendar and maps claimed to be aimed at all Latvian-speakers and reach across confessional divisions by providing information about

[127] *Altbass* (1894): back cover. [128] Ibid: 35.
[129] Karnups, 'Matīss Siliņš': 153.
[130] 'Ostseeprovinzen,' *Libauische Zeitung*, no. 209, 14 September 1891: 2.
[131] *Balss*, no. 43, 21 October 1892: 10 (advertisement).

Lutheran, Orthodox, and Catholic calendar dates. However, in the context of the early 1890s, his audience would primarily have lived in Livland and Kurland. Although Siliņš included the western part of Vitebsk province in his maps of the Latvian territory, in the 1890s the region was subject to a ban on printing and publishing in Latin script, which remained in force in the whole of the North-Western Territory through 1864–1904.[132] Consequently, during this period Vitebsk province largely remained part of a separate information space to Livland and Kurland. Moreover, literacy levels in western Vitebsk were far lower than in the neighbouring Baltic provinces and literacy among Catholic Latvian/Latgalian-speakers in 1897 was just 58 per cent.[133] Thus, while Siliņš included the Latvian (specifically, Latgalian) inhabitants of Vitebsk province within his spatial concept of Latvia, the population themselves remained largely isolated from the networks of circulation, consumption, and production of geographical knowledge about Latvia during this period. It is only in the 1910s that we find examples of intellectuals from Vitebsk province engaging with Siliņš's maps. Even then, the reception was lukewarm. Notably, the Latgalian newspaper publisher and political figure Francis Kemps (Fraņcs Kemps) criticized Siliņš for using transcribed Russian language versions of toponyms on his maps rather than using the local Latgalian place names.[134]

Matīss Siliņš's Idea of Latvia

What ideas about Latvia and Latvians did Siliņš convey through his maps? The first issue of Siliņš's calendar published in 1889 contained a roadmap of Kurland province, which was followed in 1890 by his *Map of Latvia* (Plate 14). According to the map, the territory of Latvia comprised Kurland, Livland, and western Vitebsk province. In doing so, Siliņš's map was the first cartographic depiction to unite the three separate imperial provinces under the geographical umbrella of 'Latvia'. Prior to the 1880s, the term 'Latvia' was only used by a small circle of Latvian-speaking intellectual elites in the Baltic provinces and its political and geographical meaning was the subject of much debate.[135] In the early 1880s, the concept of Latvia became increasingly politicized when members of the Riga Latvian Society presented a programme for reform in 1883 to Nikolai Manasein during his senatorial inspection of Livland and Kurland, which included the proposal to transform the three Baltic provinces into two more ethnically homogenous Estonian and Latvian provinces. The political-administrative reformist agenda of the tsarist government combined with the increasingly territorially and ethnically

[132] On the 'Russification' policies in Latgale, see Bukšs, *Die Russifizierung*: 47–80.

[133] The majority of this population spoke Latgalian, regarded either as a Latvian dialect or distinctive language. The 1897 imperial census did not recognise Latgalian as a separate language and classified these people as Latvian-speakers. Raun, 'Literacy in the Russian Empire': 71–2.

[134] 'Wietu nosaukumi Latgalê,' *Dzimtenes Vēstnesis*, no. 113, 19 May 1912: 5.

[135] Von Hirschhausen, *Die Grenzen der Gemeinsamkeit*: 341–66.

orientated nationalist ideology of the Riga Latvian Society to consolidate the concept of Latvia as a region within the empire.[136]

Over the next decade, Siliņš published further maps of Latvian settlements, cities, and regions, as well as an overview map of the coastal lands of the Baltic Sea. The topics and appearance of Siliņš's maps reflected his broad interests and passion for collecting: the maps covered multiple topics and contained numerous inset maps. Siliņš incorporated the latest findings from scientific publications, such as the research by University of Dorpat geologist Constantin von Grewingk, revealing how the boundaries between academia and popular science were becoming increasingly blurred.[137] Siliņš presented knowledge from academic-orientated publications in a more accessible visual form and in Latvian language. At the same time, Siliņš's maps assumed that the newspaper readers, teachers, and pupils viewing the map had a basic knowledge of physical geography and we can question the extent to which the intended audience would have fully comprehended their content.[138]

Siliņš's trademark cartographic style consisted of cramming his maps full of tiny details (Plates 14, 15, and 15a).[139] While the maps' cluttered appearance can make them quite hard to decipher, we can interpret Siliņš's approach as a strategy of 'mapping abundance', as described by Candace Fujikane in her study of how native Hawaiian communities used cartography to challenge the claims of colonial powers to be settling underdeveloped or empty lands.[140] In contrast to Bielenstein's representation of Latvians as an ethnolinguistic group which existed outside of contemporary politics, Siliņš created images of a thriving and vibrant modern Latvian society. He filled his maps with references to Latvian-language toponyms, places of noteworthy historical and cultural significance, and the latest scientific knowledge to make a powerful statement about the Latvian presence in the world, undoubtable in the face of the overwhelming volume of evidence presented on the maps.

Siliņš only produced one map that was explicitly entitled an 'ethnographic map', which was published much later during the interwar period.[141] However, he incorporated ethnographic elements into almost all his maps published during the imperial period too. For example, he defined the borders of 'Latvia' based on the Latvian language border. On his 1890 *Map of Latvia*, Siliņš labelled the northern border of the Latvian territory as the 'extreme border of the Latvian language', corresponding to the northern boundary marked on Bielenstein's maps of the Latvian language area (Figure 4.6). In 1891, Siliņš published his

[136] Ījabs, 'Als "Latwija"': 403.

[137] Zabele, 'Latvijas ģeoloģiskas kartēšanas aizsākumi'.

[138] Vytautas Petronis raises a similar point with respect to maps produced at the same time of the Lithuanian territory. Petronis, *Constructing Lithuania*: 233.

[139] Štrauhmanis, 'Matīsa Siliņa ietekme': 17.

[140] Fujikane, *Mapping Abundance*.

[141] Siliņš, *Latvija. Etnogrāfiski*.

Figure 4.6. Excerpt from Matīss Siliņš's *Map of Latvia* (1890) depicting the 'Latvian language extreme border' (*latweeschu walodas galezas robeschas*).

Source: Matīss Siliņš, *Latvijas karte (Kurzeme līdz ar Vidzemes un Vitebskas guberņu latviešu daļu)* (Riga: M. Siliņš, A. V. Grotusa litogrāfija, 1890). Reproduced with permission from the National Library of Latvia.

General (Political and Physical) and Geological Map of Latvia, which marked the 'borders of Latvia' in a thick red line that also snaked through Vitebsk province.[142] Siliņš's practice of marking the language border on maps of the region's physical geography was copied by other mapmakers. A popular German-language map of the Baltic provinces published by the N. Kymmel printing house in Riga, and reprinted in nine editions between 1898–1914, replicated Siliņš's practice of marking the language border between Estonian- and Latvian-speakers in Livland.[143]

Siliņš sometimes included information about the areas inhabited by different ethnolinguistic groups on his maps. The 1890 *Map of Latvia* contained an inset map of the 'Lithuanian-Latvian ethnic land', which included text indicating the regions inhabited by Estonians, Poles, 'Gudi', and Belarusians (Plate 14).[144] His 1891 *General Map…of Latvia* incorporated colour-coded lines to indicate the settlement of 'Russian colonizers before the 18th century' (i.e. Old Believers), Poles, and Estonians. Similarly, his 1901 (reissued in 1911) *Map of the Latvian Land* (Plate 15a) used a combination of colour and patterns to show the distribution of German colonies in Kurland. He depicted Latvians, Belarusians, Great Russians, and Estonian 'colonists' in western Vitebsk. Lithuanians and Poles were later added to the 1911 edition of the map (Plate 15b). By singling out the areas inhabited by non-Latvians on his maps, Siliņš presented these populations as minorities living within a majority Latvian territory.

[142] Siliņš, *Latvijas vispārīga (politiska un fiziska)*.

[143] *Dr. Henry Lange's Karte*.

[144] Gudi is a Lithuanian-language term used in this context to refer to Belarusians or the Eastern Slavic population of the former Grand Duchy of Lithuania. It is not clear how Siliņš distinguished them from Belarusians. Dziarnovič, 'Gudas as a Historical Name'.

While Siliņš positioned himself against Bielenstein and the Baltic German tradition of Latvian ethnography that he symbolized, Siliņš also borrowed extensively from and reworked previous maps. Often, the intertextuality of Siliņš's maps was explicit. The 1890 *Map of Latvia* included an inset dialect map of the Latvian language area based on the 1881 map by 'Bielenstein and Döring' (compare Figures 4.1 and 4.7). The same map also featured an inset map of Latvian 'colonies' in Vitebsk province, based on the *Ethnographic Map of Vitebsk Province* (1872) published by Aleksandr Sementovskii and the Vitebsk Statistical Committee (compare Plate 6 and Figure 4.8). Siliņš extracted information from Sementovskii's ethnographic map about the areas inhabited by Latvian 'colonies', pockets of Latvian-speakers living in predominantly non-Latvian territories, and ignored the other nationalities depicted by Sementovskii. The frame of Sementovskii's map was also cropped so that only central and eastern Vitebsk province was shown, the implication being that western Vitebsk was an integral part of the main 'Latvian' language area and not an area of Latvian 'colonies'. The way that Siliņš distilled previous maps and presented data in new contexts highlights the importance of reading his maps not only in the context of emerging ideas about Latvian territory, but as part of the interconnected information space of the Russian Empire.

Siliņš's maps became revered in the early decades of the twentieth century. Latvian literary critic Alfrēds Goba attended Ģipterāni parish school in Kurland from 1900–5 and remembered how in his early schooldays there was only a map of Palestine, an old wooden globe in German, and later a map of Russia in the classroom. However, Goba noted how subsequent generations of schoolchildren benefited from Siliņš's maps:

> There were the borders of our land and there was our Latvia! There is the Baltic sea, there are the borders with neighbouring counties. There was the Daugava, Riga, other cities, there Vidzeme, Zemgale, Kurzeme, Latgale. All such familiar names, but now you can also see them on the map and well imagine where the county is located. The pupil saw it and memorized it. The concept of the fatherland grew.[145]

Goba was writing much later in 1940 and so we need to treat his lyrical statements about the impact of Siliņš's maps on emerging Latvian national consciousness with caution. What Goba's reminisces do convey, though, is the sense that cartography was becoming a familiar everyday object of material and visual culture. By the early decades of the twentieth century, Latvian-speakers were increasingly able to access printed cartographic materials in their native language.

[145] Goba, 'Skolas devums latviešu sētai': 574.

Figure 4.7. Insert *Map of Latvian Dialects* from Matīss Siliņš's *General (Political and Physical) and Geological Map of Latvia* (1891). The map is based on the 1881 map by August Bielenstein and Julius Döring (Figure 4.1).

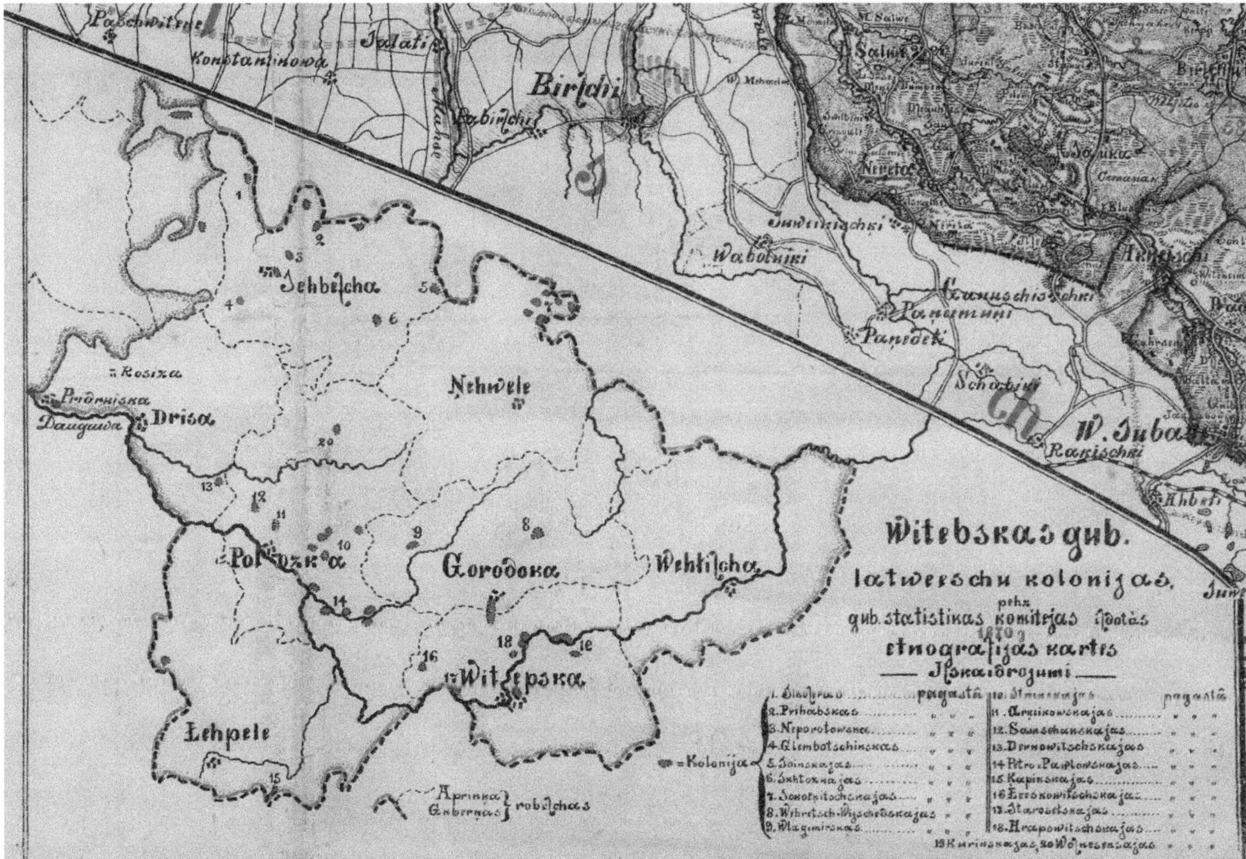

Figure 4.8. Insert map of *Latvian Colonies in Vitebsk Province* from Matīss Siliņš's *General (Political and Physical) and Geological Map of Latvia* (1891). The map is based on Aleksandr Sementovskii's *Ethnographic Map of Vitebsk Province* (1872) (Plate 6) and Latvian 'colonies' are marked in red numbered dots.

Source: Matīss Siliņš, *Latvijas vispārīga (poiltiska un fiziska) un ģeoloģijas (Inflantijas jeb poļu Vidzemes latviešu etnogrāfijas) karte* (Riga: M. Siliņš, A. Štāla litogrāfija, 1891). Reproduced with permission from the National Library of Latvia.

Maps also came to be perceived as playing an important role in the cultural work to persuade populations to think in terms of a Latvian territory spanning three imperial provinces.

From Siberia to Brazil: Locating Latvian Migrants, Colonies, and Diasporas

The Russian Empire's disastrous defeat in the Russo-Japanese War and the infamous 'Bloody Sunday' of January 1905, when imperial guards shot at peaceful demonstrators outside the Winter Palace in St. Petersburg, prompted a wave of solidarity strikes across the Baltic provinces. A flurry of small political parties and meetings were held to issue political demands. In November 1905, the first Estonian national congress and Latvian congress of teachers convened in Iur'ev (Tartu) and Riga to demand various changes, including democratic elections, abolition of censorship, cultural self-determination by ending the Russification policies in education, land reform, and a degree of local self-governance based on ethnic borders. Political fermentation led to an outbreak of revolutionary violence across the region targeting Baltic German-owned manors, businesses, and members of the clergy. The imperial authorities meted out harsh punishments to the perpetrators.[146] While Tsar Nicholas II's concessions to the demands were limited, pre-publication censorship was eliminated in 1906 and parliamentary elections to the first State Duma in St. Petersburg gave Estonian and Latvian national deputies the opportunity to gain experience of professional politics and network with representatives from across the empire.[147] As one member of the First Duma famously noted, the Duma resembled a 'live ethnographic map of Russia' in its representation of the imperial diversity.[148] These political developments contributed to a growing idea among the Latvian-speaking inhabitants of Livland, Kurland, and western Vitebsk provinces that they lived within a distinctive Latvian territory within the Russian Empire, views which Matīss Siliņš's maps gave a tangible visual and material form to.[149] However, national conceptions of space were not the only way of thinking about territory that emerged on ethnographic maps in the late imperial period.

In the months following the outbreak of the Russo-Japanese war in February 1904, Siliņš departed from his usual repertoire of maps of Latvia or plans of cities in Kurland and Livland to publish three maps of East Asia, consisting of maps of

[146] Raun, 'The All-Estonian Congress in Tartu'; Raun, 'Violence and Activism', Karjahärm, 'The Awakening of the Estonian Nation'; Ījabs, 'Another Baltic Postcolonialism'.

[147] Cadiot, 'Un empire un et indivisible"?'.

[148] Semyonov, 'The Real and Live Ethnographic Map of Russia': 191; Semyonov, 'Imperial Parliament for a Hybrid Empire'.

[149] Rosenberg, 'From National Territorial Autonomy to Independence'.

Manchuria, Korea, and a general wall map of Asia.[150] Over the late nineteenth and early twentieth centuries populations in the Baltic provinces gradually came to learn more about faraway places such as Palestine or America in their geography lessons or through the ethnographic shows that frequently performed in the region. Following the outbreak of the Russo-Japanese war 1904, attention turned to East Asia where several thousand men from the Baltic provinces were fighting for the Russian imperial army.[151] Advertisements in local newspapers announced that Siliņš's map would enable school pupils and newspaper readers 'to better follow events that took place in the Far East'.[152] Siliņš's maps of the empire's eastern borderlands indicate the broadening geographical horizons of the region's mapmakers beyond their Baltic homeland and an increasing trend to think about the Baltic provinces in relation to Russia as a Eurasian, rather than just a European, empire and its place in the wider world.[153]

Siliņš's *Map of Asia* deserves special attention as an important landmark in geographical thinking in the Baltic provinces. Consisting of one large map of Eurasia and twenty-eight insert thematic maps, it resembled a Latvian-language geographical atlas on one sheet (Plate 16). Although the map's title suggested that the main subject of the map was Asia, which was positioned at the centre of the map, the inset maps running down the left-hand side and bottom of the map all consisted of world maps. These maps provided Latvian readers with a global overview of different aspects of natural and human geography such as average summer and winter temperatures, air pressure, precipitation, flora and fauna, agriculture and livestock, languages, population density, and world cultures and religions.

The map portrayed a distinctly modern view of the world divided into territories ruled by different colonial powers. Siliņš's map shaded and labelled territories to indicate which European colonial power they were governed by. The oceans were criss-crossed with arced lines indicating shipping lanes and journey times in days. A full analysis of the richly detailed map is beyond the scope of the present discussion. However, if we look closely at the map from the perspective of ethnographic cartography, Siliņš's *Map of Asia* marked a radical departure from established practices of mapping ethnolinguistic groups based on the concentration of a population in a defined, potentionally national, territory. As indicated by the map's full title, 'Map of Asia with all of Russia and Latvian Colonies', Siliņš broadened the scope of ethnographic mapping to Latvian communities living in far-flung locations outside the Baltic provinces (Figure 4.9 and 4.10).

The term 'colony' in the Latvian historical context is often associated with several overseas territories purchased by the Duchy of Courland in the seventeenth century.

150 Siliņš, *Mandžūrijas—Korejas*; Siliņš, *D-Mandžūrija un Koreja*; Siliņš, *Āzijas karte.*

151 Jēkabsons, 'Latviešu piedalīšanās Krievijas-Japānas karā'.

152 'Kara-lauka kartes,' *Latviešu Avīzes*, no. 22, 16 March 1904: 3.

153 Siliņš's approach here resonates with contemporary historiographical discussions about Estonia's transnational past. See Brüggemann, *On Saints, Migrants and Communists.*

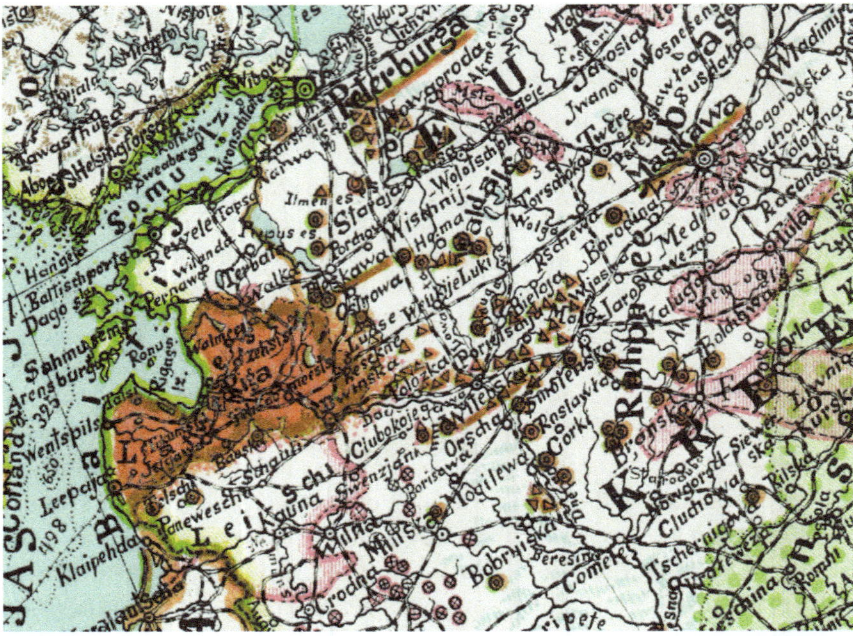

Figure 4.9. Detail of Matīss Siliņš's *Map of Asia* (1904), including the borders of Russia (outlined in green), 'Latwija' (shaded in orange), and 'Latvian agricultural colonies' (indicated by orange circles and triangles). See Plate 16 for full map.

Under the leadership of Duke Jakob von Kettler, the Duchy of Courland negotiated the rights to two overseas colonies in the Caribbean—the islands of Tobago and St. Andrews (later renamed James island)—and several islets and plots at the mouth of the river Gambia in West Africa. These overseas territories have been framed in contemporary national discourse in Latvia—and to a lesser extent also in Poland—as an early modern colonial past.[154] However, Siliņš's use of the term 'colonies' referred to Latvian-speakers who had emigrated from the Baltic provinces at various times and established settlements. From 1871–1916 more than 9 million peasants migrated from European Russia to Siberia, the Central Asian steppe, or the Far East.[155] In the case of the Baltic provinces, the abolition of serfdom, easing of passport laws, poor harvests and famines, and the prospect of land and warmer climates drove thousands of Baltic inhabitants to seek out a better life in the empire's interior.[156] By 1917 an estimated one-sixth of the

[154] Sooman et al., 'From the Port of Ventspils'; Merritt, 'The Colony of the Colonized'.
[155] Sunderland, 'The "Colonization Question"': 213.
[156] Korovushukin, *Latyshi i estontsy v zapadnoi Sibiri*; Tooming, 'Nälg ja ränne'; Jürgenson, 'The Role of Religion'.

Figure 4.10. Detail of Matīss Siliņš's Map of Asia (1904) showing Latvians 'colonies' (indicated by orange circles) near Omsk. See Plate 16 for full map.

Source: Matīss Siliņš, *Āzijas karte līdz ar visu Krieviju un latviešu kolonijām: skolām, avīžu lasītājiem* (Riga: Šnākenburga drukātava, 1904). Reproduced with permission from the National Library of Latvia.

Estonian-speaking population lived outside of the Baltic provinces.[157] As Nick Baron argues, these mass population resettlements, along with rapid industrial growth and the expansion of transport and communication networks by the turn of the twentieth century, prompted a radical rethinking of space, ideas about centres and peripheries, and questions of distance and imperial diversity.[158]

The turn of the nineteenth century was a period of growing public interest in Estonian and Latvian communities outside of the Baltic provinces.[159] Linguist and ethnographer Eduard Vol'ter (Eduards Volters) studied 'Vitebsk Latvians' (Latgalians) in the 1890s with funding from the Imperial Russian Geographical Society.[160] Estonian linguist and ethnographer Oskar Kallas conducted ethnographic expeditions to study Kraasna Estonians in Pskov province and the 'Lutsi country-people'

[157] Raun, 'Estonian Emigration within the Russian Empire'.

[158] Baron, 'The Mapping of Illiberal Modernity'. See also Bassin, 'Russia between Europe and Asia'; Bassin, *Imperial Visions*; Tolz, *Russia's Own Orient*; Schenk, *Russlands Fahrt in die Moderne*.

[159] Meomuttel, *Eesti asunikud laialises Wene riigis*.

[160] Vol'ter, *Ob etnograficheskoi poezdke*.

(*Lutsi maarahvas*) living in Liutsin (Ludza) district in Vitebsk province, who had migrated from the Baltic provinces in the sixteenth and seventeenth centuries.[161] Contemporary discourse in the Baltic press presented the Lutsis as a 'forgotten tribe of the Estonian people' and Kallas's research was met with considerable public interest.[162] Jakob Hurt, a linguist, folklorist, and pastor of St. John's Estonian Church in St. Petersburg, carried out a research expedition in 1903 to study the Seto people living in western Pskov province. The Setos were a Finno-Ugric Orthodox population living in the border region between Estonians and Russians. Hurt's research was also supported by the Imperial Russian Geographical Society and he produced an ethnographic map of the Seto-inhabited territory to accompany his published report.[163] Siliņš's map spoke to these broader trends in scholarly and public interest. The overarching message of the *Map of Asia* was to consolidate the idea of an ethnic Latvian heartland or metropole spanning Kurland, southern Livland, and western Vitebsk province, from which waves of ethnic Latvians had spread out over the centuries.

In addition to drawing readers' attention to the location of Latvian-speaking communities scattered across the Russian Empire, the *Map of Asia* also included an inset map in the lower left-hand corner depicting 'Latvian colonies' in North and South America (Figure 4.11). In the late nineteenth and early twentieth century, inhabitants from the Baltic provinces sought out new lives in the United States and Canada, part of a wider mass migration from Central and Eastern Europe during this period.[164] Brazil was an especially popular destination for Latvian migrants, who settled there in rural agricultural communities in search of a more prosperous and free life.[165] The first colony was established in 1890, when twenty-five families travelled to Santa Catarina state under the leadership of Lutheran pastor Kārlis Baložs. Another wave of 400–600 Latvian-speakers followed in 1906.[166] After the First World War, a further 2,000 migrants, mostly Baptists, made the journey across the Atlantic ocean to Brazil and established a Latvian settlement in Vārpa and later a collective farm in Palma.[167] Reports about Latvian-speaking migrant communities around the world regularly appeared in

[161] Jääts, 'Üks kuulus välitöö ja selle pikk vari': 15; Mela, *Läti eestlased*. Kallas's published his findings as Kallas, *Lutsi maarahvas*; Kallas, *Kaheksakümmend Lutsi Maarahva muinasjuttu*.

[162] Weske, 'Über die Witebskischen Esten'; 'Eestirahwa unustatud suguharu', *Postimees*, no. 202, 10 September 1892: 1. Cited in Jääts, 'Üks kuulus välitöö ja selle pikk vari': 16. Kallas spread awareness of the Lutsi-speakers among the broader Estonian-speaking public in an extended thirteen-instalment summary of his research published in *Postimees*. Kallas, 'Üks surew eesti rahwa oks', *Postimees*, nos. 206–23, 15 September–5 October 1895.

[163] *Izvestiia IRGO za 1903* (1904): 653; NA RGO f. 83, op. 1, d. 180, l. 1–29; Hurt, *Über die Pleskauer Esten*; Kalkun, 'Russkaia vera ili vizantiiskoe nasledie?'.

[164] On mass migration from Eastern Europe during this period, see Zahra, *The Great Departure*.

[165] On Polish migration to Brazil during this period, see Ureña Valerio, *Colonial Fantasies, Imperial Realities*: ch. 5. On Estonian migration to Argentina, see Jürgenson, *Ladina rahva seas*.

[166] Cornford, *Latvian Immigration to São Paulo*: 45–8; Tēraudkalns, 'Latviešu izceļošana uz Brazīliju'.

[167] Augelli, 'The Latvians of Varpa'. The experiences of early Latvian migrants to Brazil are the subject of the documentary film *Vārpa—apsolītā zeme* (2017), dir. Bruno Aščuks.

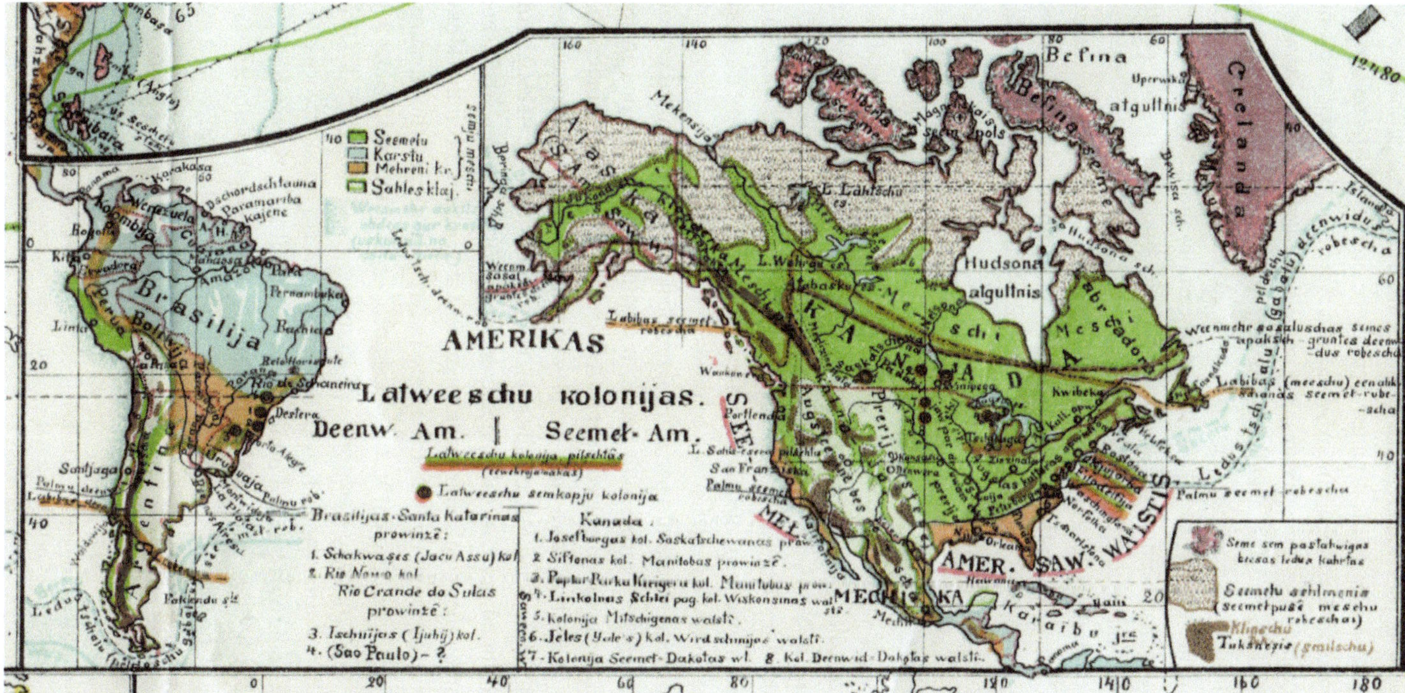

Figure 4.11 Insert map from Matīss Siliņš's *Map of Asia* (1904) depicting the location of Latvian colonies in North and South America , indicated by orange circles.

Source: Matīss Siliņš, *Āzijas karte līdz ar visu Krieviju un latviešu kolonijām: skolām, avīžu lasītājiem* (Riga: Šnākenburga drukātava, 1904). Reproduced with permission from the National Library of Latvia.

the Baltic press.[168] Siliņš's map, however, was the first to present readers with a visual overview of all these different 'colonies'. The map also distinguished between types of Latvian colonies. Siliņš depicted farming colonies in the Canadian states of Saskatchewan and Manitoba, and in Wisconsin, Michigan, Virginia, and North and South Dakota in the United States. The so-called 'Latvian colonial towns' of Boston, New York, Philadelphia, and Chicago were underlined in red. Information about some of these destinations was still vague. For example, at the time of the publication of the map in 1904, Siliņš was uncertain about whether Latvian farming colonies had been established in São Paulo, indicated by the question mark placed after the name of the city.

European colonial expansion and global migration in the late nineteenth century sparked discussions in the Baltic provinces about who should be seen as still belonging to the imagined boundaries of the nation.[169] In the Latvian case, Siliņš's *Map of Asia* consolidated the idea of a compact and ethnolinguistically homogenous Latvian homeland spanning the provinces of Kurland, southern Livland, and western Vitebsk. At the same time, Siliņš broke away from this narrow focus on the Baltic. His map encouraged viewers to connect Baltic Latvians with far-flung settlements—or 'colonies'—in Siberia and the Americas located at the edges of the map. According to Siliņš, Latvian-speakers who had departed at some point in history from the Baltic homeland were still part of the Latvian nation. Siliņš's map thus broadened the concept of Latvian ethnographic territory beyond the northwestern Russian Empire to all Latvian colonies around the globe.

In broadening the scope of Latvian geographical imagination, Siliņš's *Map of Asia* disrupted earlier cartographic narratives which had predominately portrayed ethnic Latvians as colonized peoples living either under German, Polish, or Russian yoke since the fourteenth century. As we saw in previous chapters, Estonians and Latvians had historically been mapped through the Baltic German and Russian colonial gaze as 'indigenous' and 'exotic' populations.[170] Together with the empire's other small non-Slavic peoples, Latvians had been the subjects of ethnographic mapping. By the early 1900s, the tables had turned. We can read Siliņš's map in the light of recent art history scholarship demonstrating how Baltic visual culture contributed to a growing self-awareness about the self and "Other". For example, Bart Pushaw notes how in the early twentieth century Dorpat-born artist Emilie Rosalie Saal painted hundreds of botanical illustrations during her time in the Dutch East Indies, which contributed to her self-awareness as an Estonian and of her 'newfound whiteness in the colonial south east Asia'.[171]

[168] Gūtmanis, *Kolonistu dzīve São Paulo*; Ķikuts, 'Diskusijas par migrācijas jautājumiem latviešu sabiedrībā'.

[169] For similar discussions about the impact of migration to Brazil on German and Polish perceptions of space and concepts of nationhood, see Conrad, *Globalisation and the Nation*: 275; Ureña Valerio, *Colonial Fantasies, Imperial Realities*: ch. 5.

[170] Pushaw, 'The Visual Whitening of Estonians'.

[171] Ibid.

Similarly, by visualizing the spread of ethnic Latvians across the globe, Siliņš complicated the power relationship between colonizer and colonized peoples. Imperial subjects such as Siliņš engaged in knowledge production about "Others", using the medium of Latvian-language cartography to map the Russian Empire and the territories governed by European colonial powers.[172]

The *Map of Asia* was also a symbol of something larger than itself. The size and spread of the Latvian ethnographic territory had become a measure of the Latvian nation's power. The process of inscribing Latvian settlements in Siberia and overseas 'colonies' in North and South American onto the map was used by Siliņš as a way to portray Latvians as an advanced modern nation, a historical actor, and undeniable through its sheer global presence. The spatial imagination empowered by colonialism and migration redefined the concept of Latvian territory.

Conclusion

Bielenstein's and Siliņš's maps shaped a field of study which was later taken up by nationalist activists in the early decades of the twentieth century as the scientific basis for the construction and dissemination of the idea of a collective Latvian nation forged from inhabitants spread across three imperial provinces. The process of collecting and analysing ethnographic material in preparation for the 1896 Latvian Ethnographic Exhibition held in Riga stimulated a powerful retrospective on the history of the Latvian people and their cultural distinctiveness.[173] Efforts to map the Latvian language area and ethnographic territory shaped emerging ideas about Latvian nationhood in the late nineteenth century. It is, however, important to stress that independent Latvian statehood was not on Bielenstein's and Siliņš's horizons. Instead, they constructed visions of a Latvian ethnolinguistic territory within the Russian Empire on the basis of regional studies of language, culture, and history. Emerging ideas about a Latvian national territory were explored in conjunction with visions of Latvians in a Russian imperial future, as participants in a colonial world order, and as global migrants.

Taking a comparative approach to examining Bielenstein's and Siliņš's maps demonstrates how their maps departed in striking ways from earlier renderings of the Baltic provinces on ethnographic maps. Instead of portraying a multitude of different ethnolinguistic groups living in Livland, Kurland, and western Vitebsk province, their maps communicated the area where Latvian was spoken by the majority of the population. Non-Latvians disappeared from the maps or were recast as minorities living with a majority Latvian language area. At the same

[172] On the contributions of indigenous scholars to the study of the Russian Empire in the late imperial period, see Tolz, *Russia's Own Orient*: ch. 5.

[173] Ķikuts, '1896. gada Latviešu etnogrāfiskā izstāde': 47.

time, while intellectuals such as Bielenstein and Siliņš both recognized the importance of using maps to construct a unifying narrative about the history and culture of the Latvian people, there was no clear consensus about the form that this new spatial concept of a Latvian territory should take.

Despite the fact that Bielenstein's and Siliņš's maps were displayed side-by-side at the Latvian Ethnographic Exhibition, the two mapmakers are emblematic of two different approaches to mapping Latvians. Bielenstein embodied the so-called 'Baltic German' approach to Latvian ethnography in his German-language maps and connections with the Baltic German-dominated Latvian Literary Society. Bielenstein's linguistic maps consolidated the idea of a common Latvian language area spanning three administrative provinces, but his maps also made readers aware of the considerable internal dialectal diversity distinguishing speakers across this territory. These dialectal differences were an object of curiosity for Bielenstein in his work to develop Latvian grammars and compile collections of folksongs in his parishioners' native tongues. Siliņš, by contrast, was linked to the Riga Latvian Society, whose membership comprised a younger generation of educated, upwardly socially mobile, and more nationally orientated Latvian-speakers. Siliņš's Latvian-language maps conveyed a strong message about a singular Latvian people and he used maps to communicate ideas about a Latvian territory as the basis for a shared positive identification. Speakers of different Latvian dialects from Kurland, Livland, and Vitebsk provinces were subsumed into a larger historical and cultural narrative about a unified Latvian people. In this way, nationalizing cartographic projects could end up replicating many of same colonizing and assimilating tendencies as mapmakers seeking to advance imperial interests. For example, both Bielenstein's and Siliņš's brought Latgalians into the fold of the Latvian nation, irrespective of whether the members of these groups might see themselves as different from one another.[174] Bielenstein's and Siliņš's maps enable us to pinpoint differing notions of how a Latvian space within the Russian Empire should be defined, as well as diverging ideas about who should be doing the work of spatially defining the Latvian people and in what language.

Comparing the process of constructing a Latvian ethnolinguistic space with the development of ideas about a Lithuanian ethnic territory to the south yields some crucial similarities and differences. In both cases, mapmakers carved out ethnic territories for Latvians and Lithuanians without challenging Russian imperial rule. As Vytautas Petronis argues, ethnographic cartography was used by Lithuanian-speaking intellectuals in the late nineteenth century to promote the idea of a '"Lithuania" *in/and* "the Russian Empire", rather than "Lithuania" *under* "the Russian Empire"'.[175] Moreover, mapmakers in both the Latvian and Lithuanian contexts

[174] This point resonates with Matthew Edney's argument that when it comes to cartography, there are no differences between the mapping practices of empires and modern states. Edney, 'The Irony of Imperial Mapping'.

[175] Petronis, *Constructing Lithuania*: 238.

had to confront the question of how to conceptualize ethnolinguistic spaces that spanned administrative borders. In the 1880s–1890s, mapmakers in the Baltic provinces produced maps uniting the predominantly Protestant Latvian-speaking inhabitants of Kurland and Livland with Catholic Latvian-speakers (Latgalians) in Vitebsk province. In the early twentieth century, intellectuals conceptualizing a common Lithuanian ethnolinguistic space began to map Catholic Lithuanian-speakers in the Russian Empire together with Protestant Lithuanian-speaking inhabitants in Prussia, a territory known as Prussian Lithuania or Lithuanian Minor.

At the same time, developments in Latvian geographical thinking also differed in crucial ways from the discussions taking place among Lithuanian-speaking intellectuals. Notably, Lithuanian national activists struggled to reconcile two competing visions of a 'Lithuanian' space—a historical Lithuania based on the former territory of the Grand Duchy of Lithuania and an ethnographic Lithuania corresponding to the lands inhabited by Lithuanian-speakers.[176] When it came to imagining Latvia, there was no comparable unifying historical tradition of prior statehood on which to base the borders and from which to claim continuity. The heritage of Terra Mariana (medieval Livonia) was shared with cultural promoters and national activists constructing a narrative of an Estonian national history to the north. Moreover, in contrast to the Lithuanian case, the individual provinces that were in the process of being integrated into a wider overarching concept of Latvian territory had much stronger and developed separate historiographies as a legacy of the borders of the fiefdoms of the noble corporations (*Ritterschaften*) which had existed in some form since the thirteenth century.

Finally, perhaps the most striking development to emerge from this period is the broadening of audiences for cartography, both geographically within the Baltic region and socially across different classes and linguistic groups. Mapmakers such as Siliņš sought to take advantage of the growing demand for cheap popular print culture to bring ethnographic maps to a much wider audience. The distribution of maps in school textbooks and the introduction of mapping as an educational method in schools across the Baltic provinces set the stage for widespread cartographic literacy among the general population. As I will examine in the next chapter, this emerging cartographic literacy would go on to have far-reaching consequences when it came to transforming ethnolinguistic borders into blueprints for new nation-state borders after the First World War. The acquisition of skills to read and draw maps facilitated opportunities for a broad social spectrum of society to participate in conversations about territory and nationhood. Cartographic literacy empowered local populations with the means to articulate their own proposals for drawing ethnographic borders.

[176] For the rich literature on competing visions of 'Lithuania' in the nineteenth and early twentieth century, see Snyder, *The Reconstruction of Nations*; Petronis, *Constructing Lithuania*; Staliūnas, ed. *Spatial Concepts of Lithuania*; Weeks, *Vilnius Between Nations*; Mačiulis & Staliūnas, *Lithuanian Nationalism*; Nenartovič, *Kaiserlich-russische*; Safronovas, *The Creation of National Spaces.*

5
Post-War Ethnic Boundary-Mapping from Above and Below

On 16 November 1921 the Estonian-language newspaper *Postimees* published a cartoon encapsulating the experiences of the local inhabitants during the process of demarcating the border between Estonia and Latvia (Figure 5.1).[1] The series of images depicted an Estonian man named Jüri getting out of bed one morning to discover that the border fence separating the Estonian and Latvian state had been moved by the Latvian border guards to the middle of his house during the night. Jüri's trousers are out of reach on the Latvian side of the border, blocked by a barbed wire fence. In the space of one night and without even leaving his house, the spatial ordering of Jüri's world was disrupted by the changing state border. Jüri's experience shares similarities with a story recounted by British travel writer Owen Rutter during his trip through the border town of Valka/Valga in the early 1920s. Rutter recalled hearing how the frontier 'even bisects a house, in so exact a manner that when the owner of it goes to bed his feet are in one country and his head in the other'.[2]

Laying aside the humorous framing of these border misfortunes, these allegories speak to the very real and practical difficulties faced by the inhabitants of multiethnic regions during the process of establishing new state borders after the First World War. Jüri's separation from his trousers symbolized the problems encountered by individuals whose land, property, or material possessions were now unreachable in another country. The warnings to pay attention to where border fences were being built and instructions not to move the boundary markers resonated with the confusion over the new rules and regulations restricting mobility that were rapidly introduced with the end of empire. Yet, these stories also carry strong messages about the agency of the local population in their interactions with national governments during the process of drawing new state borders. While scholars have observed how populations in twentieth-century Central Eastern Europe 'wandered' between political units throughout their lifetimes due to border shifts, not all inhabitants of pre-war imperial Livland province unquestioningly accepted their newfound status as Estonian or Latvian citizens.[3] The border-dweller in Rutter's story became embroiled in a legal confrontation

[1] 'Eesti-Läti piiri asi', *Postimees*, 16 November 1921: 5.

[2] Rutter, *The New Baltic States*: 178.

[3] Bös & Zimmer, 'Wenn Grenzen wandern'; Batt, 'Transcarpathia: Peripheral Region'; Brown, *A Biography of No Place*; Ciancia, *On Civilization's Edge*.

Geographies of Nationhood: Cartography, Science, and Society in the Russian Imperial Baltic. Catherine Gibson, Oxford University Press. © Catherine Gibson 2022. DOI: 10.1093/oso/9780192844323.003.0006

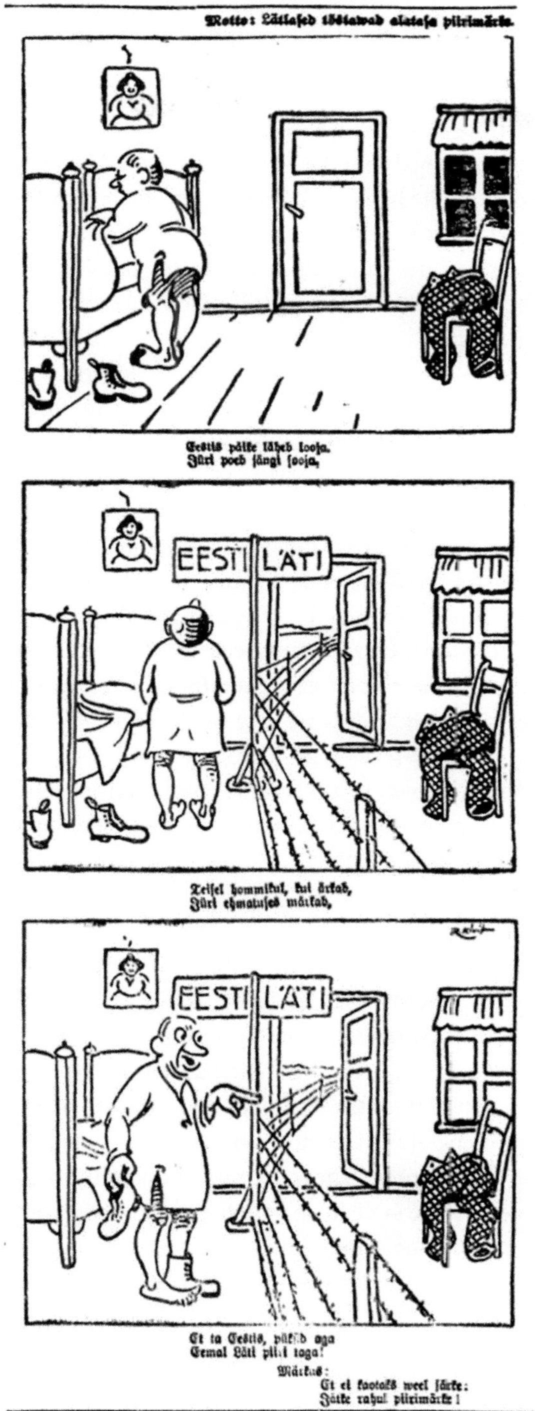

Figure 5.1. Caricature of the Estonian-Latvian border. The caption reads: 'Motto: Latvians are always lifting the boundary markers. In Estonia the sun is setting. Jüri crawls into the warm bed. Next morning when he wakes up, Jüri startlingly notices that he is in Estonia, but [his] trousers are beyond the Latvian border! Note: So as not to lose shirts as well, leave the boundary markers alone!'

Source: *Postimees*, 16 November 1921: 5. Public domain. Scan provided by the National Library of Estonia.

over which state he should pay taxes to. When charged with paying taxes in Latvia, Rutter describes how 'the gentlemen (whose sympathies were Estonian) responded by turning his bed round, so that his pillow would be in Estonia'.[4] The protagonist believed that he had a right to choose which state his sympathies lay with and took steps to exercise this right to self-determine his future.

Our understanding of the Estonian and Latvian state-building processes has been dominated by the idea that national self-determination emerged as a powerful ideology of mass mobilization after the First World War.[5] Estonian and Latvian national leaders declared independence on behalf of the territories and peoples of the former Baltic provinces on the back of the 1917 revolution, and amidst the civil war that erupted following the Bolshevik seizure of power and Germany's defeat in November 1918. While the new states continued to fight for independence against the Red Army and other military forces in the region through 1919–20, the freshly formed provisional governments began negotiating their borders, which they perceived as key steps towards realizing national sovereignty. Estonia's and Latvia's peace-making process with Bolshevik Russia, and Latvia's negotiations with Lithuania, Poland, and Belarus, have been richly documented from the perspective of diplomatic treaties and the work of the technical commissions in the 1920s to survey and demarcate the borders with posts, fences, and guard towers.[6]

My interest in the border-drawing process emerges from the question of how different stakeholders used ethnographic mapping to make territorial claims on the basis of nationality. This chapter examines an episode from the first year of the Estonian-Latvian border-making process from 1919 to 1920. It shifts the focus from the outcome of the border to the laborious processes of blueprinting the boundary line. Through an analysis of the multilevel interactions between national governments, international advisors, and local communities I address the question of why ethnolinguistic borders were so appealing, yet ultimately so difficult to enact. In chasing the ideal of national self-determination as the guiding principle of state-building, the Estonian and Latvian governments went to incredible lengths to separate out populations and draw state borders that closely matched ethnolinguistic borders. The scope of ethnographic mapping was applied on an unprecedentedly large scale to analyse the micro-geographies of nationality by zooming in on private households and farmlands. The obsession with making state and ethnolinguistic borders coincide drove the Estonian-Latvian Boundary

[4] Rutter, *The New Baltic States*: 178. [5] Prott, *The Politics of Self-Determination*: 1.

[6] On the Estonian-Latvian border, see Päts, *Eesti Vabariigi ja Läti Vabariigi vaheline piir*; Arumäe, 'Piirikonfliktist liidulepinguni'. On the establishment of other borders in the region, see Butkus, "Great Britain's Mediation'; Alston, 'James Young Simpson'; Jēkabsons, *Piesardzīgā draudzība*; Jēkabsons, *Polijas ārējā politika*; Golubev, 'Stanovlenie Belorussko-Latviiskoi granitsy'; Mattisen, *Eesti-Vene piir*; Musaev, 'Russko-estonskie pogranichnye problemy'; Pae, 'Eesti-Vene piir'.

Commission to scale up their thinking about nationhood and extend the states' panoptic surveillance over border region dwellers.

Our understanding of the uses of ethnographic cartography during the First World War and the post-war peace settlements has revolved around the use of maps by various national activists, experts, and politicians to propose a myriad of different post-war territorial designs based on ethnolinguistic or national borders.[7] Mapmakers became an in-demand resource during a period of unprecedented questioning about 'what binds people and what divides them; what is close by and what is distant; what is backward and what is modern; what is a part and what is the whole'.[8] A cohort of 'map men' laboured to put into practice Wilsonian ideas about 'national self-determination', reducing ethnicity to an objective factor that served both border-drawing and collective political mobilization.[9] Yet, for the most part, we do not hear the voices of the local inhabitants in the process of mapping new state borders. The records of the Estonian-Latvian Boundary Commission, however, allow us to do so.

At the heart of this chapter is a story of how the very real personal stakes of the border-making process in the post-war period motivated people in the Estonian-Latvian border region to produce their own maps to communicate locally and personally significant cultural, social, and economic geographies. The work of the Boundary Commission offers unrivalled evidence of the agency of the local population in the process of mapping ethnolinguistic borders. The extension of ethnographic mapping into the spaces of private lives had the unexpected effect of facilitating opportunities for local populations to intervene in meaningful ways in the border-making process. As a result, the discussion of where to draw borders in the aftermath of the First World War was not only confined to political elites, but also emerged as a topic of debate among the inhabitants of the border region.[10]

Counter-mapping as a form of participatory politics and tool to legitimize claims to land and resources has often been regarded as a late twentieth-century phenomenon.[11] However, in the early twentieth century, populations in the Baltic provinces were using mapmaking to delineate and formalize territorial claims. Non-elite populations of the Baltic provinces were the beneficiaries of social and cultural trends that had transformed maps into familiar objects of material and

[7] Crampton, 'The Cartographic Calculation of Space'; Palsky, 'Emmanuel de Martonne'; Górny, 'Futuristic Geography'; Nemes, 'Mapping the Hungarian Borderlands'; Sinanoglou, *Partitioning Palestine*.

[8] Lebow et al., 'Making Modern Social Science': 138.

[9] Seegel, *Map Men*; Prott, *The Politics of Self-Determination*; Górny, *Kreślarze ojczyzn*; Górny, *Science Embattled*; Wolff, *Woodrow Wilson*.

[10] Dunlop, 'Mapping a New Kind of European Boundary': 263.

[11] Peluso, 'Whose Woods Are These?': 400.

visual print culture, which they gained regular exposure to in school textbooks, wall maps, and newspapers. Advertisements make it clear that there was an existing market for maps even in small towns such as Valga/Valka, where the local Estonian-language local newspaper, *Rajalane*, published from 1919 to 1921, regularly contained publicity for the Tartu-based publisher Heinrich Laakmann's *Map of Estonia* (1919).[12] The inhabitants of the Baltic provinces not only internalized this geographical knowledge that they had learnt at school and through consuming popular media for 'shaping practices of self-representation and political identification'.[13] They also went a step further and used their cartographic literacy to advocate for their views and rights. By the early twentieth century there was an embedded cultural understanding in the region of mapping as a way to give voice to territorial aspirations. Cartographic literacy politically empowered disgruntled individuals to confront the state about their individual and societal concerns through the medium of maps.

Through a focus on the Estonian-Latvian border, this chapter deepens our understanding of the popular dimensions of ethnographic border-mapping in the early twentieth century. Histories of border-drawing often distinguish between 'high' and 'low' forms of cartography, contrasting official border surveys with unofficial 'popular' maps produced to sway public opinion and present alternative imagined borders.[14] However, examining only printed maps obscures the cartographic activities of those with less financial and material resources, but who nonetheless put pencil to paper and drew maps to make visual arguments about borders. This chapter deepens our understanding of the kinds of people involved in the production of 'low cartography' by examining hand-drawn maps made by farmers and villagers in the Estonian-Latvian border region. These populations were not only consumers of cartographic print culture, but also engaged in mapmaking in their efforts to influence the border outcome. This chapter also breaks down the sharp distinction often made between official and unofficial border maps by demonstrating how the maps produced by the local population were in close dialogue with the official work of the boundary commission, and often used the same cartographic symbology and colours. As the maps were submitted by the local population to accompany petitions, they formed part of the official communication channels for how local populations voiced their objections to the authorities and attempted to influence the border outcome in their favour.[15]

[12] For instance, *Rajalane*, 23 October 1920: 4 (advertisement).

[13] On the development of mass geographic literacy and its impact on identity formation in Early America, see Brückner, *The Geographical Revolution*: 139.

[14] Dunlop, *Cartophilia*: 8.

[15] Ibid: 6.

Occupation, Population Displacement, and Administrative Restructuring

The multiple military conflicts fought on the territory of the Baltic provinces between 1914–20 produced new and long-lasting ways of thinking about territories, populations, and borders.[16] Between July and September 1915 German forces took control of Kurland. For the next two years the stabilized front line ran from Riga, where the Russian imperial army held its position on the west bank, along the river Daugava following the boundary between Kurland and Livland, before running south from Dvinsk to Pinsk. Kurland was incorporated into the German occupied administrative territory of *Ober Ost*, while Estland and Livland remained under the control of Russian imperial troops.[17] In the second half of 1917, the German military began to push northwards, capturing Riga in August, the islands of Saaremaa, Hiiumaa, and Muhu in October, and Tallinn in February 1918. By the time the Treaty of Brest-Litovsk was signed in March 1918, the whole territory of the former Baltic provinces was under German control. Between 1918 and 1920, the region became embroiled in the Russian Civil War, Polish-Soviet War, and the Estonian and Latvian Wars of Independence.[18]

These years of conflict led to massive population displacements across the Russian Empire's western borderlands.[19] In response to the German advance, on 2 February 1915 the tsarist government issued an imperial decree establishing a 160 km-wide and 6,000 km-long border zone along the front line. Ethnic Germans, branded as unreliable and dangerous, were forced to register and sell their properties and were displaced to the Russian interior. Many Latvians and Jews with German-sounding surnames faced the same suspicion and discrimination.[20] The front also absorbed the Pale of Settlement and in spring 1915 military commanders ordered the expulsion of Jews from Kovna and Kurland provinces, as well as the Military District of Dvinsk.[21] Over the course of the conflict, the army deported between half a million and one million Jews and 250,000 ethnic Germans from areas under military rule.[22] Altogether, more than a million forcibly displaced people and refugees fled the western borderlands and were evacuated to the empire's heartlands.[23] In the process, Kurland province lost 54.4 per cent of its pre-war population.[24]

[16] On the impact of the First World War on the reshaping of borders in the Baltic, see Richter, *Fragmentation in East Central Europe*.

[17] Liulevicius, *War Land on the Eastern Front*: 19–21.

[18] Minnik, 'The Cycle of Terror in Estonia'; Minins, 'Latvia, 1918–1920'.

[19] Baron & Gatrell, *Homelands*.

[20] Gatrell, *A Whole Empire Walking*: 149, 159.

[21] Lohr, 'The Russian Army': 410–11.

[22] Ibid: 404–6.

[23] Purs, 'Orphaned Testimonies'; Purs, 'Working towards "An Unforeseen Miracle"'; Balkelis, 'Nation-Building and World War I'.

[24] Liulevicius, *War Land on the Eastern Front*: 30.

The massive population displacement made a fiction of pre-war statistical data. War disrupted traditional methods of counting, categorizing, and keeping records about populations. The work of the provincial statistical committees was suspended during the war and the second Russian imperial census scheduled for December 1915 was cancelled. Forms of identification were also in flux. The experiences of war, occupation, and displacement shattered social identity boundaries and created new economic, political, and social reconfigurations. As Peter Gatrell writes, the experience of refugeedom 'made nonsense of official attempts to maintain the fiction of social stability in wartime... established taxonomies of social description—whether ascribed "estate" (*soslovie*) or class—were subverted and redefined'.[25] Moreover, scholars have shown how the German *Ober Ost* administration provided partial support for certain kinds of national self-awareness among local Latvians, Lithuanians, and Belarusians as part of a strategy to weaken Russian and Polish influence in the region.[26] The German occupation thus functioned as a powerful catalyst for national activism by institutionalizing national categorizes as a means of ordering the inhabitants of the western borderlands of the Russian Empire and for promoting the idea that nationality was something physical, visible, and measurable. Meanwhile, the wartime purging of 'enemy aliens' helped to nationalize border regions, setting the stage for both the national movements and Soviet nationalities policy.[27]

By 1917, the developing nationalist discourse about the incongruence of ethnolinguistic and administrative borders in the Baltic provinces paved the way for a series of administrative reforms to allow a greater degree of self-government.[28] Following the February revolution, the Provisional Government in Petrograd (formerly, St. Petersburg) made concessions to regional aspirations and granted provisional autonomy to Estonia and Latvia.[29] The Estonian-speaking districts of Livland province were transferred to Estland in order to create two more ethnolinguistically homogenous and autonomous Estonian and Latvian provinces. The proposal to create a South Estonian province, with its provincial capital in Iur'ev/Tartu, was also briefly considered but the idea was eventually dismissed. The precise border was determined by a special commission, comprising representatives from Estland and Livland.[30]

The process of redrawing administrative borders in the Baltic along ethnolinguistic lines inspired a wave of bottom-up demands for border changes that

[25] Gatrell, *A Whole Empire Walking*: 3–4.

[26] Rudling, *The Rise and Fall*: 72–5; Safronovas, 'Zum Wandel des räumlichen Begriffs "Litauen"'; Zariņš, 'Das Land Ober Ost'; Kasmach, 'Forgotten Occupation'; Balkelis, *War, Revolution, and Nation-Making*.

[27] Lohr, *Nationalizing the Russian Empire*.

[28] Karjahärm, 'The Problem of Reorganization'; Rosenberg, 'From National Territorial Autonomy to Independence'; Brüggemann, '"Foreign Rule" during the Estonian War of Independence': 212.

[29] Brüggemann, *Die Gründung der Republik Estland*: 52–3.

[30] Bakhturina, 'Izmenenie administrativnykh granits': 179.

reverberated from south to north along the former eastern boundary of the Baltic provinces. The Latgalian War Victims Relief Society in Petrograd organized a meeting of Latgalian intellectuals, workers, and soldiers to discuss the idea of separating the districts of Dvinsk/Daugavpils, Liutsin/Ludza, and Rezhitsa/Rēzekne from Vitebsk province and joining them to the other Latvian provinces of Livland and Kurland. At the First Latgale Latvian Congress held in Rezhitsa/Rēzekne in April 1917, Latgalian representatives voted in favour of joining the other Latvian territories to the west. At the time, the representatives did not envisage that they would become part of an independent Latvian state, but rather a Latvian ethnic region with a reformed and federalist Russia. The decision was not unanimous and a minority of those present at the meeting opposed the idea of an outright unification of Latvians and Latgalians. Opposition coalesced around the Latgalian writer, political figure, and publisher Fraņcs (Francis) Kemps. In 1916, Kemps had written a forceful article outlining the basis for Latgalian nationhood and advocated that Lithuanians, Latvians, and Latgalians should all receive national autonomy within the Russian Empire.[31] Meanwhile, demands for territorial restructuring along national lines continued to reverberate northward along the eastern borders of the Baltic provinces in the summer of 1917. On 1 July 1917, seventy-four Seto representatives petitioned the Estonian Provisional Assembly asking for Seto-inhabited areas of Pechory/Petseri district in Pskov province to be transferred to Estland.[32] On 2 July 1917, the Narva Public Committee issued a proclamation to join Estland, which was followed by a plebiscite on 17 December 1917 where the residents of Narva voted in favour of transferring the city from Petrograd province to Estland.[33]

To keep pace with this rapidly unfolding administrative restructuring to accommodate political demands for national territorial autonomy, the Academy of Sciences established a Commission for Studying the Tribal Composition of the Population of Russia (KIPS) in the spring of 1917. KIPS was tasked with providing expert recommendations for demarcating the boundaries of western Russia (Lithuania, Galicia, Bukovina, Bessarabia, Poland) and the northern part of Asia Minor, Azerbaijan, and Northern Persia. Its members wrote articles on the peoples living in Russia, produced ethnographic maps, and created list of official nationalities.[34] Two experts on Belarusian and Latvian-Lithuanian ethnography and linguistics, Efim Karskii and Eduard Vol'ter, prepared an ethnographic map and report describing ethnic borders in the

[31] Kemps, 'Latgalijas nokutne': 1–3.

[32] Raun, 'The Petseri Region'; Mezhevich, 'Rossiisko-Estonskaia granitsa': 139.

[33] Veiss-Vendt, 'Komu prinadlezhit Narva?': 39.

[34] Psianchin, *Komissiia po izucheniiu plemennogo sostava naseleniia*; Francine Hirsch, *Empire of Nations*: 7.

Latvian-Belarusian-Russian boundary region for KIPS in 1917 and early 1918.[35] As specialist consultants on nationalities questions, Karskii and Vol'ter provided expert recommendations on where future borders should be drawn in the run up to the Treaty of Brest Litovsk signed between Germany and Bolshevik Russia in March 1918. While topographic maps were closely connected to the military, during the First World War ethnographic mapping also came to be perceived as a strategic resource for military intelligence and territorial administration.[36] The process of restructuring internal administrative borders in the Baltic laid the groundwork for the political territories of the future Estonian and Latvian states as part of a gradual process of fragmentation during the war years.[37]

Uncertain Futures: Cartographic Propaganda for Estonian and Latvian Independence

In the spring of 1919, Sir Henry Wilson, Chief of the British Imperial General Staff, discussed the future of Estonia and Latvia with fellow British delegate Sir Esmé Howard. Wilson pointed to a map of Russia and said: 'Look at those two little plots on the map and look at all that enormous country beside them. How can they hope to avoid being gobbled up?'[38] The Estonian Provincial Assembly declared independence for Estonia on 24 February 1918 and the People's Council of Latvia proclaimed Latvian independence on 18 November 1918. However, the territories claimed as part of Estonia and Latvia were variously controlled by Bolshevik, German, Polish, and Lithuanian troops. In the eyes of the international community, both self-proclaimed states continued to occupy an ambiguous status in relation to their bigger and more powerful neighbours. While the Allied powers were generally sympathetic to the Baltic claims for independence, the uncertain situation of Russia and the multiple armed conflicts ongoing in the region meant that they were hesitant to move to *de jure* recognition. They remained doubtful about the geopolitical and economic viability of these small states nestled between Russia and Germany.[39] The majority of Allied diplomats and policymakers also had very limited knowledge of who Estonians and Latvians were. For instance,

[35] Karskii, *Etnograficheskaia karta*: 7. Vol'ters's report can be found at: SPbF ARAN f. 135, op. 2, d. 85, l. 1–8. A transcription and discussion of the report is presented in: Galiona, 'Etnicheskaia prigranichnogo regiona Pytalovo/Abrene'.

[36] Heffernan, 'Geography, Cartography and Military Intelligence'; Brown, *A Biography of No Place*: 18–51.

[37] On the gradual process of reconfirguring territories during the First World War, see Richter, *Fragmentation in East Central Europe*; Baron, 'Nature, Nationalism and Revolutionary Regionalism'; Smith, ed. *Beyond the Limits*: 241–58; Hirsch, 'Towards an Empire of Nations'.

[38] Howard, *Theatre of Life*: 260. Cited in Alston, *Makers of the Modern World*: 85.

[39] Alston, *Makers of the Modern World*; Alston, 'The Suggested Basis for a Russian Federal Republic'. On the narrative of states in East Central Europe as small and lacking in performance and capacities, see Richter & Hein-Kircher, 'Too Small to Succeed?'

French-readers' knowledge of the Baltic provinces was largely based on Jules Verne's famous *Drama in Livonia* (1904), a thrilling tale of courageous Russians who fall victim the cruel machinations of Baltic Germans from Riga, but which had very little to say about the region's Estonian- and Latvian-speakers.[40]

The viability of Estonia and Latvia as independent states was further cast into doubt by the flurry of ethnographic maps produced during this period which presented parts of the region as belonging to other state-building projects on the basis of sweeping claims about populations' ethnolinguistic identities.[41] Nineteenth-century discussions about the idea of the Baltic provinces as part of a German sphere of influence, examined in Chapter 3, were strengthened after 1915 when Kurland fell under German control. Discourse about the Germanness of the Baltic provinces spread to popular travelogues and reports published in scientific journals, which rendered the Baltic provinces as German territories.[42] Meanwhile, both Kurland and Vitebsk lay within the territories claimed by Poland based on visions of recreating the former borders of the Polish-Lithuanian Commonwealth expressed in Eugeniusz Romer's famous *Geographical Statistical Atlas of Poland* (1916), which was widely consulted by experts at the Paris Peace Conference (Figure 3.4).[43] Belarusian and Lithuanian claims to the border territories of Kurland and western Vitebsk province, including the city of Dvinsk/Daugavpils, further destabilized any firm sense of where Latvia's borders lay in the south and east.[44] Finally, proposals for alternatives to nation-states, such as Jonas Šliūpas's vision of a Lithuanian-Latvian confederation, were also thrown into this complex mix of competing imagined geographies.[45]

Even though the Estonian and Latvian governments were not formally invited to send a delegation to the Paris Peace Conference, both sent representatives to garner support for their independence claims and lobby for international recognition. Faced with the threat of being 'gobbled up' by their neighbours, scepticism about the economic and political viability of these small states, and lack of knowledge about the region, the Estonian and Latvian delegations turned to cartography to represent their territories to the world from an Estonian and Latvian national perspective. An avalanche of short books, pamphlets, and brochures

[40] Gueslin, *La France et les petits états Baltes*: 34.

[41] Sanborn, 'War of Decolonization'; Sanborn, *Imperial Apocalypse*; Nelson, 'The Baltics as Colonial Playground'.

[42] Rohrback, *Das Baltenbuch*; Rosen, 'Die etnographischen Verhältnissen den baltischen Provinzen'.

[43] Seegel, *Map Men*: 48; Labbé, 'Eugene Romer's 1916 Atlas'; Osowska & Przybytek, 'Thematic Maps'. On Latvian-Polish border disputes, see Maliszewski, *Polacy na Łotwie*; Jēkabsons, *Piesardzīgā draudzība*; Jēkabsons, *Polijas ārējā politika*; Jēkabsons, 'Polish-Latvian Border'.

[44] Dounar-Zapolski, *Les bases de l'Etat de la Ruthénie Blanche*. On symbolic and spatial concepts of the BNR, see Rudling, *The Rise and Fall*: 86–96; Michaluk, *Belaruskaia Narodnaia Respublika*: 207–33. On the Latvian-Lithuanian border dispute, see Surgailis, 'Boi Litovskoi armii'; Alston, 'James Young Simpson'; Butkus, 'Great Britain's Mediation'.

[45] Šliupas, *Lithuania in Retospect and Prospect.*

appeared in French, English, and German to imprint the idea of Estonia and Latvia on the mental map of the international diplomatic community as deserving a place in the future of Europe.[46] In 1920, Johan Laidoner, Commander of the Estonian Army, described one of these maps of Estonia as 'serving as the first geographic calling card of the Republic of Estonia', printed in forty copies for distribution to foreign governments.[47] In the original Estonian quotation, Laidoner played on the pun of 'map' (*kaart*) and 'business card' (*visiit kaart*) to highlight the close relationship between the production and circulation of geographical knowledge and diplomacy. The image of the Estonian delegation handing out maps as business cards vividly conveys a sense of how the delegations used cartography to raise awareness on the international stage of their claims to independent statehood and to manage misinformation.[48]

The map of Estonia circulated by the Estonian delegation in their memorandum for the Paris Peace Conference defined the territory of Estonia based on ethnographic principles, mapping the territory where 'Estonians—the autochthonous inhabitants of the country—constitute more than 90% of the total population' (Figure 5.2).[49] At the same time, the map also proposed an alternative, more maximalist version of Estonia's 'ethnographic borders'. A grey dashed line depicted the 'limits of continuous colonization' extending to the east. In a throwback to Peter Köppen's ethnographic map of St. Petersburg province from 1849 (Figure 1.2), which had depicted the Finno-Ugric presence in this region, the Estonian delegation's map now proposed that large swathes of Petrograd and Pskov provinces be included within a vision of a greater Estonian ethnographic space. Defining the Estonian ethnographic territory in maximalist terms enabled the Estonian delegation to proclaim that the total number of Estonians worldwide was 2 million, and thus they were not such a small nation. Moreover, they explained that this figure had been reached by 'correspondingly augmenting' the number of Estonian-language speakers that had been recorded in the 1897 All-Russian census to reflect the 'national renaissance' and 'increasing political influence', meaning those Estonians who formerly 'impersonated Germans or Russians'.[50] The map of Estonia circulated by the Estonian representatives extrapolated data that was more than twenty years old to project aspirations of what a future Estonian state could look like.

The equivalent memorandum prepared by the Latvian delegation made similar use of cartographic arguments. The Latvian memorandum contained two colour maps of Latvia. Tellingly, the title of the first was *Map of Lines of Communication of Latvia*, and it depicted navigable rivers and railways that

[46] *Mémoire sur l'indépendance de l'Esthonie*; Martna, *Estland, die Esten und die estnische Frage*; *La Revue Baltique*, no. 1 (13), 15 September–1 October 1919.
[47] *Eesti Kaart/Carte d'Estonie* (Estonian Embassy in Paris, 1920). RA, ERA.495.10.28. Sheet 1.
[48] Laidoner's map is discussed in Pae, *Eesti Rahvusatlas*: 13–15.
[49] *Mémoire sur l'indépendance de l'Esthonie*: 3.
[50] Ibid.

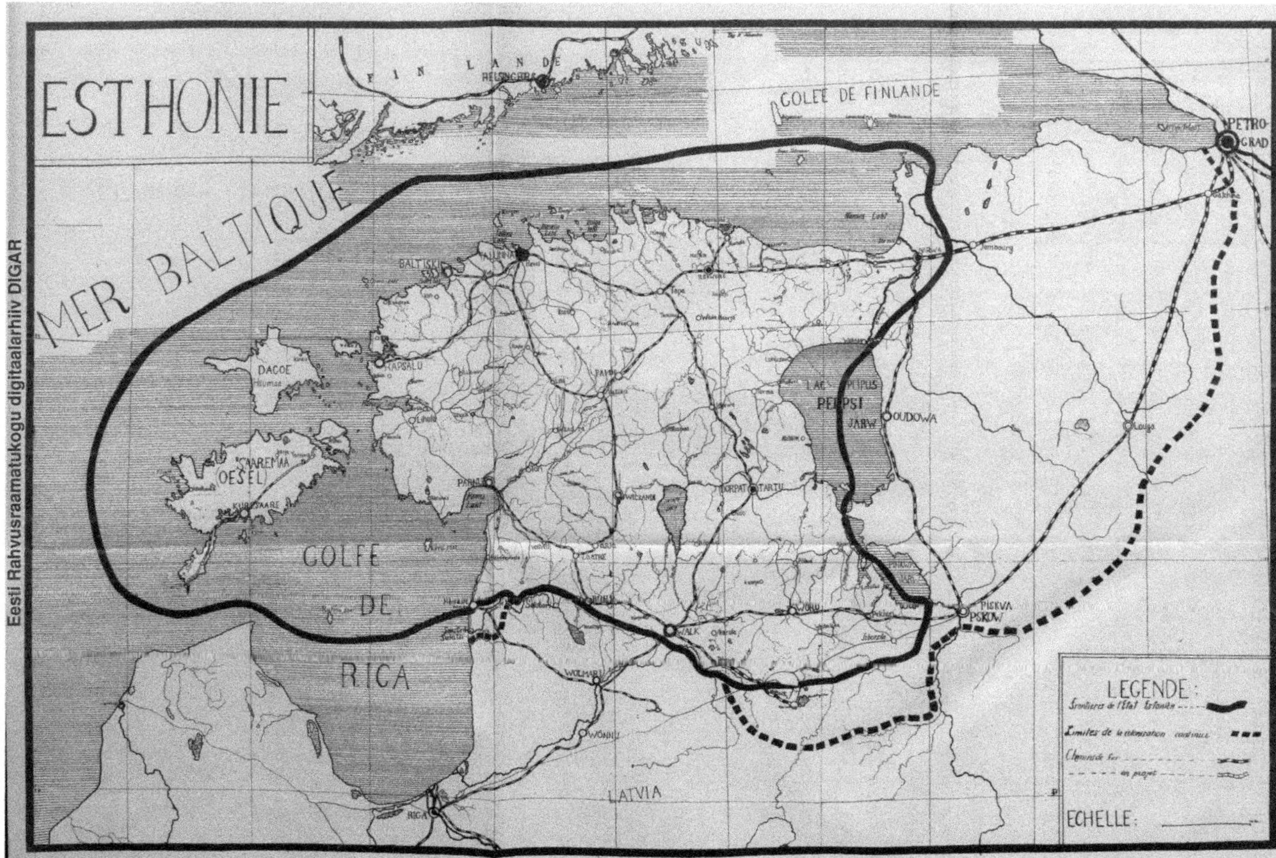

Figure 5.2. Map of Estonia from the Estonian delegation's memorandum for the Paris Peace Conference. The map depicts state borders (solid black line), the borders of Estonian settlement (dashed line), as well as railways and lines planned for construction.

Source: *Mémoire sur l'indépendance de l'Esthonie présenté a la Conférence de la Paix par la délégation Esthonienne* (Paris: Imprimerie de la Bourse de Commerce, 1919). Scan provided by the National Library of Estonia.

criss-crossed the country (Figure 5.3). The map also marked key ports and shipping channels, thereby presenting Latvia as a maritime state and important trading partner. Similar to its Estonian counterpart, the map of Latvia tackled the perception of Latvia's smallness by comparing the surface area of Latvia with the familiar Western European countries of Switzerland, Denmark, the Netherlands, and Belgium. In this way, the map attempted to connect what was known and unknown in readers' minds. Despite the map's title, it also depicted the borders as a close overlap of administrative and ethnographic frontiers. The map shaded 'regions with mixed population' which directly corresponded to the areas being disputed by Latvia and its neighbours in the ongoing border negotiations with representatives of Lithuania, Estonia, and the Belarusian People's Republic.

The Latvian delegation also produced a second map entitled *The Population of Latvia According to Nationalities* (Figure 5.4). Latvians were symbolically shaded in the dark red of the national flag. Lithuanians and Belarusians were shaded in a similar shade of red, which meant that taken at a quick glance, the map conveyed the impression that the whole territory claimed by Latvia was inhabited by a majority Latvian population. The map also made a persuasive argument about the socio-economic character of the proposed Latvian state. Unlike nineteenth-century ethnographic maps which had focused on shading the 'geo-body' inhabited by Latvians, this map concentrated instead on showing the ethnographic composition of towns and cities in pie charts and waffle charts (squarified pie charts).[51] The top righthand corner of the map contained a graph showing the percentage of Latvians, Russians, and Germans in Riga between 1867–1917, highlighting the increasing Latvian population. In this way, the map supported the argument outlined in the Latvian delegation's memorandum which challenged the stereotypes and negative discourse about Latvians as a 'peasant-nation' and predominantly rural population. As in the Estonian map, we find the same entanglements of ethnographic and socioeconomic arguments.

On the one hand, the Estonian and Latvian representatives passed out maps like business cards. In the hands of politicians and diplomats, these outline maps functioned as logos, conveying a memorable impression of the shape of these states from a quick glance based on the ethnographic territories inhabited by Estonians and Latvians.[52] The maps used simple elements of cartographic design and data visualization to present Estonia and Latvia as socioeconomically developed and modern nation-states that had a legitimate place on the future map of Europe at a time when these newly emerged state-building projects in the Baltic region still faced considerable international scepticism. These maps demonstrate the importance of visual culture in bringing into being an idea of a territory that

[51] On the growing popularity of this cartographic techique during this period, see Palsky, 'Emmanuel de Martonne'.

[52] Batuman, 'The Shape of the Nation'; Anderson, *Imagined Communities*: 170–8.

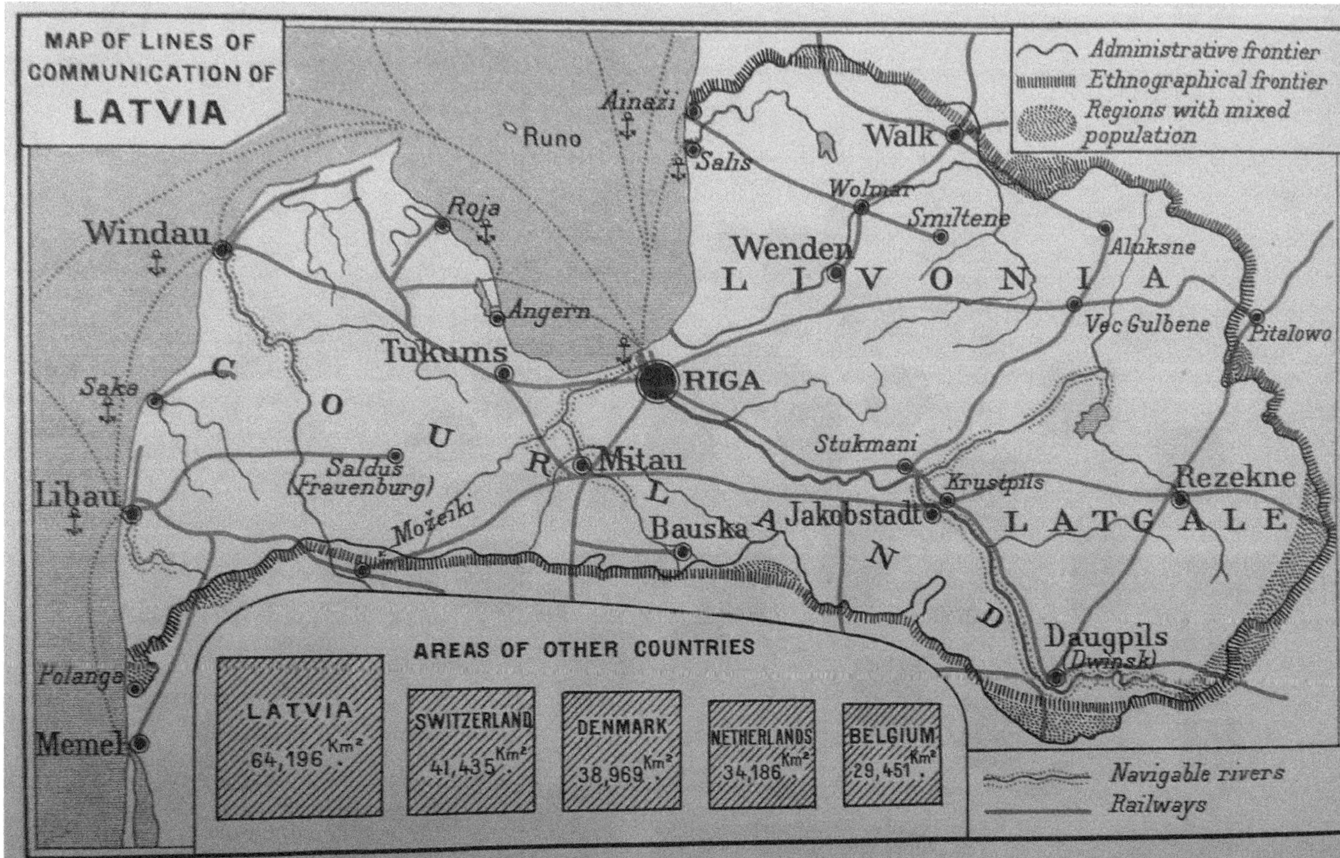

Figure 5.3. *Map of Lines of Communication of Latvia* published in the Latvian delegation's memorandum for the Paris Peace Conference in 1919.

Source: *Memorandum on Latvia Addressed to the Peace Conference by the Lettish Delegation.* ELMAL Balt Book IV 84. Reproduced with permission from the Estonian Literary Museum Archival Library.

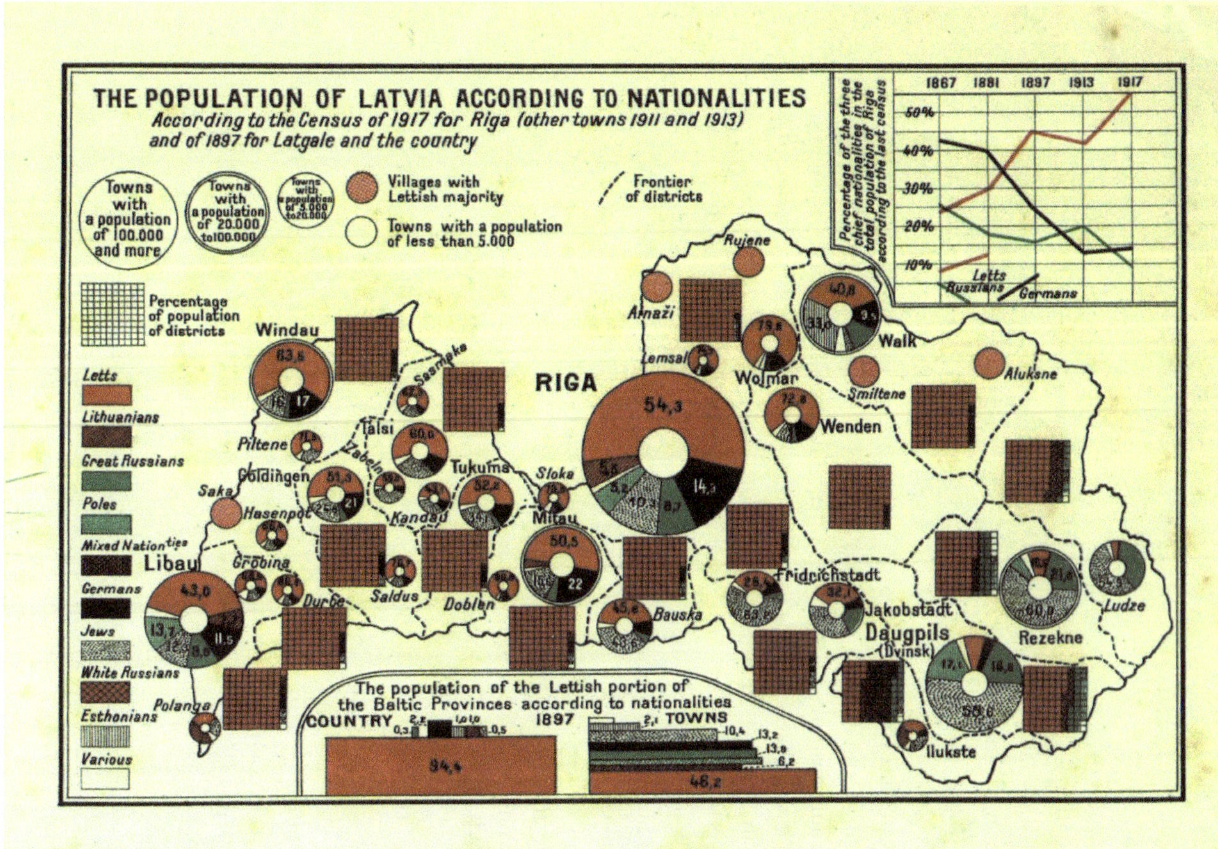

Figure 5.4. *Map of the Population of Latvia According to Nationalities* published in Latvian delegation's Memorandum for the Paris Peace Conference in 1919. National Library of Latvia.

Source: *Memorandum on Latvia Addressed to the Peace Conference by the Lettish Delegation*. ELMAL Balt Book IV 84. Reproduced with permission from the Estonian Literary Museum Archival Library.

was not yet consolidated in geopolitical terms.[53] On the other hand, these maps also demonstrate how ethnographic mapping was used in the summer of 1919 to spread ideas about the ambiguous location of Estonia and Latvia's future borders. Dashed lines hinted at the possibility that borders could be interpreted in various ways.[54] Hatched areas denoted mixed border regions that were still up for grabs. The flexibility of ethnographic mapping thus allowed the Estonian and Latvian representatives in Paris to sketch out multiple possible future versions of their national 'geo-body'.[55]

The way in maps were used in 1918–19 by new governments to lay down on the negotiation table which territories were still open for discussion is clearly illustrated by the meetings in June 1919 in Paris between the Latvian representatives and the government-in-exile of the Belarusian People's Republic (BNR). Both sides discussed how the future border between their states should be based on the boundaries of the former Russian imperial administrative districts of Dvinsk, Lutsin, and Rezhitsa. However, the Belarusian representatives emphasized that these administrative borders did not align with ethnographic borders. Anton Lutskevitch, who had served as Prime Minister during the BNR's short period of existence in 1918, referred to Karskii's *Ethnographic Map of the Belarusian Tribe* (1903) as evidence of how the eastern part of Ilukste, the area north of Dwinsk, and Lutzin (as these places were called by their Belarusian names) formed part of the Belarusian language area. Iauhen Ladnou (Evgenii Ladnov), Minister of Foreign Affairs and chairman of the Belarusian delegation in Paris was born in Dvinsk/Dwinsk, so the Belarusian connection to the region also had a personal resonance for him. As the meeting drew to a close, both delegations agreed that 'the population in certain parts of the contested districts is very mixed' and that there was a lack of precise statistics on the matter on which to base a decision.[56]

The Estonian-Latvian Boundary Commission

In the summer of 1919, the Estonian-Latvian Boundary Commission convened to begin the process of transforming these uncertain blueprints of national territorial sovereignty into more concrete plans for state borders.[57] Both parties agreed that the 1917 administrative boundary between Estland and Livland should be used as the 'source line' for the new international state border.[58] However, both sides agreed that this line needed to be 'adjusted' and 'corrected' so as to align the border with 'ethnographic principles', as well as other economic, transport, and

[53] Pourchier-Plesseraud, *Arts and a Nation*: 1.

[54] On the use of dashed lines to communiate uncertainty, see Lotz, Gestrichelte Linien und Schattierte Flächen: 58.

[55] Winichakul, *Siam Mapped*; Monmonier, *How to Lie with Maps*: 110.

[56] LVVA f. 2574, apr. 3, l. 7, lp. 214.

[57] LVVA f. 2574, apr. 3, l. 42, lp. 3.

[58] LVVAf. 2574, apr. 3, l. 42, lp. 4.

political considerations.[59] The framing of the boundary commission's remit is revealing. From the outset the boundary commission's work was underpinned by the assumption that ethnolinguistic borders were fundamental to national self-determination. It was not enough to have military control over a territory and governments needed an ethnolinguistic rationale to legitimize their rule. At the same time, tensions quickly emerged between the lofty rhetoric of self-determination and economic and political priorities, a clash of interests that would encumber the border-making process over the coming months. The boundary commission turned to ethnographic mapping as a technology to inform the border-drawing process. Important for the argument developed in this book—namely how spatial thinking came to dominate ways of thinking about nationhood—is that there was no interrogation of why ethnographic mapping was needed. The unquestioned use of ethnographic cartography reveals the extent to which ethnographic maps had come to be instrumentalized in the early twentieth century as a form of useful knowledge for solving territorial disputes. As we shall see, using ethnographic maps to solve border questions, as John Agnew argues, also 'trap[ped] thinking about and acting in the world in territorial terms'. Mapping conventions '*limit*[ed] the exercise of intellect, imagination, and political will' when it came to thinking about the possible options for constructing new national territories.[60]

Unsurprisingly, both national governments took maximalist stances when it came to mapping the Estonian and Latvian ethnographic territories. The *National Language Border* map produced by the Estonian Military Topography for the Estonian-Latvian Boundary Commission in 1919 formulated the Estonian position on the border question (Plate 17). The map marked the 1917 administrative border (in orange, perhaps originally red) and Estonian language border (in blue). Mapping in this case fulfilled an analytical purpose. The juxtaposition of the two lines made visible hitherto hidden patterns about exactly where the two states' approaches to border drawing aligned and where they diverged. The blue line of the Estonian language border is clearly visible snaking below the administrative border used as the base line. Notice how the blue line only departs from the base line to claim territory to the south, never to concede land to the north. The areas created by deviating lines defined the parameters of uncertainty in the border region, an effect not dissimilar to the error bars found on contemporary graphs. The map implied that the eventual border would likely fall somewhere within this zone. This cartography-driven approach to understanding the issues at stake in the border question both enabled the Estonian government to identify areas in the border region that would form the focus of future negotiations and, even more significantly, set the scope for how far they were prepared to negotiate. The converging

[59] LVVA f. 2574, apr. 3, l. 42, lp. 4 & 9.

[60] Agnew, 'Borders on the Mind': 176.

lines pre-empted Latvian territorial claims pushing northward by suggesting which areas—from the Estonian perspective—were not up for discussion. A reverse map was produced by the Latvian delegation marking the reach of Latvian-inhabited territories north of the administrative border line.[61] The map also included dashed lines to indicate the border guard line and green hatching indicating Estonian 'occupied territories', a provocative statement that only heightened tensions between the two sides. The Estonian representatives objected in the strongest terms to being designated as an 'occupying' power.

It is worth pausing here to note how ethnolinguistic groups were represented on these maps in light of the developments in cartographic thinking traced over the preceding four chapters. As I have argued, for most of the nineteenth century mapmakers were primarily driven to make ethnographic maps to construct different spatial communities, using coloured shading to define areas characterised by linguistic commonality and affinity. By 1919, the focus noticeably shifted to conceptualizing Estonians and Latvians solely in terms of the dividing lines between national groups. The shift in understandings of what the methods and purposes ethnographic mapping entailed—from areal community-building to delineating difference—reveals how the use of the 'ethnographic principle' as a driver of border-making was mirrored in cartographic practices.

One of the major sticking points in the Estonian-Latvian border negotiations was a town located in the centre of pre-1917 Livland province known in Russian as Valk, in German as Walk, in Estonian as Valga, and Latvian as Valka. Nineteenth-century ethnographic maps traditionally presented the ethnolinguistic border between Estonians and Latvians in the vicinity of the town. While mapmakers had historically jostled the town back and forth between the Estonian and Latvian ethnolinguistic territories, it only assumed a political significance when the town became the site of competing territorial claims in 1919. The town had been taken by Estonian forces in January 1919 as they pushed out the Latvian regiments fighting for the Bolsheviks (known as Latvian Riflemen) during the War of Independence. In autumn 1919 the town remained under Estonian control, which gave the Estonian representatives an upper hand in the negotiations. The Estonian position was further strengthened as during these months the Latvian government had their attention trained on the south and east where they were preoccupied until February 1920 with fighting against the Bolsheviks and the mercenary army of Pavel Bermondt-Avalov.[62]

The Estonian and Latvian delegations agreed that it was in the economic interests of the town to belong to just one state. Yet both sides outlined arguments

[61] Päts, *Eesti Vabariigi ja Läti Vabariigi vaheline piir*. Appendix 1. Eesti-Läti sõjaaegne piir (Latvijas-Igaunijas robežu josla) [kaart 1:84000].

[62] For an overview of the complex matrix of conflicts on the territory of Latvia during this period, see Minins, 'Latvia, 1918–1920'.

about why the town was either Estonian or Latvian on ethnolinguistic grounds. The Estonian representatives presented nationality statistics gathered by the town administration in August 1919 showing that the population of the town they called Valga comprised more than 7,000 Estonians and 5,600 Latvians. They explained that the number of Latvians also included 900 refugees who had not been resident in Valga prior to the war and thus, in their view, should be excluded from the nationality calculations.[63] The report listed further supporting arguments, including how more than 400 of the 750 houses in Valga belong to Estonians, how the economic life of the town was 'organically' linked 'with pure-Estonian places' to the north, and, looking to the south from Tallinn, the importance of Valga as a railway 'node' for the Estonian state. In addition, the representatives listed geographical and environmental arguments about how Valga was surrounded by forests and swamps to the south, and thus 'merges with Estonia from a natural geographical position'. Finally, they emphasized Valga's historical importance for the Estonian state and Estonian soldiers who had died in the offensive to capture the town in January 1919.[64]

Latvian claims to the town they called Valka rested on diametrically opposed arguments. Echoing the points put forward by the Estonians, the Latvians likewise argued for the importance of Valka as a railway junction and important regional economic centre for the Latvian state from the vantage point of Riga.[65] When it came to the ethnic statistics cited by the Estonian representatives, the Latvians claimed that as the data had been collected during 'unusual times' of conflict it could neither be verified nor taken as an accurate reflection of the population. The Latvian delegation countered by reaching back for statistics from the 1881 Livland census, one of the rare cases in the Russian Empire where a specific question about nationality was asked, showing that the town had historically been inhabited by a Latvian majority.[66] Reinterpreting these results forty years later through the lens of nationalist politics, the Latvian delegation pointed to how, of the town's 4,200 inhabitants, 1,604 had reported their nationality in 1881 as Latvian and only 978 as Estonian. By the All-Russian imperial census of 1897, the population had risen to 4,453 Latvians and 3,594 Estonians, although this data was based on language as no specific question about nationality was asked.[67]

To settle the question of Valga's/Valka's ethnolinguistic composition, the boundary commission organized a census of the town's population between 12 and 17 October 1919.[68] Counters were sent to every household to record individuals' self-proclaimed nationality and language spoken in the family. The questionnaire was printed in Estonian and Russian, as the town was under the control of

[63] LVVA f. 2574, apr. 3, l. 42, lp. 19.
[64] LVVA f. 2574, apr. 3, l. 42, lp. 19–20. [65] LVVA f. 2574, apr. 3, l. 42, lp. 15.
[66] Plakans & Wetherell, 'The 1881 Census'; Gibson, 'Experiencing Enumeration'.
[67] LVVA f. 2574, apr. 3, l. 42, lp. 60. [68] LVVA f. 2574, apr. 3, l. 42, lp. 53, 58.

the Estonian authorities at the time.[69] On this basis the Latvian delegation later contested the results, claiming that it did not reflect 'true figures' about the Latvian population as they had been 'influenced by the administrative authority exercised by the Estonian occupation'.[70] The need to solicit the opinions of the local population about the border question was also discussed in the sessions of the Estonian-Latvian Boundary Commission on 26–9 February 1920, when the issue of organizing a plebiscite was proposed. However, no agreement was reached about whether to organize a vote solely in Valga/Valka or to give the inhabitants of all the disputed border regions a vote.[71] By the middle of March, tensions were mounting and threats of conflict breaking out were in the air.[72] These exchanges highlight how the First World War introduced whole new levels of uncertainty into the already shady terrain of nationality statistics. The case of Valga/Valka is a case in point of how population displacement and the refugee crisis made both pre-war and current nationality statistics highly contentious. Yet, governments were limited in their ability to collect new data to support their claims in territories controlled by another regime.[73]

British Intervention in the Border Arbitration

When it became clear in early spring 1920 that no border agreement could be reached without a mediator the British Foreign Office stepped in to offer their help. The British had maintained a strong presence in the region since 1918–19, when they had carried out a naval operation as part of the Allied intervention in the Russian Civil War and in support of the Estonian and Latvian forces in the respective independence wars. The British government sympathized with the Estonian and Latvian independence movements, but they also had their eye on wider geopolitical issues in the region. Russia was one of Britain's major competitors and would be considerably weakened if it lost the Baltic littoral. Britain therefore had multiple interests in the stability and mutual cordiality of the Estonian, Latvian, and Lithuanian governments.[74] On 22 March 1920, the Estonian and Latvian provisional governments agreed to appoint Stephen Tallents, former British Commissioner to the Baltic who had also acted as civil governor of Riga during the summer of 1919, as chair of the Estonian-Latvian Boundary Commission. Captain Bentley and Major Caledon oversaw the Eastern and Western Sub-Commissions, respectively, while Colonel Rowan-Robinson

[69] LVVA f. 2574, apr. 3, l. 42, lp. 70; RA, ERA.1604.1.83.2.
[70] LVVA f. 2574, apr. 3, l. 46, lp. 10.
[71] LVVA f. 2574, apr. 3, l. 42, lp. 97.
[72] Arumäe, 'Piirikonfliktist liidulepinguni': 1121.
[73] Gatrell, *A Whole Empire Walking*: 3–4.
[74] Fletcher, 'The British Navy in the Baltic'; Butkus, 'Great Britain's Mediation'.

chaired the Central Walk Sub-Commission and acted as overall president of the three Sub-Commissions.

To untangle the complex mesh of ethnolinguistic, social, and economic relations in the border region the Estonian-Latvian Boundary Commission heavily relied on cartographic methods over the following months. Mapping enabled the Boundary Commission to aggregate a bigger picture of the issues and offer up solutions for how to settle the border dispute. The Sub-Commissions reused cadastral maps of estates from before the war, onto which they shaded the nationality of the owners of farmsteads and households using coloured pencils based on information gathered through inspections and interviews with proprietors. The maps created a powerful visual iconography of the process of nationalizing territory by shading Estonian land in blue and Latvian in red, consolidating and extending land-based territorial claims. As one report by the Latvian delegation noted with exasperation, from their perspective, the solution became evident just from glancing at a map: 'If [Valka] was awarded to ESTONIA, it would form a wedge in Latvian territory, and from a strategical standpoint such a wedge should not exist. Without studying rows of figures, it was apparent to anybody regarding the map that it was impossible for ESTONIA to provide the town of WALK with food.'[75] This 'wedge' was not readily apparent in rows and columns of data tables. Confrontations over space could only be resolved using spatial methods.

The maps of the Estonian-Latvian border region built on methodological innovations within the genre of ethnographic mapping examined in previous chapters. Mapmakers in the second half of the nineteenth century began to probe questions of nationality in urban spaces and used statistics on households and property ownership to understand the ethnolinguistic composition and socioeconomic inequalities of cities in the Baltic provinces. When applying cartography to address the question of border-drawing in 1919–20, property and landownership once again gained relevance as a spatial unit for dividing territory by nationality. Both the Estonian and Latvian maps categorized land in the border region according to the nationality of the landowners. Estonian-owned property was shaded in blue, while their Latvian counterparts were depicted in orange (or likely faded red). The result was a border proposal that twisted and turned around individual plots of land. This meandering border is the very antithesis of the borders drawn during the late nineteenth century by colonial administrators in Africa as straight lines on maps, which appeared to arbitrarily slice up territory with little or no regard for local geographies or the interests of the region's inhabitants.

By choosing to prioritize land and property ownership as the spatial framework for thinking about nationality, thereby ethnicizing property boundaries between neighbours, the creators of the boundary maps were likely interested in

[75] LVVA f. 2574, apr. 3, l. 46, lp. 13.

conveying a deep connection to the land. The impression of national rootedness, however, was illusory and deeply misleading. Peasant laws in the Baltic provinces in 1856–65 had facilitated the individual purchase of land in designated areas, often on the large estates owned by Baltic Germans. Thus, this land owned by Estonian- and Latvian-speakers had only been purchased in the last generation or two.[76] The logic of basing national territorial claims on land ownership was also destabilized in the light of the land reforms being enacted in both states at this time. On 10 October 1919, exactly around the time that the aforementioned Estonian *National Language Border* map was created, the Estonian Constituent Assembly passed the Land Reform Act. The reform expropriated large estates from the Baltic German elite and redistributed it to small farmers, a move designed to give peasants a stake in the new democratic state.[77] The Latvian Constitutional Assembly passed similar legislation on 16 September 1920 in the form of the Agrarian Reform Law. The cartographic ethnicization of property thus created a false sense of stability in a landscape of rapidly changing land ownership.

The emphasis on land and property ownership also raised important questions about whose rights should be considered when determining the placement of the border. What status did renters have in the process of ethnicizing attachment to the land? How should hired labourers be accounted for? The mobile lifestyles of seasonal labourers, as well as the presumed transience of renters, did not easily fit into the boundary commission's vision of the tightly overlapping relationship between territory and nationality. Those farming on rented land were listed in the maps' legends under the separate category of 'renters', whose presence was perceived as having less influence on the determination of national borders. The border-mapping process created hierarchies among the inhabitants of the border regions and weighted the claims of various groups differently.

After years of military conflict, maps remained shrouded in secrecy and were perceived as valuable intelligence sources. The Estonian authorities complained that the archives of former Livland province were located in Riga, which gave the Latvian delegation an unfair advantage when it came to gathering evidence for their border claims.[78] Meanwhile, the Latvian delegation refused to allow the Estonian representatives to keep their maps 'for military reasons' and British officers had to supervise the Estonian delegates while they consulted the maps.[79] On 30 May 1920, Colonel Rowan-Robinson sent the boundary commission chairman Tallents a map depicting the proposal for dividing the station: 'I attach a plan of VALK Station (for Mr. Tallents only) showing a Division suggested by the Lettish railway expert for the common working of the Station'.[80] The value

[76] Roasto, 'The Political Debate about the Land Question'.
[77] Maandi, 'Land Reforms and Territorial Integration'.
[78] Päts, *Eesti Vabariigi ja Läti Vabariigi vaheline piir*: 9.
[79] RA, ERA.1604.1.82.18.
[80] RA, ERA.1604.1.21.5.

placed on maps is clear from a report from the Valga/Valka Sub-Commission to Tallents on 17 May, which noted with urgency how: 'No satisfactory result can be obtained until a good map of Neu Rosen Commune is obtained. The original Commune map was taken by the Bolsheviks. A wire has been sent to Reval asking if another is available, and a draughtsman is being sent to the Commune to prepare a new map on the spot.'[81]

The maps of land and property ownership by nationality are evidence of the huge effort undertaken by the Estonian-Latvian Boundary Commission to fully explore the possibilities of using the 'ethnographic principle' as a basis for drawing state borders. They used mapping to aggregate data, identify outlying farms, and make recommendations about where the boundary should be drawn as a line of 'best fit' through the mix of farmsteads owned by Estonians and Latvians. The remarkable attention to detail on these maps, which sought to account for every individual plot of land and household, took ethnolinguistic border-drawing to the extreme. Their approach to mapping the border was founded on the belief that more data was better and, like many other boundary commissions, they became obsessed with 'the devil in the detail'.[82] The logic governing the Estonian-Latvian Boundary Commission's work was that if only they had more data, then a clear solution to the border question would present itself. However, the boundary commission quickly came to realize that this assumption was flawed. As Rowan-Robinson frustratedly wrote to Tallents on 2 May: 'There is no system in the distribution of Estonian and Lettish farms. It would, therefore, be impossible to find a line of division that would satisfy everyone.'[83] Paradoxically, cartography ended up disproving the very arguments that the mapmakers set out to make in their maps. Maps cast doubts on whether creating an ethnographic border was possible.

Zooming in: Estonian Valga and Latvian Valka

The most remarkable demonstration of large-scale ethnographic mapping emerged from an effort to resolve the Valga/Valka dispute (Plate 18). Unfortunately, like many of the maps encountered in this book, we know little about the circumstances behind the creation of the ethnographic town plan of Valga/Valka creation. The map was made by tracing over a pre-war town plan and then colouring addresses as either Estonian or Latvian using data from the October 1919 town survey.[84] By shading individual households in red or blue, the boundary commission hoped to shed light on the overall proportion and distribution of Latvians and Estonians in the town

[81] LVVA f. 2574, apr. 3, l. 46, lp. 116.
[82] Donaldson, 'Politics and Scale in Boundary-Making': 393.
[83] LVVA f. 2574, apr. 3, l. 46, lp. 129.
[84] LVVA f. 6862, apr. 2, l. 463, lp. l.

based on the respective ground occupied by property.[85] This attempt to analyse the spatial syntax of nationality and property ownership scaled up ethnographic cartography to an unprecedented level of detail. The plan of Valga/Valka bears a strong resemblance to urban plans of nationalities in large British and North American cities produced in the late nineteenth century as tools of social inquiry to analyse ethnic clustering among immigrant communities and social segregation.[86] Of importance to the argument in this chapter is how the same ideological stakes were amplified in pursuit of border-drawing in the case of a relatively small town in the Estonian-Latvian borderlands. By zooming in on Valga/Valka's residents, the map demonstrates how the drive to implement national self-determination led governments to carry out extensive surveillance of borderland populations.

The decision to map the data from the October 1919 census of Valga/Valka represented an effort to make what was counted count in the wider border debate. Devolved from a similar set of circumstances as the urban statistical maps examined in Chapter 3, the Valga/Valka ethnolinguistic town plan is a good example of how maps can be exploratory, created with the specific purpose of uncovering the spatial syntax of nationalities living together in high-density urban space. The map also made a clear statement about whose rights were being prioritized in the border-making process and whose were not. Although statistical data on 'foreigners' (e.g. Germans, Russian, Jews) had been gathered in October 1919, property owned by these nationalities was left blank on the map as they were deemed by the Boundary Commission to have no bearing on the outcome of the border. Building on the earlier observation about the lower weighting given to renters and hired labourers when linking nationality to territory, the silencing of certain groups from the map of Valga/Valka—and thus as a factor in the wider border debate—indicates how the mapping process required governments to set their priorities about who they defined as belonging in the borderlands. The ethnographic town plan of Valga/Valka is a potent example of what happens when border maps are created by political elites to perpetuate a certain nationalizing worldview, and reminds us of who stands to benefit and who is overlooked in this process.

The ethnographic town plan of Valga/Valka did not provide a clear resolution to the problem of how to divide the town along national lines. In fact, the map once again worked to challenge arguments for drawing ethnolinguistic borders by showing how Estonians and Latvians lived alongside one another all mixed up. The map did not convey a sense of the social uses of space in the town, how the townsfolk interacted with space in their daily lives, nor geographies of mobility, such as where the inhabitants worked and shopped, or which church they

[85] LVVA f. 2574, apr. 3, l. 46, lp. 117.

[86] Vaughan, *Mapping Society*: 129–67.

frequented.[87] As the Latvian delegate Voldemārs Zāmuēls, who would go on to become Prime Minister of Latvia in 1924, argued at a meeting of the boundary commission on 23 June, it did not make sense to debate who built the electricity substation or schools as 'They were built by the Letts and Esthonians, when the town was one administrative whole, and the question of Lett or Esthonian did not exist'.[88] The two-dimensional surface of the map also did not account for the verticality of the urban space, where families lived in multistorey buildings in apartments stacked on top of one another.[89]

Ultimately, the map only served to confirm doubts present from the beginning that the ideal of an ethnolinguistic border was incompatible with strategic interests and local realities on the ground. As expected, the Estonian and Latvian representatives levied a host of complaints against the map. The Latvian delegation objected that the data 'was apparently not wholly reliable' and pointed out how 'exactly opposite conclusions could be arrived at from figures, according to the way in which they were arranged'.[90] They cited cases of inhabitants who had changed their nationality since the statistics had been collected, giving the example of 'one woman who in October declared herself a German and now called herself Lettish'.[91] After studying the map, Rowan-Robinson remarked that 'It was clear that both races and their properties are mixed up together in a way that renders a division of the town very difficult'.[92] Zāmuēls concurred that 'The one clear fact was that the town was Esthonian and Latvian'.[93] The ethnographic town plan of Valga/Valka highlighted how any attempt at drawing an ethnolinguistic line of 'best fit' to divide the town between Estonians and Latvians clashed with economic and infrastructural concerns. As Tallents concluded, 'it is physically impossible to reach a completely just or satisfactory solution'.[94]

Public Participation in the Boundary Discussion

'We also have the right to self-determination', wrote Gustav Lillemann, the elder of Heinaste/Ainaži parish, in a petition he penned on behalf of the local community to the Estonian-Latvian Boundary Commission in March 1920.[95] Lillemann's pronouncement reveals how the declarations of Estonian and Latvian independence created expectations on the part of the local population that they could have a voice in shaping the form this state took, including its borders. The boundary commission received hundreds of petitions like Lillemann's from local

87 Vaughan, *Mapping Society*: 163; Dennis, *Cities in Modernity*: 69.
88 LVVA f. 2574, apr. 3, l. 46, lp. 17.
89 On the often-overlooked vertical dimensions of territoriality, see Elden, 'Secure the Volume'.
90 LVVA f. 2574, apr. 3, l. 46, lp. 16.
91 LVVA f. 2574, apr. 3, l. 46, lp.11.
92 LVVA f. 2574, apr. 3, l. 46, lp. 38.
93 LVVA f. 2574, apr. 3, l. 46, lp.16.
94 LVVA f. 2574, apr. 3, l. 46, lp. 33.
95 RA, ERA.1604.1.53.11.

men and women seeking to insert themselves into the discussion of the border question. Petitioning was a long-standing and widespread tradition in the Russian Empire as a legitimate means to lay out individual or collective grievances and appeal to the tsarist authorities.[96] In 1920, petitioners in the Estonian-Latvian border region used these same strategies to communicate with the new state authorities. The petitions give us important insights into public opinion on the border issue, revealing the local inhabitants' emotional responses to the bordering process and worries about how the proposed frontier would impact their everyday lives.

The local population followed the border developments with great interest. They talked to members of the boundary commission who visited their villages and homesteads to carry out inspections. Local newspapers republished the protocols of the boundary commission and printed outpourings by nationalistic journalists criticizing the border proposals and spreading rumours. For instance, in February 1920 Leena Schults explained her shock at discovering the news that her farm would fall on the Latvian side of the border:

> I cherished the hope that my farm at the fixing of the Esthonian-Latvian frontier line will remain within the frontiers of the Esthonian Republic, but I have now read in the newspaper that my farm has been left within the Lettish boundaries. This is not just, neither from the logical nor from the geographical point of view, for the fixed frontier line leaves my farm like in a sack—between the farms of Tolla and Kassi.[97]

Schults heard about the plans for her farm before the border proposal was formally announced, revealing the extent to which information was leaked and rumours circulated among the local population via the press.

Petitioners wrote to the boundary commission to articulate their own judgements about the ethnolinguistic character of their local community. In contrast to many of the examples encountered in previous chapters, the arguments put forward in the petitions demonstrate how most of the population were not 'indifferent' to the concept of nationhood.[98] The establishment of the state border according to 'ethnographic principles' made questions of nationhood directly relevant to border communities. The bordering process was a situation in which it was meaningful for the local inhabitants of the border region to articulate their sense of self in national terms.[99] To prove the national character of their local

[96] On the petitioning tradition in the Russian Empire, see Verner, 'Discursive Strategies in the 1905 Revolution'; Pyle, 'Peasant Strategies for Obtaining State Aid'; Hearne, 'To Denounce or Defend'.

[97] RA, ERA.1604.1.52.40.

[98] Zahra, 'Imagined Noncommunities'.

[99] On national commitment as 'situational' and 'event-driven', see Judson, *The Habsburg Empire*: 274.

community as either Estonian or Latvian, they provided evidence of local language use and behaviours, such as which church or school a person attended. Petrs Tilts, the elder of Neu Rosen (Vaaste-Roosa) parish, presented himself as 'a representative of the Latvian people [and writing] in the name of the Latvian people'.[100] He backed up his claim that he was part of a Latvian community by explaining that the local population spoke Latvian and used Latvian in the parish court and church services. Tilts juxtaposed these visible demonstrations of Latvianness with the weak 'Estonian element' in the area, noting how the Estonian language was much less prevalent.[101]

A similar rhetorical formulation can be found in the petition by Jaan Karrm from Kööma, whose farm lay on the proposed Latvian side of the border in Valmiera district. Karrm asserted: 'I am also completely an Esthonian, have been to an Esthonian school and am registered in an Esthonian parish for which reason I am regarded by the Letts as a stranger.'[102] Karrm constructed himself as a loyal member of the Estonian national community based not only on language use, but also through his behaviour and actions in the areas of education and religion. The rhetorical technique of role-playing as 'completely an Esthonian', who was perceived as a victimized "Other" by his Latvian neighbours, strengthened Karrm's appeal to the Estonian authorities to listen to his grievances. The act of writing a petition was as much transformative as it was affirmative, requiring petitioners to define on paper what being Latvian or Estonian meant to them.[103] In contrast to the boundary commission's fixation on property ownership for determining the national character of a locality, the petitions highlighted a range of different social practices that were perceived as locally significant manifestations of nationhood.

Military service performed during the recent wars was frequently used as proof of patriotic loyalty and service to the state, but also as a contract which required the state to fulfil certain obligations towards his citizens. Schults, the previously mentioned owner of Katkine farm, proclaimed 'I am . . . an Esthonian', but she also elaborated that 'my son Jaan Schults serves at present in the frontier guards' battalion and I and my children desire to belong with our farm to the Esthonian Republic'.[104] Gustav Arskalu, leaseholder of the Hoppenhof/Ape watermill, wrote to the boundary commission to claim compensation for the early termination of his contract on the watermill. In addition to requesting permission to move to Latvia, Arskalu tried to convince the authorities that he deserved financial compensation by describing how his only son, 17-year-old Kaube, had volunteered to join the Estonian National Army and 'fought, on the front line, for one year and

[100] LVVA f. 2574, apr. 3, l. 46, lp. 85.
[101] LVVA f. 2574, apr. 3, l. 46, lp.75.
[102] RA, ERA.1604.1.52.40.
[103] Verner, 'Discursive Strategies in the 1905 Revolution': 90.
[104] RA, ERA.1604.1.52.40.

three months against the Bolsheviks. Also in this time ca. four months on Estonian territory.'[105] The petitions are evidence that many of the local inhabitants acted on the expectations that the state would fulfil its obligation to listen to the interests of the local population. When they perceived the state as failing in its duty, individuals sometimes reacted in forceful ways. Johann Mihkli, a farm-owner from Pajo near Valga/Valka, objected that his 'farm, without my consent, is unexpectedly in the Latvian border'. Mihkli viewed the lack of consultation about the fate of his private property as an infringement of his rights as an Estonian citizen and opposed the decision in the strongest terms.[106]

While the majority of petitions contained forceful proclamations of national identification, not all the inhabitants wanted the borders to be drawn along national lines. Among the petitions are numerous examples of individuals who challenged nationality as the primary basis for determining political territories. Among the thirty-seven signatories of a petition from Heinaste/Ainaži were two people who expressly self-identified as 'Latvians who wanted to stay in Estonia'.[107] The community council members of Hummuli/Hummelshof outlined arguments for keeping their community intact. They explained how 'the inhabitants of the Hummuli community of Estonian as well as of Lettish nationality have till up to the present lived together in complete concord' and how dividing the community would leave some of the inhabitants without access to a mill. They prioritized economic interests and connections to markets over ideas about national self-determination and explicitly objected to a state border based on nationality: 'if the frontier will be fixed according to ethnographical principles it will be too sinuous, which is not suitable for a state frontier.' The petitioners' emotive language was designed to elicit sympathy for their cause: an ethnolinguistic border would be a 'calamity' that would cause many to 'suffer'.[108]

The decision to prioritize economic interests over national identification was not unique to the Estonian-Latvian border region. Similar concerns were voiced by borderland communities across Central and Eastern Europe in the wake of the First World War. Peter Haslinger's work on the Czechoslovak-Hungarian borderlands reveals how 'the population in the area alongside the demarcation line communicated concerns that did not exclusively follow a purely national patriotic logic'.[109] Following years of conflict and rapidly unfolding geopolitical changes across the region, it is quite reasonable that some border dwellers valued the territorial integrity of their local socioeconomic community. They prioritized spatial relations that were familiar and proximate, feeling less emotional attachment to the abstract concept of whatever larger political polity their community was embedded in. The petitions giving voice to these viewpoints constituted small-scale

[105] RA, ERA.1604.1.81.15r. [106] RA, ERA.957.2.16.6. [107] RA, ERA.1604.1.53.4.
[108] LVVA f. 2574, apr. 3, l. 46, lp. 96–7. [109] Haslinger, 'Dilemmas of Security': 194.

pushbacks against the Estonian-Latvian Boundary Commission and the larger ideal of ethnographic border-making. The various ways in which the petitioners articulated their relationship to the border invited the boundary commission to consider how the border might look if they prioritized different goals and factors.

Counter-Mapping

The border region dwellers not only voiced their opinion on the boundary question in the form of written petitions, but also mapped these ideas into visual narratives about territory and belonging. In 1920, the very high and personal stakes of the boundary-drawing process motivated populations in the Estonian-Latvian border region to put pencil to paper. The spread of cartographic literacy in the second half of the nineteenth century meant that these people did not need to raise funds or go through the logistical process of working with a printer. Being equipped with an elementary education, pencil, and paper was enough for one to become a mapmaker.

The maps sent by the petitioners to the Estonian-Latvian Boundary Commission can be grouped into two types. The first consisted of plans of property boundaries produced prior to the war which the inhabitants enclosed with their petitions as proof of their claims and to inform the authorities about information they might have overlooked in the course of their investigations. The use of plans as evidence to settle land disputes was a common practice dating back to seventeenth-century Muscovite lawsuits.[110] Landowners in Neu Rosen (Vaaste-Roosa) parish enclosed maps with their petition to the boundary commission for the same purpose, writing: 'Here we add the maps of the farms, or the plans of the land, to support the fact that our request is reasonably justified!'[111] Jaan Karrm complained that the border split his farm, explaining that the land owned by this wife had been assigned to Estonia, while the adjoining farmlands that he owned had been placed on the Latvian side of the border. The division of his lands would have dire long-term economic consequences for Jaan and his family: 'All my land products I have exported till up to the present to Esthonia, this ceases only in the last time owing to the new frontier line.'[112] To support his claim, he included a map with his petition showing the 'approximate scheme of the boundaries of my property'.[113] In both these cases, maps were exchanged as proof to substantiate individual claims.

The second type consisted of maps specifically drawn by petitioners to protest the authorities' decision about the placement of the border and present alternative arguments for where the boundary should be drawn. Boundary-mapping not

[110] Kivelson, 'Early Mapping': 30–1. [111] RA, ERA.1604.1.81.7r.
[112] RA, ERA.1604.1.52.40. [113] RA, ERA.1604.1.52.43.

only served as a state instrument to categorize, rationalize, and control people and territory, but was appropriated by local populations as a dialogic tool for protest and popular resistance. The petition maps colour-coded farms using the same blue and red colour scheme to depict Estonian and Latvian property ownership as on the maps made by the boundary commission. By incorporating the same 'map language' into their own petition maps, local farmers subverted hierarchies of knowledge to intervene meaningfully in the border debate. They challenged the boundary commission on its own terms, framing their local knowledge of space within a cartographic language that was acceptable in the context of the legal state system. This citizen-led border-mapping initiative allowed farmers who felt that their lands had been assigned to the wrong state to show how these actions held very real consequences and posed practical problems impacting their everyday lives.

Landowners and renters in Neu Rosen (Vastse-Roosa) were especially active in attaching hand-drawn maps to their collective petitions to communicate visual narratives of how the proposed border separated their farms from woodlands that they owned to the south and which the boundary commission proposed to give to Latvia (Figure 5.5 and Plate 19). The farmers perceived these woods as crucial for their livelihood and they topped the map with the inscription, 'If you do not change the border, it is impossible to live'.[114] The petitioners depicted the boundary commission's proposed border in a black-and-red dashed line, which was then surrounded by other information showing why the border should be changed based on the needs of the inhabitants and local patterns of land use. Whereas the boundary commission approached the border as a dividing force that would bring about the desired ethnic unmixing, the petitioners' map told alternative spatial stories about a border that could be positioned in such a way so as to function as a unifying force that maintained access to farmland and other environmental resources. In contrast to the thin lines drawn on maps by the members of the boundary commission, the local inhabitants mapped a frontier region that was 'thick' and malleable.[115] Juxtaposing these two approaches to border-mapping highlights the clash that emerged between ethnolinguistic border-making and socioeconomic relations to land that had been consolidated over many decades.

The use of petition mapping by the local population in the Baltic to challenge borders was not unique to the Estonian-Latvian border region. For example, the Latvian-Russian Boundary Commission received similar petitions accompanied by hand-drawn maps from locals protesting the placement of the border with Soviet Russia. In former Vitebsk province, however, the much lower literacy rates meant that these maps tended to be produced by estate owners rather than

[114] RA, ERA.1604.1.81.50.

[115] On the concept of the Soviet border as 'thick', see Dullin, *La frontière épaisse*.

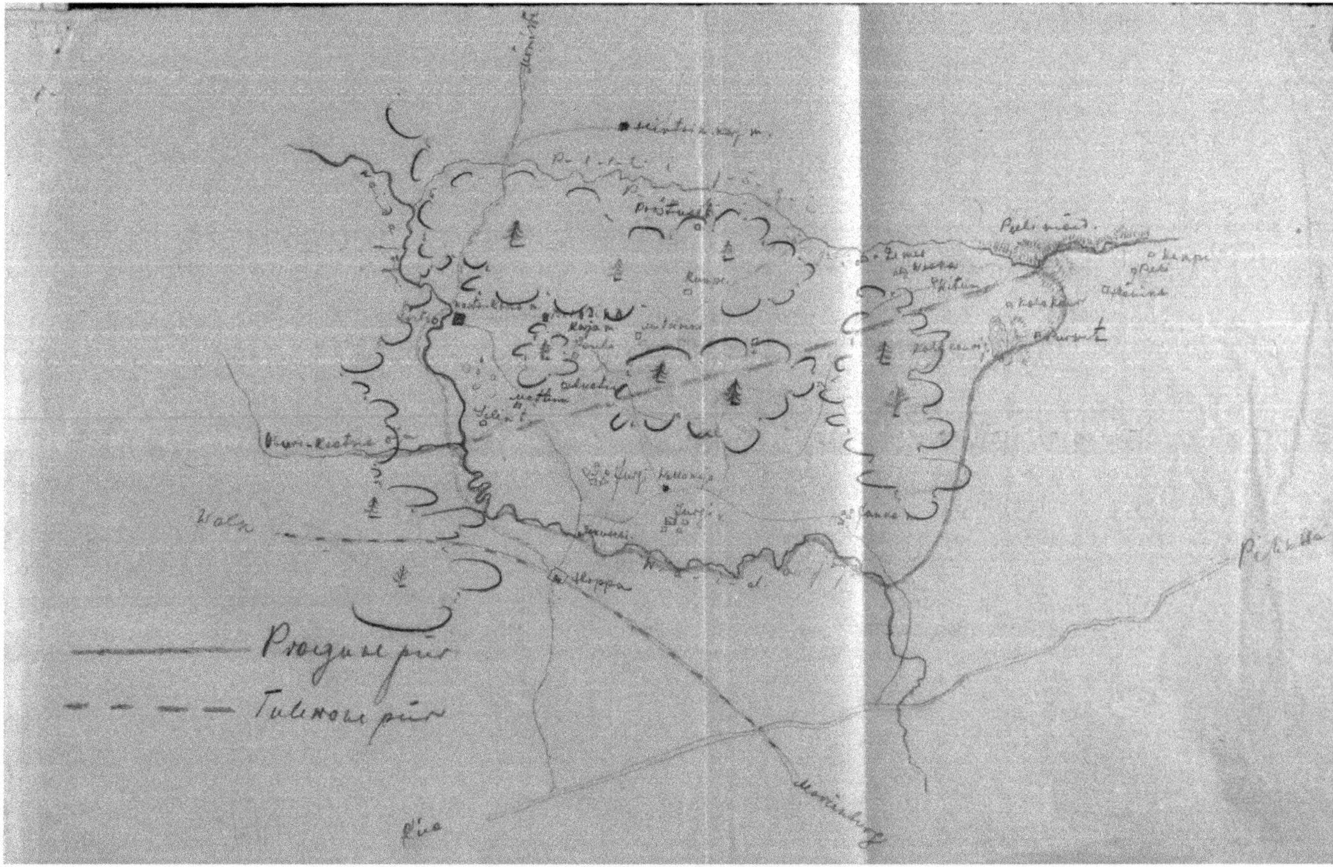

Figure 5.5. Petition map of border region in Vastse-Roosa sent to the Estonian-Latvian Boundary Commission in May 1920. The solid red line denotes the 'current border' proposed by the boundary commission, while the dashed red line marks the 'future border' desired by the local farmers.

Source: RA, ERA.1604.1.81 sheet 18.

lower-class farm owners and renters. One such map was sent in March 1921 by V. K. Tepffer, owner of the Domanovo estate on the left bank of the river Siniukha (Figure 5.6). Although Tepfer wrote his petition and annotated his map in Russian, he presented himself to the Latvian-Russian Boundary Commission as a 'native of Latvia'. He complained that the proposed border with Soviet Russia left his property and fertile farmlands, which are 'perhaps singular in Latgale', on the Russian side. He referred the boundary commission to a map he had drawn to explain to them how previously his land had been farmed by Latvian peasants from the left bank of the river and those living on church lands on the right bank

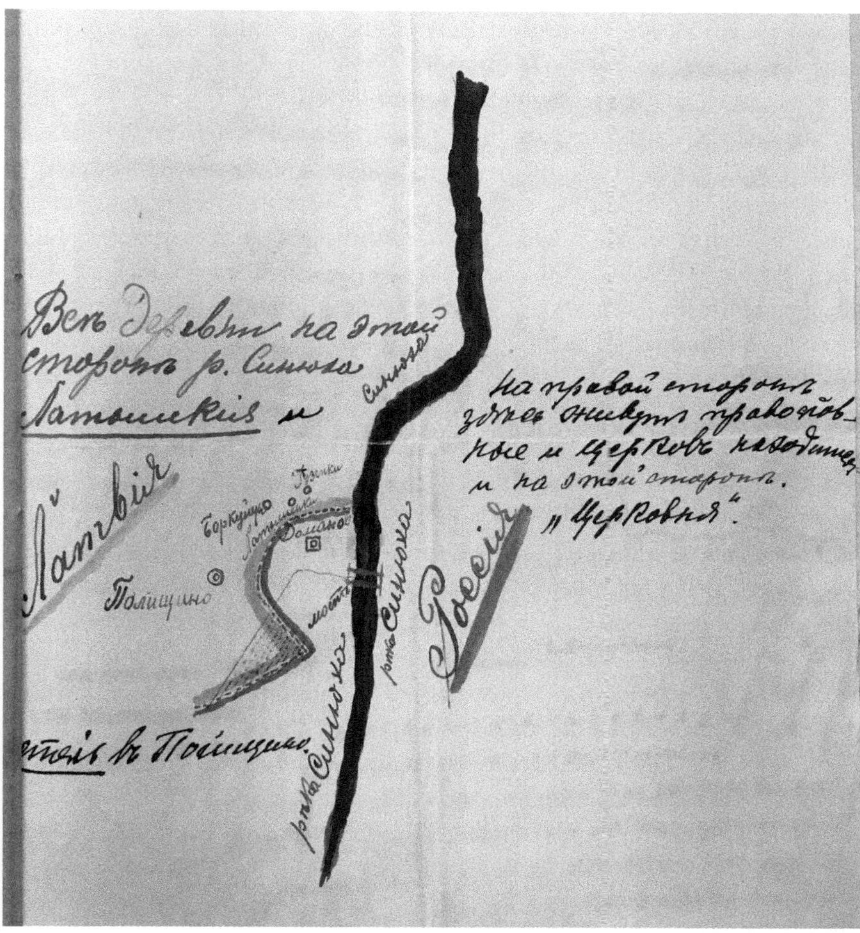

Figure 5.6. Petition map sent by the owner of the Domanovo manor, V. F. Tepffer, to the Latvian-Russian Boundary Commission in March 1921.

Source: LVVA f. 1313, apr. 2, l. 792, lp. 146.

had no connections to the estate. He argued that 'on all ethnographic, topographic, and economic conditions the "Domanovo" estate ought to belong to Latvia'. He requested that the boundary commission change the international border around the perimeter of his estate so that it followed the river and left his property in Latvia. Tepfer also marked an adjacent area of infertile land of an equivalent area on his map, which he proposed could be given to Russia in compensation.[116]

Much of the discussion of cartography in this book has centred on the ways in which ethnographic maps worked to reduce complexity, translating complex intersectional forms of identification into mutually exclusive and impermeable identity categories. However, in the case of the petitioners' maps, cartography was put to use for the reverse aim. By mapping the landscape from their own home vantage point, the petition maps challenged the binary classifications of farms along ethnolinguistic lines and instead communicated a more complex and nuanced picture of local geography. The maps incorporated pictorial vocabulary to indicate locally significant landmarks, such as a curvy blue line to indicate small streams, tree symbols to show forests, and inscriptions specifying the types of crops sown in different fields (Plate 19).[117] As Valerie Kivelson argues in her work on the uses of 'sketch' maps (*chertezhi*) in seventeenth-century legal disputes, it was often precisely this pictorial vocabulary used to communicate small details about the local topography, land use, and environmental relationships that would determine a judge's ruling.[118]

Petitioners harnessed the compelling nature of cartographic storytelling to enable the viewer to see the problem from the perspective of the local inhabitants and provide a glimpse into local ways of life. The map in Plate 19 was orientated so that the south was at the top. The viewer was positioned as if they were standing on the Estonian side of the border and looking south towards Latvia, thereby deconstructing the standard Cartesian north-south view that we are accustomed to on maps. Further information was given about the direction of the rising sun—'morning' (*hommik*)—marked on the left-hand side of the map, while the setting sun—'evening' (*õhtu*)—was indicated on the right-hand side. The petitioners disassembled cartographic conventions to reflect the positionality of the mapmaker, make the boundary commission understand the existential challenge they faced, and present their local knowledge to new audiences. Mapmakers harnessed these subjective and emotional responses to border maps, pushing the boundaries of what affective cartographic visual communication could look like.[119]

[116] LVVA f. 1313, apr. 2, l. 792, lp. 144–6.

[117] The use of pictorial cartographic symbols also appears on other petition maps, e.g. RA, ERA.1604.1.81.47.

[118] Kivelson, 'Early Mapping': 37.

[119] On 'affective' data visualization, see D'Ignazio & Klein, *Data Feminism*: 84–8.

Figure 5.7. Photograph of the Estonian-Latvian 'ethnographic' border near Valga/Valka in the early 1920s.

Source: Eesti-Läti piir Valga juures, ERM TM Fk 205, Eesti Rahva Muuseum, http://www.muis.ee/en_GB/museaalview/1532276.

While population centres such as Valga/Valka polarized national governments and became the focal points of international inventions to draw borders to enact ethnic unmixing, for the local population a border drawn on 'ethnographic principles' looked rather different. The Estonian-Latvian Boundary Commission based their survey of the border region on the idea that there were towns, villages, and farmsteads which needed to be divided by a border line based on ethnolinguistic principles. For the inhabitants of the border region, the space between these population points comprised a rolling landscape of fields, meadows, and forests (Figure 5.7). The process of joining up population points led to the arbitrary appropriation and division of these natural spaces, and 'dissonant concepts of "nature" and "nation"'.[120] By contrast, the petitioners' maps gave meaning to these spaces by showing the spatial syntax of relations between people, land, and natural resources. Their maps projected ideas about individual identity and belonging onto territory, presenting 'territorial spaces as "dwelling" rather than national spaces'.[121]

[120] Baron, 'Nature, Nationalism and Revolutionary Regionalism': 566.
[121] Agnew, 'Borders on the Mind': 176.

Tallent's Boundary Line

Not all maps are equally impactful, a theme repeatedly encountered throughout this book. Despite the efforts undertaken by the local population to communicate their own stories through mapping, the Estonian-Latvian Boundary Commission only paid selective attention to the requests of local inhabitants. A handful of petitions were translated into English for the benefit of the British arbitrators and submitted as part of the evidence gathered by the sub-commissions. On some occasions, the boundary commission took note of the petitioners' wishes, especially when the requests aligned with their own proposals of where to adjust the border, such as in the case of the southern part of Laura district which was later transferred from Estonia to Latvia in November 1923.[122] Yet, the interactions between the multiple stakeholders on the border question were characterized by deep-rooted power imbalances. Officials were often very wary of taking at face value these statements from the local population that problematized straightforward notions of national identification and loyalty. As Haslinger notes, authorities often found it 'difficult to separate genuine patriotic enthusiasm and staged or orchestrated opinions from each other in the records of the local visits'.[123] Similar sentiments littered the reports of the Estonian-Latvian Boundary Commission too. In response to petitions received from inhabitants of the coastal commune of Ainaži/Heinaste, Bentley cautioned that 'these petitions are not a reliable basis on which to work as the private and public feeling of the people vary to a considerable extent'.[124]

The Estonian and Latvian representatives also accused one another of manipulating the outpouring of testimonies from the local population for political ends. Bentley criticized the Estonian mayor of the town of Ainaži/Heinaste as 'hopelessly incompetent and unfit for the position he occupied', having supplied him with 'erroneous' figures that exaggerated the number of Estonians.[125] The Estonian authorities complained that the registers of inhabitants and their properties submitted by local pro-Latvian authorities 'contain evidently false statements' and accused the Latvian authorities of 'partiality... as well as a lawless and incorrect fulfilment of their duties'.[126] Meanwhile, the Latvian representatives accused Jaan Pommer (Raudma), editor of the local Estonian-language newspaper *Rajalane* in Valga/Valka, of publishing testimonies by the local population in his newspaper that were deliberately designed to 'embitter feelings between the two countries'. Pommer defended his actions by replying that 'some of the statements made by Lettish farmers to the Commission were so ridiculously untrue that he felt bound to report on them'. The Latvians later conceded that 'many of them

[122] Ščerbinskis, 'Lauri Latvians'.
[123] Haslinger, 'Dilemmas of Security': 194.
[124] LVVA f. 2574, apr. 3, l. 46, lp. 46.
[125] LVVA f. 2574, apr. 3, l. 46, lp. 67.
[126] RA, ERA.1604.1.82.30.

were somewhat exaggerated but that was natural and not perhaps sufficient grounds for an article headed "Perversion of Truth"'.[127]

By the middle of May, after several months of intensive work, the British officers began to lose patience. By now the 'Walk Question' had become synonymous with the whole border issue between Estonia and Latvia.[128] As Tallents noted in his memoire: 'I had known, when I first intervened in this dispute, that its solution was likely to be a disagreeable and difficult job; but I had not taken the full measure of its difficulties or of the bitter feeling which it would arouse.'[129] The everyday grind of the work of the boundary commission began to take its toll. As we have seen in earlier chapters, the process of gathering data to make maps was cumbersome. Delays in receiving maps, motorcar breakdowns, illness, the unwillingness of the Estonian and Latvian delegates to cooperate, and the slow pace of work wore down the patience of the British mediators.[130] Rowan-Robinson's exasperation is palpable in the report he sent to Tallents on 19 May 1920:

> In my opinion the whole frontier question has been thoroughly thrashed out and nothing is to be gained by further discussion. Except in the most obvious cases, agreement is impossible to obtain. It remains therefore for me as President of the Sub-Commissions to recommend a line, and that I am prepared to do.[131]

Ultimately, Tallents was forced to use his prerogative as chair of the Estonian-Latvian Boundary Commission to make a decision and draw the deliberations to a close. Based on recommendations by the sub-commissions, on 1 July 1920 Tallents divided Valga/Valka along the small Konnaoja/Varžupīte river, a tributary of the Pedeli/Pedele river. The town centre and main station to the east were given to Estonia, while Latvia was awarded the western part of the town and the second railway station on the Marienburg-Walk-Pernau line. The northern half of the town henceforth was officially known by its Estonian-language name—Valga, while the southern half of the town went by its Latvian-language name—Valka. On 3 July, Tallents outlined a proposal for the rest of the border, stretching from the coast to the boundary with Soviet Russia in the east. As compensation for Latvia's smaller share of Valka, it was awarded the coastal parish of Heinaste/Ainaži. Contrary to the initial plan to draw a border based on ethnographic principles, economic factors played the decisive role in shaping Tallent's decision. He felt that to draw the border solely on an ethnolinguistic rationale and 'run a frontier line through the main body of the town of Walk would cripple its activity and might endanger its life'.[132]

[127] LVVA f. 2574, apr. 3, l. 46, lp. 32.
[128] LVVA f. 2574, apr. 3, l. 46, lp. 27.
[129] Tallents, *Man and Boy*: 373.
[130] LVVA f. 2574, apr. 3, l. 46, lp. 76, 111, 116.
[131] LVVA f. 2574, apr. 3, l. 46, lp. 120.
[132] LVVA f. 2574, apr. 3, l. 46, lp. 33.

Members of the boundary commission marked the border, which came to be known colloquially as the 'Tallents boundary line', on an updated border map together with ongoing disputed regions.[133] Individual sections of the border were also referred to according to the British officer who had chaired the sub-commission responsible for delimiting that part of the boundary, such as the 'Bentley line'.[134] In an explicit manifestation of the palimpsest-like layering of different forms of spatial organization in the transition from empire to nation-state, the Estonian-Latvian border map was made from a repurposed Russian imperial administrative map. The new Estonian-Latvian border was literally inscribed over the top of former imperial administrative borders.

Tallent's recommendation to split Valga/Valka and give the majority of the town to Estonia, while leaving a small part to Latvia, was met with fierce objections from the two governments and was heavily criticized by the local press.[135] On 11 July, a Russian-language newspaper in Riga published a caricature depicting Tallents as a Solomon-like figure brandishing a sword to divide the town, depicted as an innocent baby (Figure 5.8). Satirizing the Old Testament story of two mothers fighting over a baby, Estonia (on the right) was depicted as continuing to tug at the baby's feet, while Latvia (on the left) chose to renounce the child rather than letting it be killed as a result of the dispute. Tallents was deeply offended by this popular portrayal of his actions, describing the cartoon as a 'scurrilous attack by the gutter press of the Baltic'.[136] As a result of his experience arbitrating the Estonian-Latvian border negotiations, Tallents turned down the invitation to mediate the Latvian-Lithuanian Boundary Commission in the autumn of 1920. In his place, the British Foreign Office appointed James Young Simpson, a Scottish zoologist and theologian from the University of Edinburgh. Simpson had worked in the Russian Section of the Foreign Office and spent six weeks in Paris in 1919 with the British Delegation to the Peace Conference as an expert on Russia's borderlands.[137] In a twist of fate, Tallents later served as secretary to Ireland's Lord Lieutenant and then Imperial Secretary for Northern Ireland from 1922–6. Drawing on his experiences of diplomacy, propaganda, and research gained during his civil service career, Tallents became a pioneering figure in the emergence of the concept of public relations.[138]

Despite the considerable disagreements over the boundary line, war did not break out between Estonia and Latvia. Both governments agreed to sign a border treaty on the basis of Tallent's line on 19 October 1920 and over the following years the boundary commission only made minor adjustments to the border as they worked to settle the matter of the maritime border, the question of Ruhnu/Runö

[133] RA, ERA.T-6.3.1142 sheet 1. [134] LVVA f. 2574, apr. 3, l. 46, lp. 202.
[135] Päts, *Eesti Vabariigi ja Läti Vabariigi vaheline piir*, appendix 9.
[136] Tallents, *Man and Boy*: 378.
[137] Simpson, *The Garment of the Living God*; Alston, 'James Young Simpson'.
[138] Anthony, *Public Relations and the Making of Modern Britain*.

Figure 5.8. 'Solomon of our days': a caricature of Stephen Tallent's dividing the town of Valga/Valka between Estonia and Latvia. Zāmuēls, on the left, is saying: 'As for me, Tallents, give her the baby alive and do not destroy my town by death!'

Source: *Segodnia*, 11 July 1920: 1. Reproduced with permission from the National Library of Latvia.

island, and the border to the east. The technical commission began work on the lengthy process of surveying and physically constructing the border fences, guard posts, and crossing points along the frontier between Estonia and Latvia, which was completed in April 1927.[139] The incredible lengths undertaken by the boundary commission in 1919 to 1920 to map the border according to the 'ethnographic principle' resulted in a state border that was remarkably close to the ideal of a nation-state border. The small number of Estonians and Latvians left on the "wrong" side of the border found themselves recast in the new category of 'national minorities'.[140]

Conclusion

'A state of affairs…which is almost Gilbertian' was how Owen Rutter described the division of Valga/Valka (Figure 5.9). Rutter's account of his visit to the town in the

[139] The process is detailed in: Päts, *Eesti Vabariigi ja Läti Vabariigi vaheline piir*.
[140] Mela, *Eesti lätlased*.

early 1920s conveys a sense of his uncanny encounter with the border running through the town:

> I had walked some distance through the rather uninteresting town and was just beginning to think that the police station was a long way when, looking up to see what street I was in, I saw the strange word *tän* instead of the familiar Lettish *iela*—street. It seemed that I was in Estonia .... Standing in the middle of the cobbled roadway, with one foot in Latvia and the other in Estonia, I took a photograph of that crooked street and then hurried on, keeping to the Latvian pavement.[141]

A century later, the impact of the border dividing Valga/Valka has been somewhat lessened, especially since Estonia and Latvia entered the Schengen zone in 2007 and with the adoption of the common Euro currency in Estonia in 2011 and Latvia in 2014. Today, one has to look carefully to spot the international border signposts within the town. Nevertheless, the division of the town stands as a potent reminder of the early twentieth-century paradox of national self-determination: the massive amounts of labour invested by people across all segments of society into blueprinting borders based on the 'ethnographic principle', yet the practical impossibility of enacting such an ideal. Not only did the boundary commission's experience of border-drawing in Valga/Valka highlight the difficulties of spatially unmixing ethnolinguistic groups in former imperial spaces, but it also demonstrated the impossibility of harmonizing an ethnolinguistic logic of border-drawing with other economic and social interests, as well as transport and communication infrastructures.

Most of the maps analysed in this chapter do not conform to our prevailing expectations of ethnographic cartography. The application of ethnographic mapping to border-drawing meant that the genre took on new meanings and approaches in the turbulent aftermath of the First World War and the collapse of the Russian Empire. The ethnographic maps produced by the Boundary Commission zoomed in on the nationality of landowners to convey the micro-geographies of individual farms and houses. Ethnographic mapmakers also cemented a major methodological shift by using property and land ownership, rather than administrative divisions, as the dominant spatial framework for conceptualizing the relationship between nationality and space. At the same time, there are many recognizable tropes in these border maps which we already encountered in our examination of the development of the ethnographic map genre in the nineteenth-century. Not least, mapping continued to be used by those in positions of power to define who mattered in the discussions of where to

[141] Rutter, *The New Baltic States*: 178–9.

Figure 5.9. Estonian-Latvian border checkpoint in Valka/Valga *c.*1930.
Source: RA, EAA.2073.2.197.1.

draw ethnolinguistic borders and to create hierarchies about who had the right to belong in different places.

The most remarkable development during this period was how the process of drawing state borders brought ethnographic mapping to the realm of the local population in an unprecedented way. Barriers to participation in mapmaking encountered in earlier chapters, such as cartographic literacy, money, and access to a printing press, were gradually eroded as local populations picked up paper and pencils to engage the authorities in discussion about the geopolitics of border-drawing. Cartographic narratives about territory and national belonging were not only debated on the pages of newspapers and in cheap mass-produced books, but on scraps of paper by people with very little material means. The petitions and maps produced by the region's inhabitants to communicate with the authorities and contest the proposals of the boundary commission marked a reversal, and deliberate subversion, of established practices about who could define the conversations around nationhood and territory through the medium of ethnographic mapping. The subjects of ethnographic maps became mapmakers themselves, while governments and officials became the recipients and intended audiences of ethnographic maps. The declarations of independence raised expectations on the part of the local population that borders should no longer be solely decided by high-level decision-makers, but that they too could have a voice in shaping the border outcome. Ethnographic maps acquired new functions and were used to attribute new meanings to the territories and peoples they mapped.

In previous chapters, we have seen how people saw lines on maps and made judgements about the linguistic differences that they heard. However, the process of physically demarcating a state border line solidified ethnolinguistic borders into tangible phenomena that could be touched, traversed, and embodied. The process of deciding which house or field should belong on which side of the border line intimately linked ethnolinguistic borders to bodies, land, and soil. Reading these maps as complex visual arguments about spatial relations enables us to examine how border dwellers perceived, and sought to influence, the relationship between ethnicity and territory during this period of dramatic historical change. The border inhabitants used maps to hold the states accountable for the disruption to their economic, social, and everyday life caused by the pursuit of national self-determination and used cartographic language to assert their perceived right to have a say in shaping the course of national borders. The process of establishing state borders, therefore, lets us see how vocal and self-determining local populations could be during this period when borders in the region were being dramatically redrawn. Ethnographic cartography was not only a tool of imperial governments, scholars, national activists, and bureaucrats, but by the early twentieth century the cultures and practices of ethnographic mapping had permeated across all levels of Baltic society.

Epilogue

Afterlives of Maps

During a routine inspection of Belarusian minority schools in the Latgale region of eastern Latvia in March 1924, the Daugavpils district school inspector Ozoliņš discovered a map of the Belarusian People's Republic on a classroom wall of Kapiņu Belarusian primary school. The map depicted the borders of the truncated attempt to establish a Belarusian state on the German-occupied territory following the Treaty of Brest Litovsk in 1918, a project which shortly collapsed with the invasion of Polish and Bolshevik forces in 1919.[1] Glancing at the map of this ephemeral state on the classroom wall, the school inspector was shocked to discover that the map depicted parts of the territory of Latvia within the borders of this imagined Belarusian nation-state and in Russia. The inspector uncovered the same map in other Belarusian schools in the region, as well as on numerous other smaller postcards in school libraries and in the hands of schoolchildren throughout Latgale. Moreover, teachers in the region's Belarusian minority schools were using a Belarusian-language geography textbook published by Arkadz Smolich, the famous Belarusian soil scientist and Soviet Belarus's first university chair in geography, as supplementary teaching material.[2] Machnovskis, the headteacher of the Belarusian primary school in Kapiņu, denied malicious intent and explained to Ozoliņš that the map had been sent to him by the Belarusian Education Department within the Latvian Ministry of Education. However, it later emerged that Smolich's geography textbook had not been approved for use in Belarusian schools.

Eight Belarusian schoolteachers from Latgale were arrested, including the head of the Belarusian school department and founder of the "Bakkovshchina" Belarusian cultural society Kaustus' Ezavitau (known in Latvian as Konstantīns Jezovitovs), and headteachers and prominent figures in the Belarusian community in Latgale, Uladzimir Pihuleuski/Vladimirs Piguļevskis and Ivans Krasovskis.[3] They were charged with inciting separatism among the Belarusian minority in Latvia by communicating with Belarusian organizations abroad and distributing geographical maps, textbooks, and periodicals among pupils that showed parts of Latvia incorporated within the Belarusian territory. The trial of the

[1] Rudling, *The Rise and Fall*: 78–96. [2] Smolich, *Heagrafiia Belarusi*.

[3] 'Baltkrievu sazvērestības lietu sāka tiesāt 2. aprilī, Latgales apgabaltiesā', *Daugavas Vēstnesis*, no. 14, 3 April 1925: 3.

Geographies of Nationhood: Cartography, Science, and Society in the Russian Imperial Baltic. Catherine Gibson, Oxford University Press. © Catherine Gibson 2022. DOI: 10.1093/oso/9780192844323.003.0007

Belarusian schoolteachers from 2–5 April 1925 at the Latgale Regional Court uncovered no evidence of undercover activity to promote secessionist ideas among the Belarusian-speaking community in Latgale. The court ruled that the offending maps merely depicted the ethnographic distribution of Belarusians rather than the political borders of the Belarusian People's Republic, and accepted the arguments put forward by the defendants that the geography textbooks were being used to teach Belarusian language due to a lack of other pedagogical materials.[4] Although the accused were eventually acquitted, the so-called 'Belarusian Trial' ignited a public debate about the political loyalty of minorities to the Latvian state, policies towards minority education, and relations between the new Latvian state and its neighbours.[5] The panic sparked by the prospect of maps being used in Belarusian minority schools to promote alternative ways of thinking about ethnolinguistic territories and borders reveals how by the early twentieth century ethnographic maps were perceived not only as powerful shapers of elite discourses about space, but could exert a powerful influence over children's hearts and minds.[6]

The controversy encapsulates many of the key questions and themes which have driven this book. Over the course of the nineteenth century, mapmakers experimented with different cartographic approaches, techniques, and scales to map people. They studied the distribution of populations through fieldwork and statistics, divided them into national groups, sorted them into taxonomies based on relative similarity and difference to one another, colour-coded them, and inscribed them onto the natural landscape. The reduction of the complex intersection of religious, linguistic, and socioeconomic forms of identification into a two-dimensional rationalized image evolved into a powerful way of perceiving the human world. Maps streamlined hybrid, multifaceted, and fluid forms of identification into a neat mosaic of national spaces. With the collapse of empires in Central and Eastern Europe at the end of the First World War, this territorialized way of thinking about nationality encouraged nationalists to declare independence on behalf of nation-states, arguing that—in theory at least—the mapped spaces and borders of ethnolinguistically-defined national populaces should determine state sovereignty.

In the preceding five chapters, I have shown how mapmakers explored different cartographic techniques to examine ideas about ethnolinguistic communities in the Russian Empire's Baltic provinces and neighbouring regions. Specifically, this book has examined the role of cartography in embedding a particular map-based view of nationality, which strongly linked national groups to particular

[4] Kazachenok, '"Belorusskii protsess" v Latvii'.

[5] 'Balss no latvju skolotāju vidus', *Daugavas Vēstnesis*, no. 15, 9 April 1925: 4; Ezavitau, *Belarusy u Latvii*; Apine, *Baltkrievi Latvijā*; Jēkabsons, 'Baltkrievi Latvijā'; Purs, 'The Price of Free Lunches'; Purs, 'Unsatisfactory National Identity'.

[6] On borderland children as targets of nationalizing policies in the interwar period, see Venken, *Peripheries at the Centre*.

territories. The case studies highlight the plurality of ideas, often conflicting, about how nationality should be represented in map form. Mapmakers engaged in methodological discussions about best practices for collecting data, how to translate numbers or ethnographic descriptions into cartographic form, and the scales to which this space-based way of thinking about nationality should be applied. The nineteenth century witnessed an exponential growth in the amount of statistical data collected about populations, necessitating new ways to process and manage information. Cartography provided a visual method to synthesize and abstract this unwieldy data and communicate it in a clear and accessible way. However, in doing so, maps profoundly impacted understandings of language use, ethnicity, and nationhood.

This book has placed particular emphasis on the production of maps as objects of material and visual culture. The broader discussion of ethnographic mapmaking not only encompassed different ideological viewpoints but was also shaped by technological and economic factors related to gathering data, financing, printing, publishing, and selling maps. More often than not, the process of making an ethnographic map was hampered by financial difficulties and wrought with logistical challenges. The available printing technology shaped how mapmakers designed and created their maps, such as the number of colours, level of detail, and kinds of patterns that could be engraved. Nineteenth-century mapmakers were very self-reflective about their role as craftsmen and viewed their work in the context of a broader period of experimentation with different methods of data visualization in the second half of the nineteenth century.

The use of ethnographic mapmaking as a tool to shape ways of thinking about nationality and territory did not end with the creation of the new independent states in the Baltic after the First World War. Within the newly independent states of Estonia and Latvia in the 1920s, State Statistical Bureaus were established and immediately set to work on counting and mapping the population of their countries. The first Estonian statistical atlas was published in three volumes from 1925 to 1928 and the first Latvian atlas in 1938, the publication dates coinciding with the anniversaries of independence.[7] The maps of nationalities and religions in the atlas reinforced the idea that these states belonged to the titular nation. In both cases, the atlases were printed in the official state languages and in French, revealing how the atlases were directed at international audience and presented a portrait of Estonia and Latvia for their political allies and trade partners.[8]

In addition to the traces of Belarusian maps found in schools in eastern Latvia, the 'phantom borders' of German and Polish influence in the Baltic provinces continued to reverberate in the interwar years. In December 1919, the Polish government laid claim to several parishes of former south-west Kurland and the

[7] *Eesti—Statistiline Album*; *Latvijas Statistikas Atlass*.

[8] Schulz, 'The Statistical Atlases of the Baltic States'.

town of Gryva/Griva, which had been rendered as Polish in Eugeniusz Romer's *Geographical-Statistical Atlas of Poland*. Only in 1929 did Poland formally recognize the inclusion of these territories into Latvia and a Latvian-Polish commission began to survey the border region in 1933.[9] Estonia and Latvia also continued to occupy an important place in the minds of Weimar economic and foreign policy-makers.[10] In the second half of the 1930s, German annexationist thinking towards the region underwent a resurgence when numerous statistical maps were published in Weimar Germany depicting Estonia's and Latvia's Baltic Germans population as 'Germans abroad' (*Auslandsdeutsche*).[11] The construction of spatial narratives about the Baltic and the region's historical, cultural, political, and economic relationship to Europe and Russia carried on through mapping into the twentieth and twenty-first centuries. Changing approaches to mapping Estonian history in school textbooks, for instance, reveal how the Baltic continues to be shaped by shifting 'centripedal forces' and political attempts to reposition the region in relation to different 'natural' geographical and historical orientations to the east, west, and north.[12]

A major aim of this book has been to show how ethnographic mapmaking was a far more widespread activity than has previously been thought. Rather than being the sole purvey of states or intellectual elites located in places traditionally thought of as centres of scientific knowledge production, a wide range of actors engaged in mapmaking in the course of their daily lives or everyday work and in locations often overlooked as peripheral to scientific activity in imperial Russia. Approaching mapmaking as a broader participatory activity that occurred across diverse sites and settings in the Russian Empire foregrounds how ethnographic cartography also had a strong subversive tendency. Maps were wielded by local populations as instruments of enlightened public education, to assert local agency, and even to directly protest and challenge the state. Taking this multilevel approach to the history of ethnographic cartography is crucial for explaining the permeation of ethnographic maps across multiple social and political spheres in the second half of the nineteenth century. It is only by considering the spread of cartographic literacy among the general population that we can understand how maps came to resonate with audiences and hold such a powerful sway over popular understandings of nationhood.

[9] Jēkabsons, 'Sešu pagastu un Grīvas pilsētas problēma'; Maliszewski, *Polacy na Łotwie*; Maliszewski, *La Pologne d'aujourd'hui*: 52–3.

[10] Hiden, *The Baltic States and Weimar Ostpolitik*.

[11] Handrack, *Karte der Deutschen*; Von Maydell, *Die Bevölkerung Estlands und Lettlands*; Von Maydell & von Bornkam, *Die Verbreitung der Deutschen*.

[12] Pääbo, 'Constructing Historical Space'.

Bibliography

Archives and Libraries

Estonian Literary Museum Archival Library
Francis Skaryna Belarusian Library and Museum
Herder Institute Library, Document Collection and Archives of the Baltic Knighthoods
Latvian State Historical Archive
National Archives of Estonia
National Historical Archives of Belarus
National Library of Estonia
National Library of Finland
National Library of Latvia
National Library of Russia
Russian State Library
Russian State Historical Archives
Scientific Archive of the Russian Geographical Society
St. Petersburg Branch of the Archive of the Russian Academy of Science
Tartu University Library Manuscripts Collection

Maps (Excluding Works Listed in Figures and Plates)

Dr. Henry Lange's Karte von Liv-, Est- und Kurland. Nach den neuesten Quellen bearbeitet von K. v. Lowis of Menar und F. A. Brochkaus' Geographisch-Artistischer Anstalt in Leipzig, 9th edn (Riga: N. Kymmel, 1914).

Eesti—Statistiline Album/Estonie—Album statistique 3 vols (Tallinn: Riigi Statistika Keskbüroo, 1925–8).

Erkert [Erckert], R. F. *Etnograficheskii atlas Zapadno-Russkikh gubernii i sosednikh oblastei* (St. Petersburg: 1863).

Grewingk, Constantin. *Geognostische Karte der Ostseeprovinzen Liv-, Est- und Kurland* (Dorpat: Naturforscher Gesellschaft zu Dorpat, 1861).

Handrack, H. *Karte der Deutschen in Lettland nach Städten und Gemeinden (nach Angaben der lettischen Volkszählung 1935, überprüft durch die Deutsche Volksgruppe in Lettland)* (Dorpat: Volksdeutsche Forschungsgemeinschaften, 1936).

Jakobson, Carl Robert. *Kooli atlas ehk maa kaardi-raamat* (Tartu: H. Laakmann, 1873).

Karskii. E. F., *Etnograficheskaia karta belorusskogo plemeni (s ob'ianitel'noi zapiskoi)*. Ser: Trudy Komissii po izucheniyu plemennogo sostava Rossii, vol. 2 (Petrograd: Tipografiia Rossiiskoi Academii Nauk, 1917).

Kentmann, Woldemar Friedrich. *Geograahwia kaardid. Kooliaste Geograahwia-raamatu lilkas* (Tallinn: Eestimaa Sinod, 1884).

Khoziaistvenno-statisticheskii atlas Evropeiskoi Rossii, sostavlennyi pri Departamente sel'skogo khoziaistva MGI (St. Petersburg: Departament sel'skogo-khoziaistva MGI 1851).

Keograhwi, ehk õppetus Ma-ilma surussest ja Ma-ilma madest (Tartu: H. Laakmann, 1854).

Kiepert, Heinrich. *Übersichtskarte der Verbreitung der Deutschen in Europa* (Berlin: Dietrich Reimer, 1887).

Köppen, Peter. *Ethnographische Karte des St. Petersburgischen Gouvernements* (St. Petersburg: Kaiserliche Akademie der Wissenschaften, 1849).

Lantkharte po Latweeschu semmes (Tehrbatā: G. A. Reyhers; Mitau: C. Schulz, 1859).

Langhans, Paul. *Deutscher Kolonial-Atlas* (Gotha: Justus Perthes, 1897).

Langhans, Paul. *Sprachen und Religionen in Europa und die Grenzen zwischen west- und osteurpäischer Kultur* (Gotha: Justus Perthes, 1917).

Latvijas Statistikas Atlass/Atlas statistique de la Lettonie (Riga: Valsts Statistika Parvalde, 1938).

Von Maydell, Kurt. *Die Bevölkerung Estlands und Lettlands dargestellt in ihrem nationalen Gefüge nach dem Stande von 1934 (Estland) und 1935 (Lettland)* (Leipzig: Hirzel, 1940).

Von Maydell, Kurt and G. von Bornkam, *Die Verbreitung der Deutschen in Estland und Lettland auf Grund der amtlichen Volkzählungen von 1934 (Estland) und 1935 (Lettland)* (Berlin: Leopold Kraatz, 1940).

Meiershof, Seidlitz. *Hypsometrische Karte Ehstland's and Nord Livland's* (Dorpat: G. Schulz, 1876).

Mellin, Ludwig August. *Atlas von Liefland, oder von den beyden Gouvernementern u. Herzogthümern Lief- und Ehstland, und der Provinz Oesel* (Riga: Johann Friedrich Hartknoch, 1798).

Mirkovich, M. F. and A. F. Rittikh. *Ethnograficheskaia karta slavianskikh narodnostei* (St. Petersburg: Slavianskoe blagotvoritel'noe obshchestvo, 1875).

Plan Sankt-Petersburga. Raspredelenie ploshchadi goroda na odnogo zhitelia (St. Petersburg: Kartografiheskoe zavedenie A. Il'ina, 1868).

Rittikh, Aleksandr F. and Pompei N. Batiushkov. *Atlas Narodonaseleniia Zapadno-Russkago Kraia po Veroispovedaniiam* (St. Petersburg: Tipografiia V. Bellinga 1864).

Sbornyi list spetsial'noi karty Zapadnoi chasti Rossii (St. Petersburg: Voenno-topograficheskoe delo, 1844).

Schultz, Ernst Wilhelm Woldemar. *Maa kaardi-ramat kus sees 16 Maa Kaarti* (Tartu: R. Jakobi et Komp, 1859).

Siliņš, Matīss. *Rīga līdz ar apvidu* (Riga: M. Siliņš; A. Štāla litogrāfijā, 1896).

Siliņš, Matīss. *Latwijas skolu karte* (Riga: M. Siliņš; A. Štāla litogrāfijā, 1896).

Sitska, Jaan. *Archäologische Karte von Liv, Est- und Kurland* (Jurjew/Dorpat: H. Laakmann, 1896).

Siliņš, Matīss. *Mandžūrijas—Korejas karalauku speciālkarte* (Riga: Schnakenburg, 1904).

Siliņš, Matīss. *D-Mandžūrija un Koreja: Tālie Austrumi* (Riga: Schnakenburg, 1904).

Siliņš, Matīss. *Latvija. Etnogrāfiski pēc tautisko uzvalku un rotaslietu līdz šim ievāktiem materiāliem* (Riga: Aleksandra Štāla litogrāfija, 1930).

Tomasik, M. I. *Nagliadnaia karta Evropeiskoi Rossii*, 2nd edn (Warsaw: Aktsionernoe obshchestvo S. O. Orgel'branda, 1903).

Zuev, N. I. *Podrobnyi Atlas Rossiiskoi Imperii* (St. Petersburg: Tipografiia I. I. Glazunova i Komp., 1860).

Newspapers and Periodicals (Only Multiple References)

Altbass kalendārs

Balss

Baltijas Vēstnesis

Baltische Monatsschrift
Das Inland
Daugavas Vēstnesis
Düna Zeitung
Geograficheskie izvestiia
Latviešu Avīzes
Libauische Zeitung
Lietuvēns
Otchet RGO
Pamiatnaia knizhki Vitebskogo gubernii
Postimees
Rigascher Zeitung
Segodnia
Vestnik IRGO
Vilenskii Vestnik
Zapiski IRGO
Zeitung fur Stadt und Land

Published Primary Sources

Aksakov, Ivan. *Sochineniia*, vol. 3 (Moscow: 1886).

Aksakov, I. S. *Polnoe sobranie sochinenii*, vol. 6, Pribaltiiskii vopros vnutrenniia dela Rossii (Moscow: M. G. Bolchanov, 1887).

Andreev, E. and K. Stepanov, *Kratkii uchebnik geografii Rossii dlia nachal'nykh narodnykh uchilishch* (Riga: K. G. Zikhman, 1912).

Arbeiten und Ergebnisse des Ersten Deutsch-Baltischen Lehrertages am 3. und 4. August 1907 in Riga (Riga: Verlag von G. Löffler, 1907).

Ausgewählte Werke aus dem Verlage von Franz Kluge in Reval (Reval: Aug. Mickwis, 1918).

Ber [Baer], K. M. 'Ob etnograficheskikh isledovaniiakh voobshche i v Rossii v osobennosti'. *Zapiski RGO* 1-2 (1849): pp. 73–90.

Bericht über eine am 22. October 1864 auf dem Gute Jensel probeweise ausgeführte Volkszählung (Dorpat: Druck Karow, 1865).

Berkholz, Georg. 'Geschichte des Wortes "baltisch"'. *Baltische Monatsschrift* 29 (1882), pp. 519–30.

Bielenstein, August. *Die lettische Sprache: nach ihren Lauten und Formen erklärend und vergleichend dargestellt* (Berlin: Ferd. Dümmler's Verlagsbuchhandlung, 1863).

Bielenstein, August. *Handbuch der lettischen Sprache* (Mitau: Fr. Lucas' Buchhandlung, 1863).

Bielenstein, August. *Die Elemente der lettischen Sprache* (Mitau: J. F. Steffenhagen und Sohn, 1866).

Bielenstein, August. 'Reiseskizzen aus dem Oberlande'. *Baltische Monatsschrift* 29, nos. 7–9 (1882), pp. 569–90, 611–44, 707–43.

Bielenstein, August. *Die Grenzen des lettischen Volkstammes und der lettischen Sprache in der Gegenwart und im 13. Jahrhundert. Ein Beitrag zur ethnologischen Geographie und Geschichte Russlands* (St. Petersburg: Eggers; Kaiserliche Akademie der Wissenschaften, 1892).

Bielenstein, August. *Die Holzbauten und Holzgeräte der Letten. Ein Beitrag zur Ethnographie, Culturgeschichte und Archaeologie der Völker Russlands im Westgebiet*, 2 vols (St. Petersburg: Kaiserliche Akademie der Wissenschaften, 1907).

Bielenstein, August. *Ein glückliches Leben: Autobiographie, 1826–1907* (Michelstadt: Neuthor Verlag, 2002).

Bielenstein, August, Emil Bielenstein, and Hans Bielenstein. *Studien aus dem Gebiete der lettischen Archäologie, Ethnographie und Mythologie.* (Riga: Hoerschelmann, 1896).

Blumberg, Gustav. *Heimathskunde: Stofflich begrenzt und methodisch bearbeitet* (Dorpat: W. Glasers Verlag, 1869).

Von Bock, Waldemar. *Der deutsch-russische Konflikt an der Ostsee* (Leipzig: Dunder und Humblot, 1869).

Buchholtz, Arend. *Geschichte der Buchdruckerkunst in Riga, 1588–1888: Festschrift der Buchdrucker Rigas zur Erinnerung an die vor 300 Jahren erfolgte Einfuhrung der Buckdrunkerkunst in Riga* (Riga: Müllersche Buchdruckerei, 1890).

Danilevskii, Nikolai. *Rossiia i Evropa* (St. Petersburg: Izdanie tovarishchestva obshchestvennaia 1871).

Dolgorukova, V. M. *Vitebskaia guberniia: istoriko-geograficheskii i statisticheskii obzor. Vyp. 1. Istoriia. Priroda. Naselenie. Prosveshchenie* (Vitebsk: Gubernskaia Tipografiia, 1890).

Doŭnar-Zapol'ski, Mitrafan. *Les bases de l'etat de la Ruthénie Blanche* (Grodno: Ministère des Affaires etrangères de la République Démocratique Blanche-Ruthénienne, 1919).

Döring, Julius. 'Über die Herkunft der kurländischen Letten'. *Sitzungs-Berichte der kurländischen Gesellschaft für Literatur und Kunst nebst Veröffentlichungen des kurländischen Provinzial-Museums, aus dem Jahre 1880. Mit 2 Karten* (Mitau: J. F. Steffenhagen und Sohn, 1881), pp. 47–118.

Döring, Julius. *Was ich nicht gern vergessen möchte oder Erinnerungen aus meinem Leben* (Riga: Lettisches Nationalarchiv, 2016).

Dvatsatipiatiletie Kartograficheskogo zavedaniia A. Il'ina, 1859–1884 (St. Petersburg: A. Venke, 1884).

Eastlake, Elizabeth. *A Residence on the Shores of the Baltic. Described in a Series of Letters* (London: John Murray, 1841).

Erkert [Erckert], Rodrig F. *Vzgliad na istoriiu i etnografiiu zapadnykh gubernii Rossii* (St. Petersburg: Dom pizreniia maloletnykh bednykh, 1864).

Ethnograficheskii atlas Evropeiskoi Rossii sostavlennyi Petrom Keppenom (St. Petersburg: Al'faret, 2008).

Ezavitau, Kaustus'. *Belarusy ŭ Latvii* (Riga: Vydan'ne Belaruskaga Vydavetstva ŭ Latvii, 1927).

Francev, Vladimir Andrejevič. *Korespondence Pavla Josefa Šafaříka*, vol. 1, Vzájemné dopisy P. J. Šafaříka s ruskými učenci (1825–1861) (Prague: Česká Akademie Věd a Umění, 1927).

Gatsinskii, Aleksandr S. *Sbornik v pamiat' pervogo russkogo statisticheskogo s'ezda 1870 goda* (Nizhnii Novgorod: Tipografiia Nizhgorodskago Gubernskago Pravleniia, 1875).

Goba, Alfrēds. 'Skolas devums latviešu sētai'. *Sējējs*, no. 6, 1 June 1940, pp. 565–76.

Gūtmanis, Jānis. *Kolonistu dzīve São Paulo valstī Brazilijā* (Riga: L. Jakavičiaus, 1906).

Von Heyking [Hanking], Alfons Baron. *Statistische Studien über die ländlichen Zustände Kürlands* (Mitau: Fr. Lucas, 1862).

Howard, Esme. *Theatre of Life: Life Seen from the Stalls, 1905–1936* (London: Hodder and Stoughton, 1936).

Hurt, Jakob. *Über die Pleskauer Esten oder die sogenannten Setukesen* (Helsingfors: Finnische Literaturgesellschaft, 1904).

Jordan, Paul Eduard. *Die Resultate der Volkszählung der Stadt Reval am 16. November* 1871 (Reval: Kluge and Ströhm, 1874).

Jordan, Paul Eduard. *Die Resultate der ehstländischen Volkszählung vom 29. December 1881 in textlicher Beleuchtung* (Reval: Lindfors Erben, 1886).

Von Jung-Stilling, Friedrich. *Beitrag zur Statistik der evangelisch-lutherisch Landvolk-Schulen in Livland* (Riga: A. Stahl, 1879).

Von Jung-Stilling, Friedrich. *Ein Beitrag zur Livländischen Agrarstatistik: aus dem Material des livl. Landraths-Collegiums* (Riga: Müllersche Buchdruckerei, 1881).

Von Jung-Stilling, Friedrich and Wilfried Anders. *Ergebnisse der baltischen Volkszahlung vom 29. December 1881. Kartographische Darstellung der Bevolkerungdichtigkeit in der Stadt Riga* (Riga: Stahl'sche Buchdruckerei, 1883).

Kallas, Oskar. 'Üks surew eesti rahwa oks'. *Postimees*, nos. 206–23, 15 September–5 October 1895.

Kallas, Oskar. *Lutsi maarahvas* (Helsinki: Suomalaisen Kirjallisuuden Seura, 1894).

Kallas, Oskar. *Kaheksakümmend Lutsi maarahva muinasjuttu* (Jurjev/Tartu: Schnakenburg, 1900).

Katalog der lettischen ethnographischen Ausstellung, verfasst bei Gelegenheit des X. ärchäologischen Congresses in Riga 1896: aus dem Lettischen übertragen (Riga: Kalnin & Deutschmann, 1896).

Kemps [Skoborga], F. 'Latgalijas nokutne'. *Drywa*, no. 30, iss. 327, 12 October 1916, pp. 1–3.

Keppen, Petr. *O proiskhozhdenii, iazyke i literature litovskikh narodov* (St. Petersburg: Tipografiia Karla Kraiia, 1827).

Keppen, Petr. *Ukazatel' k karte iuzhnogo Kryma, prinadlezhashchei k Krymskomu sborniku Petra Keppena* (St. Petersburg: Imperatorskaia Academia Nauk, 1836).

Keppen, Petr. *O sushchnosti statistiki* (St. Petersburg: Tipografiia Imperatorskoi Akademii Nauk, 1840).

Keppen, Petr. *O chisle zhitelei v Rossii, v 1838 gody* (St. Petersburg: Tipografiia Imperatorskoi Akademii Nauk, 1840).

Keppen, Petr. *Vod' s S. Peterburgskoi gubernii: Otryvok iz poiasnitel'nogo teksta k etnograficheskoi karte S. Peterburgskoi gubernii* (St. Petersburg: Tipografiia Imperatorskoi Akademii Nauk, 1851).

Keppen, Petr. *Ob etnograficheskoi karte Evropeiskoi Rossii* (St. Petersburg: Tipografiia imperatorskoi Akademii nauk, 1852).

Keppen, Petr. *Predvaritel' nyie svedeniia o chisle zhitelei v Rossii, po guberniiam i uezdam v 1851 godu* (St. Petersburg: Tipografiia Imperatorskoi Akademii Nauk, 1854).

Keppen, Petr. *O tret'em izdanii Etnograficheskoi karty Evropeiskoi Rossii* (St. Petersburg: Tipografiia imperatorskoi Akademii nauk, 1856).

Keppen, Petr. *Khronologicheskii ukazatel' materialov dlia istorii inorodtsev evropeiskoi Rossii* (St. Petersburg: Imperatorskaia Akademiia Nauk, 1861).

Keppen, Fedor Petrovich. 'Biografiia P. I. Keppena'. *Odeleniia russkogo iazyka i slovesnosti imperatorskoi akademii nauk* 89, no. 5 (1911), pp. 1–170.

Köppen, Peter. *Über den Ursprung, die Sprache und Literatur litauschen Volkerschaften* (Mitau: J. F. Steffenhagen und Sohn, 1829).

Köppen, Peter. 'Die Bewohner Kur- und Livlands im Allgemeinen und die Liven insbesondere'. *Bulletin de la Classe des sciences historiques, philologiques et politiques de l'Académie impériale des sciences de Saint-Pétersbourg* (1846), pp. 257–69.

Köppen, Peter. 'Die Bewohner Estlands'. *Bulletin de la Classe des sciences historiques, philologiques et politiques de l'Académie impériale des sciences de Saint-Pétersbourg* (1846), pp. 346–7.

Köppen, Peter. 'Finnland in ethnographischer Beziehung (avec une carte coloriée)'. *Mémoires de l'Académie Impériale des Sciences de St. Pétersbourg. Sixième Série. Sciences politiques, histoire, philologie* 7 (1848), pp. 452–64.

Köppen, Peter. *Über die Deutschen im St Petersburgischen Gouvernement. Mit einem Vorworte über die Ethnographische Karte des genannten Gouvernements und einem Anhange über die auf derselben vorkommenden Dorfnammen* (St. Petersburg: Buchdruckerei der Kaiserlichen Akademie der Wissenschaften, 1850).

Köppen, Peter. 'Über die Anfertigung der ethnographischen Karte des europäischen Russlands'. *Bulletin de la Classe des sciences historiques, philologiques et politiques de l'Académie impériale des sciences de Saint-Pétersbourg* 9 (1852), pp. 137–64.

Körber, Bernhard. *Die Stadt Dorpat (Jurjew) in statistischer und hygienischer Beziehung* (Jurjew: C. Mattiesen, 1902).

Lieven, Hugo. *Die Revision des Lehrplanes der Gymnasien des Dorpater Lehrbezirks nach Ihren allgemeinen Gesichtspunkten erörtert* (Riga: N. Kymmel, 1880).

Luts, Oskar. *Spring*, trans. Aino Jõgi and Melanie Rauk (Tallinn: Perioodika, 1983).

Maliszewski, Edward. *Polacy na Łotwie* (Warsaw: Wydawnictwo T-wa Straży Kresowej, 1922).

Maliszewski, Edward. *La Pologne d'aujourd'hui* (Paris: Gebethner et Wolff, 1926).

Manteuffel, Gustaw. *Polnisch-Livland* (Riga: N. Kymmel, 1869).

Manteuffel, Gustaw. *Inflanty Polskie poprzedzone ogólnymrzutem oka na siedmiowiekową przeszłość całych Inflant* (Poznań: Księgarnia Jana Konstantego Żupańskiego, 1879).

Manteuffel, Gustaw. *Zarys z dziejów krain dawnych inflanckich czyli Inflant właściwych (tak szwedzkich jako i polskich), Estonii z Ozylią, Kurlandii i Ziemi Piltyńskiej*, ed. Kryzsztof Zajas (Kraków: Universitas, 2007).

Martna, Mikhel. *Estland, die Esten und die estnische Frage* (Olten: Trösch, 1919).

Mayr, Georg. *Gutachten über die Anwendung der graphischen und geographischen Methode in der Statistik* (München: Gotteswinter & Mössl, 1874).

Mémoire sur l'indépendance de l'Esthonie présenté a la Conférence de la Paix par la délégation Esthonienne (April 1919).

Memorandum on Latvia Addressed to the Peace Conference by the Lettish Delegation (Paris: Imprimerie de la Bourse de Commerce, 1919).

Meomuttel, Jüri. *Eesti asunikud laialises Wene riigis: Esimene katse sõnumid kõikide Eesti asunduste üle tuua* (Jurjew: 'Postimehe' trükikoja kirjastus ja trükk, 1900).

Merkel, Garlieb. *Die Letten, vorzüglich in Liefland, am Ende des philosophischen Jahrhunderts* (Leipzig: Heinrich Gräff,1796).

Nadezhin, Nikolai. 'Ob etnograficheskom izychenii narodnost russkoi'. *Zapiski RGO* 2 (1847), pp. 61–115.

Nesselmann, Georg Heinrich Friedrich. *Die Sprache der alten Preussen: an ihren Ueberresten erläutert* (Berlin: 1845).

Piatidesiatiletie Kartograficheskogo zavedeniia A. Il'ina (St. Petersburg: Tipografiia Pervoi Sb. Trudovoi Arteli, 1909).

Report of the Official Delegates to the International Statistical Congress Held at St. Petersburg in August, 1872 (Washington: US Government Printing Office, 1875).

Rittikh, A. F. *Prilozhenie k materialam dlia etnografii Tsarstva Pol'skogo: Gubernii Liublinskaia i Avgustovskaia* (St. Petersburg: 1864).

Rittikh, A. F. *Materialy dlia etnografii Rossii: Kazanskaia guberniia* (Kazan: Tipografiia Imperatorskogo Kazanskogo Universiteta, 1870).

Rittikh, Aleksandr. *Materialy dlia ethnografii Rossii: Pribaltiiskii krai* (St. Petersburg: Kartograficheskoe Zavedenie A. Il'ina, 1873).

Rittikh, Aleksandr. 'Pribaltiiskii krai i ego naselenie do pribytiia nemtsev'. In *Sbornik materialov i statei po istorii Pribaltiiskago kraia*, vol. 1, edited by E. V. Cheshkhin (Riga: A. Lipinskii, 1876), pp. 1–35.

Rittikh, Aleksandr. *Slavianskii mir: Istoriko-geograficheskoe i etnograficheskoe issledovanie* (Warsaw: Izdanie V. M. Istomina, 1885).

Rittikh, A. F. *Russkaia torgovlia i morekhodstvo na Baltiiskom more* (St. Petersburg: Leshtukovskaia Parovaia Skoropechatnaia P. O. Iablonskago, 1896).

Rittikh, Alexander. *Obizhennyi krai* (St. Petersburg: Elektro-Tipografiia N. Ia. Stoikovni, 1911).

Rohrback, Paul. *Das Baltenbuch: die baltischen Provinzen und ihre deutsche Kultur; mit Beiträgen hervorragender Balten und vielen Bildern* (Dachau: Der Gelbe Verl. Blumtritt, 1916).

Rosen, H. 'Die etnographischen Verhältnissen den baltischen Provinzen und in Litauen'. *Petermanns Geographische Mitteilungen*, 1 September 1915.

Russwurm, Carl Friedrich Wilhelm. *Eibofolke oder die Schweden an den Küsten Ehstlands und auf Runö: Eine historisch-ethnographische von der Kaiserlichen Akademie der Wissenschaften zu St. Petersburg mit einem Amidoneschen Preise gekrönte Untersuchung. Mit Urkunden, Tabellen und lithographirten Beilagen* (Reval: Fleischer, 1855).

Rutter, Owen. *The New Baltic States and Their Future: An Account of Lithuania, Latvia and Estonia* (London: Meuthuen & Co., 1925).

Šafařík, Pavel Josef. *Slovanský Narodopis* (Prague: Verf. Nákladem wydawatele 1842).

Schaffarik, Paul Joseph. *Geschichte der slawischen Sprache und Literatur nach allen Mundarten* (Ofen: Mit kön.ung. Universitäts Schriften, 1826).

Shafarik, P. I. *Slavianskoe narodopisanie*, trans. I. Bodianskii (Moscow: Universitetskaia Tipografiia, 1843).

Samarin, Iurii F. *Sochineniia*, vol. 7, Pis'ma iz Rigi i istoriia Rigi (Moscow: D. Samarin, 1889).

Samarin, Iurii F. *Sochineniia*, vol. 8, Okrainy Rossii: Vyp. 1–3 (Moscow: D. Samarin, 1890).

Samarin, Iurii. *Otvet G. G. f. Bokku i Shirrenu po povodu 'Okrainy Rossii'* (Berlin: B. Behr's Buchhandlung, 1870).

Schirren, Carl. *Livländische Antwort an Herrn Juri Samarin* (Leipzig: Duncker and Humblot, 1869).

Semenov, Petr P. *Istoriia poluvekovoi deiatel'nosti Imperatorskago Russkago Geograficheskago Obschestva, 1845–1895*, 3 vols (St. Petersburg, 1896).

Sementovskii, Aleksandr. *Otchet o deistviiakh i zaniatiakh Vitebskago gubernskogo statistiheskogo komiteta za 1869 god.* (Vitebsk: Tipografiia Gubernskago pravleniia, 1870).

Sementovskii, Aleksandr. *Etnograficheskii obzor Vitebskoi gubernii* (St. Petersburg: M. Khan, 1872).

Sementovskii, Aleksandr, ed. *Sbornik v pamiat' pervago Russkago statistiheskago s'ezda 1870 g.*, vol. 1 (St. Petersburg: M. Khan, 1872).

Simpson, James Young. *The Garment of the Living God* (New York: Charles Scribner's Sons, 1935).

Sjögren, Anders Johan. *Über die finnische Bevölkerung des St. Petersburgischen Gouvernements und über den Ursprung des Namens Ingermannland* (St. Petersburg: Buchdruckerei der Kaiserlichen Akademie der Wissenschaften, 1833).

Sjögren, A. J. 'Bericht über eine im Auftrage der russischen geographischen Gesellschaft wärend der Sommermonate des Jahres 1846 nach den Gouvernements Livland und Kurland unternommene Reise zur genauen Untersuchung der Reste der Liwen und Krewingen'. In *Denkschiften der Russischen geographischen Gesellschaft zu St. Petersburg* 1 (Weimar: Druck und Verlag des Landes-Industrie-Comptoirs, 1849), pp. 453–505.

Šliupas, Jonas. *Lithuania in Retrospect and Prospect* (New York, Lithuanian Press Association of America, 1915).

Smolich, Arkadz. *Heagrafiia Belarusi* (Vilnius: Promen', 1919).

Sreznevskii, Ismail I. 'Zamechaniia o materialakh dlia geografii russkago iazyka'. *Vestnik IRGO* 1, no. 5 (1851), pp. 1–25.

Sreznevskii, Ismail I. 'Etnograficheskaia karta Evropeiskoi Rossii'. *Izvestiia IAN po otdeleniiu russkago iazyka i slovesnosti* 1, no. 1 (1852), pp. 68–70.

Tallents, Stephen. *Man and Boy* (London: Faber & Faber, 1943).

T-vo skoropechatni A. A. Levenson: Ist. ocherk i opisanie masterskikh 1881–1903 (Moscow: A. A. Levenson, 1903).

Vol'ter, E. A. *Ob etnograficheskoi poezdke po Litve i Zhmudi letom 1887 goda* (St. Petersburg: Tipografiia Imperatorskoi Akademii Nauk, 1887).

Vos'maia sessiia mezhdunarodnogo statisticheskogo kongressa: Proekt programmy (St. Petersburg: Tipografiia Maikova, 1872).

Werden und Gedeihen der graphischen Kunstanstalt H. Laakmann, Dorpat, 1837–1927 (Dorpat: H. Laakmann, 1927).

Z okolic Dźwiny. Księga zbiorowa na dochód Czytelni Polskiej w Witebsku (Vitebsk: Zarząd Towarzystwa Dobroczynności w Witebsku; Odbito w drukarni Józefa Zawadzkiego, 1912).

Secondary Sources

Ackermann, Felix. *Palimpsest Grodno: Nationalisierung, Nivellierung und Sowjetisierung einer mitteleuropaischen Stadt 1919–1991* (Wiesbaden: Harrassowitz, 2010).

Afinogenov, Gregory. *Scholars and Spies: Chinese Secrets and Imperial Russia's Quest for World Power* (Cambridge, MA: Harvard University Press, 2020).

Agafonov, N. T., A. G. Isachenko, S. B. Lavrov, E. M. Murazaev, and Iu. P. Seliverstov. *Russkoe Geograficheskoe Obschestvo: 150 let* (Moscow: Izdatel'stvo Rossiiskoi natsional'noi biblioteki, 1995).

Agnew, John. 'Borders on the Mind: Re-framing Border Thinking'. *Ethics and Global Politics* 1, no. 4 (2008), pp. 175–91.

Aizelwood, Robin. 'Revisiting Russian Identity in Russian Thought: From Chaadaev to the Early Twentieth Century'. *Slavonic and East European Review* 78, no. 1 (2000), pp. 20–43.

Ajewski, Konrad. 'Wacław Fedorowicz—Zapomniany Kolekcjoner z Witebska'. *Muzealnictwo* 55 (2014), pp. 118–30.

Akerman, James R., ed. *The Imperial Map: Cartography and the Mastery of Empire* (Chicago: University of Chicago Press, 2009).

Akerman, James R., ed. *Decolonising the Map: Cartography from Colony to Nation* (Chicago: University of Chicago Press, 2017).

Alston, Charlotte. 'James Young Simpson and the Latvian-Lithuanian Border Settlement 1920–1921: The Papers in the Archive of the Royal Scottish Geographical Society'. *Scottish Geographical Journal* 118, no. 2 (2002), pp. 87–100.

Alston, Charlotte. '"The Suggested Basis for a Russian Federal Republic": Britain, Anti-Bolshevik Russia and the Border States at the Paris Peace Conference, 1919'. *History* 91, no. 301 (2006), pp. 24–44.

Alston, Charlotte. *Makers of the Modern World: The Peace Conferences 1919–23 and Their Aftermath. Antonius Piip, Zigfrīds Meierovics and Augustinus Voldemaras: The Baltic States* (London: Haus, 2010).

Andersen, Casper, Jakob Bek-Thomsen, and Peter C. Kjærgaard. 'The Money Trail: A New Historiography for Networks, Patronage, and Scientific Careers'. *Isis* 103, no. 2 (2012), pp. 310–15.

Anderson, Benedict. *Imagined Communities: Reflections on the Origin and Spread of Nationalism*, rev. edn (London; New York: Verso, 2006).

Andresen, A. 'Formal Stipulation and Practical Implementation of Religious Privileges in Estland, Livland, and Courland under Russian Supremacy: Researching the Core of Baltic Regional Identity'. *Ajalooline Ajakiri* 139–40, nos. 1–2 (2012), pp. 33–54.

Anderson, David G., Dmitry V. Arzyutov, and Sergei S. Alymov, eds. *Life Histories of Etnos Theory in Russia and Beyond* (Cambridge: Open Book Publishers, 2019).

Andreeva, Natalia S. 'Die "baltische Frage" und die Reformpolitik"'. In *Russland an der Ostsee: imperiale Strategien der Macht und kulturelle Wahrnehmungsmuster (16. bis 20. Jahrhundert)/Russia on the Baltic: Imperial Strategies of Power and Cultural Patterns of Perception (16th–20th Centuries)*, edited by Karsten Brüggemann and Bradley D. Woodworth (Vienna: Böhlau, 2012), pp. 243–85.

Ante, Kristine. 'Popytki ukoreneniia pravoslaviia v liuteranskoi srede Kurliandskoi gubernii', in *Pravoslavie v Pribaltike: religiia, politika, obrazovaniie 1840-e-1930e gg.*, edited by Irina Paert (Tartu: University of Tartu Press, 2018), pp. 147–77.

Anthony, Scott. *Public Relations and the Making of Modern Britain: Stephen Tallents and the Birth of a Progressive Media Profession* (Oxford: Oxford University Press, 2018).

Annus, Endel. *Eesti kalendrikirjandus 1720–1900* (Tallinn: Eesti Akadeemiline Raamatukogu, 2000).

Apine, Ilga. *Baltkrievi Latvijā* (Riga: Latvijas Zinātņu Akadēmijas Filozofijas un Socioloģijas Institūts, 1995).

Apīnis, Art. 'Latviešu valodā sarakstīto ģeogrāfijas mācības grāmatu un ģeogrāfijas karšu attīstība līdz ar viņu nozīmi no 1854.–1918. gadam'. *Izglītības Ministrijas Mēnešraksts*, no. 7 (1 July 1935): pp. 13–25.

Applegate, Celia. *A Nation of Provincials: The German Idea of Heimat* (Berkeley; Los Angeles; Oxford: University of California Press, 1990).

Applegate, Celia. 'A Europe of Regions: Reflections on the Historiography of Sub-National Places in Modern Times'. *The American Historical Review* 104, no. 4 (1999), pp. 1157–82.

Armstrong, John A. 'Mobilized Diaspora in Tsarist Russia: The Case of the Baltic Germans'. In *Soviet Nationality Policies and Practices*, edited by J. Azrael (New York: Praeger, 1978), pp. 63–104.

Armstrong, John A. 'Acculturation to the Russian Bureaucratic Elite: The Case of the Baltic Germans'. *Journal of Baltic Studies* 15, nos. 2–3 (1984), pp. 119–29.

Arumäe, Heino. 'Piirikonfliktist liidulepinguni'. *Looming*, no. 8 (1995), pp. 1114–37.

Ash, Mitchell G., and Jan Surman, eds. *The Nationalization of Scientific Knowledge in the Habsburg Empire, 1848–1918* (New York: Palgrave Macmillan, 2012).

Augelli, John P. 'The Latvians of Varpa: A Foreign Colony on the Brazilian Pioneer Fringe'. *Geographical Review* 48, no. 3 (1958), pp. 365–87.

Avrutin, Eugene M. 'The Power of Documentation: Vital Statistics and Jewish Accommodation in Tsarist Russia'. *Ab Imperio* no. 4 (2003), pp. 271–300.

Baár, Monika. *Historians and Nationalism: East-Central Europe in the Nineteenth Century* (Oxford: Oxford University Press, 2010).

Bagrow, Leo and Henry W. Castner. *A History of Russian Cartography up to 1800* (Wolfe Island: Walker Press, 1975).

Bakhturina, Aleksandra Iur'evna. 'Izmenenie administrativnykh granits pribaltiiskikh gubernii vesnoi-letom 1917 g.'. *Vestnik RGGU. Seriia 'Politologiia. Istoriia. Mezhdunarodnye otnosheniia. Zarubezhnoe regionovedenie. Vostokovedenie'* 10, nos. 2–4 (2017), pp. 177–85.

Balkelis, Tomas. 'Nation-Building and World War I Refugees in Lithuania, 1918–1924'. *Journal of Baltic Studies* 34, no. 4 (2003), pp. 432–56.

Balkelis, Tomas. *War, Revolution, and Nation-Making in Lithuania, 1914–1923* (Oxford: Oxford University Press, 2018).

Bandarchyk, Vasil. *Historyia belaruskai etnahrafii XIX st.* (Minsk: Vydavetstva 'Navuka i tekhnika', 1964).

Baranskii, N. N. *Istoricheskii obzor uchebnikov geografii (1876–1934)* (Moscow: Geografgiz, 1954).

Barbans, Jānis, ed. *100 gadi 100 kartēs* (Riga: Veikals, 2018).

Baron, Nick. 'Nature, Nationalism and Revolutionary Regionalism: Constructing Soviet Karelia, 1920–1923'. *Journal of Historical Geography* 33, no. 3 (2007), pp. 565–95.

Baron, Nick. 'New Spatial Histories of 20th-Century Russia and the Soviet Union: Exploring the Terrain'. *Kritika: Explorations in Russian and Eurasian History* 9, no. 2 (2008), pp. 433–47.

Baron, Nick. '"Ot grekha podalshe...": tsenzura i kontrol' nad topograficheskim znaniem v Sovetskoi Rossii (1918–1925)'. *Istoriko-biologicheskie issledovaniia* 2, no. 4 (2010), pp. 84–92.

Baron, Nick. 'The Mapping of Illiberal Modernity: Spatial Science, Ideology and the State in Early Twentieth-Century Russia'. In *Empire De/Centered: New Spatial Histories of Russia and the Soviet Union*, edited by Sanna Turoma and Maxim Waldstein (Farnham: Ashgate, 2013), pp. 75–95.

Baron, Nick and Peter Gatrell, eds. *Homelands: War, Population, and Statehood in Eastern Europe and Russia, 1918–1924* (London: Anthem Press, 2004).

Barons, Krišjānis. *Mūsu tēvzemes aprakstīšana un daži pielikumi īsumā saņemti: grāmatiņa prieks skolām un mājām* (Tērbata: G. A. Reyhers; Jelgava: C. Schulz, 1859).

Bassin, Mark. 'Russia between Europe and Asia: The Ideological Construction of Geographical Space'. *Slavic Review* 50, no. 1 (1991), pp. 1–17.

Bassin, Mark. *Imperial Visions: Nationalist Imagination and Geographical Expansion in the Russian Far East, 1840–1865* (Cambridge: Cambridge University Press, 1999).

Batt, Judy. 'Transcarpathia: Peripheral Region at the "Centre of Europe"'. *Regional and Federal Studies* 12, no. 2 (2002), pp. 155–77.

Battle-Baptiste, Whitney and Britt Rusert, eds. *W. E. B. Du Bois's Data Portraits: Visualizing Black America. The Color Line at the Turn of the Turn of the Twentieth Century* (Amherst, MA: The W. E. B. Du Bois Center at the University of Massachusetts Amherst; Hudson, NY: Princeton Architectural Press, 2018).

Batuman, Bülent. 'The Shape of the Nation: Visual Production of Nationalism through Maps in Turkey'. *Political Geography* 29, iss. 4 (2010), pp. 220–34.

Baycroft, Timothy and David Hopkin, eds. *Folklore and Nationalism in Europe during the Long Nineteenth Century* (Leiden: Brill, 2012).

Bayly, Christopher Alan. *An Empire of Information: Intelligence Gathering and Social Communication in India, 1780–1870* (New York: Cambridge University Press, 1996).

Beall, Karen F. 'The Interdependence of Printer & Printmaker in Early 19th-Century Lithography'. *Art Journal* 39, no. 3 (1980), pp. 195–201.

Becker, Elisa Marielle. *Medicine, Law, and the State in Imperial Russia* (Budapest: Central European University Press, 2011).

Beecher, David Ilmar. *Ivory Tower of Babel: Tartu University and the Languages of Two Empires, a Nation-State, and the Soviet Union*. PhD dissertation, UC Berkeley, 2014.

Berelowitch, Wladimir. 'Aux origines de l'ethnographie russe: la société de géographie dans les années 1840–1850'. *Cahiers du monde russe et soviétique* 31, nos. 2–3 (1990), pp. 265–73.

Berendsen, Veiko and Margus Maiste. *Esimene ülevenemaaline rahvaloendus Tartus 28. jaanuaril 1897* (Tartu: Eesti Ajalooarhiiv, 1999).

Benz, Ernst. 'Zwischen konfessioneller, regionaler und nationaler Identität: Die Katholiken in Lettgallen und Lettland im 19. und 20. Jahrhundert'. *Nordost-Archiv* 7 (1998), pp. 443–95.

Berg, L. S. *Vsesoiuznoe Geograficheskoe Obschestvo za sto let* (Moscow: Izdatel'stvo Akademii Nauk SSSR, 1946).

Bertash, Aleksandr. 'Pravoslavnoe tserkovnoe stroitel'stvo s Lifliandskoi gubernii vtoroi poloviny 1840-kh–1860-kh gg. i arkhitektor N. E. Efremov'. *Pravoslavie v Baltii* 17, no. 8 (2018), pp. 35–68.

Bittel, Carla, Elaine Leong, and Christine von Oertzen, eds. *Working with Paper: Gendered Practices in the History of Knowledge* (Pittsburgh: University of Pittsburgh Press, 2019).

Blackbourn, David and James Retallack, eds. *Localism, Landscape, and the Ambiguities of Place: German-Speaking Central Europe, 1860–1930* (Toronto: University of Toronto Press, 2015).

Blakesley, Rosalind P. 'Women and the Visual Arts'. In *Women in Nineteenth-Century Russia: Lives and Culture*, edited by Wendy Rosslyn and Alessandra Tosi (Cambridge: Open Book Publishers, 2012), pp. 91–117.

Blair, Ann. 'Note Taking as an Art of Transmission'. *Critical Inquiry* 31, no. 1 (2004), pp. 85–107.

Blumberga, Renāte. 'A. J. Sjögren's Expeditions to the Livs 1846–1852'. In *Defining Self: Essays on Emergent Identities in Russia Seventeenth to Nineteenth Centuries*, edited by Michael Branch (Helsinki: Finnish Literature Society, 2009), pp. 519–33.

Blumbergs, Andrew James. *The Nationalization of Latvians and the Issue of Serfdom: The Baltic German Literary Contribution in the 1780s and 1790s* (Amherst; New York: Cambria Press, 2008).

Boldāne-Zeļenkova, Ilze. 'Others Among Others: Latvians' View of Members of Ethnographic Shows'. *East Central Europe* 47, nos. 2–3 (2020), pp. 261–84.

Bös, Mathias and Kerstin Zimmer. 'Wenn Grenzen wandern: Zur Dynamik von Grenzverschiebungen im Osten Europas'. In *Grenzsoziologie: Die politische Strukturierung des Raumes*, edited by Monika Eigmüller and Georg Vobruba (Wiesbaden: VS Verlag für Sozialwissenschaften, 2006), pp. 157–84.

Bradley, Joseph. 'Pictures at an Exhibition: Science, Patriotism, and Civil Society in Imperial Russia'. *Slavic Review* 67, no. 4 (2008), pp. 934–66.

Bradley, Joseph. *Voluntary Associations in Tsarist Russia: Science, Patriotism, and Civil Society* (Cambridge, MA: Harvard University Press, 2009).

Branch, Michael. *A. J. Sjögren: Studies of the North* (Helsinki: Suomalais-Ugrialainen Seura, 1973).

Brendecke, Arndt. *Empirical Empire: Spanish Colonial Rule and the Politics of Knowledge* (Berlin: De Gruyter Oldenbourg, 2016).

Brenna, Brita. 'Clergymen Abiding in the Fields: The Making of the Naturalist Observer in Eighteenth-Century Norwegian Natural History'. *Science in Context* 24, no. 2 (2011), pp. 143–66.

Brian, Eric. 'Statistique administrative et internationalisme statistique pendant la seconde moitié du XIXe siècle'. *Histoire & Mesure* 4, nos. 3–4 (1989), pp. 201–24.

Bridge, Carl and Kent Fedorowich. 'Mapping the British World'. *The Journal of Imperial and Commonwealth History* 31, no. 2 (2003): pp. 1–15.

Bridges, Roy C. 'Nineteenth-Century East African Travel Records with an Appendix on "Armchair Geographers" and Cartography.' *Paideuma: Mitteilungen zur Kulturkunde* 33 (1987), pp. 179–196.

Brooks, Jeffrey. *When Russia Learned to Read: Literacy and Popular Literature, 1861–1917* (Princeton: Princeton University Press, 1985).

Brown, Kate. *A Biography of No Place: From Ethnic Borderland to Soviet Heartland* (Cambridge, MA: Harvard University Press, 2009).

Brückner, Martin. *The Geographic Revolution in Early America: Maps, Literacy, and National Identity* (Chapel Hill: University of North Carolina Press, 2006).

Brüggemann, Karsten. *Die Gründung der Republik Estland und das Ende des 'Einen und unteilbaren Russland', Die Petrograder Front des Russischen Bürgerkrieges 1918–1920* (Wiesbaden: Harrassowitz, 2002).

Brüggemann, Karsten. '"Foreign Rule" during the Estonian War of Independence 1918–1920: The Bolshevik Experiment of the "Estonian Workers" Commune'. *Journal of Baltic Studies* 37, no. 2 (2006), pp. 210–26.

Brüggemann, Karsten. 'Foreign Rule and Collaboration in the Baltic Countries, 1860–1920. New Directions in Research'. *Journal of Baltic Studies* 37, no. 2 (2006), pp. 155–62.

Brüggemann, Karsten. 'Als Land und Leute "Russisch" werden sollten: Zum Verständnis des Phänomens der "Russifizierung" am Biespiel der Ostseeprovinzen des Zahrenreichs'. In *Kampf um Wort und Schrift: Russifizierung in Osteuropa im 19.–20. Jahrhundert*, edited by Zaur Gasimov (Göttingen: Vandenhoeck and Ruprecht, 2012), pp. 27–49.

Brüggemann, Karsten. 'Fremde im eigenen Imperium: Russische ethnografische Beschreibungen finnischer und baltischer Völker an der Ostsee im späten Zarenreich'. In *Nation und Sprach in Nordosteuropa*, edited by Konrad Maier (Wiesbaden: Harrassowitz, 2012), pp. 49–73.

Brüggemann, Karsten. 'The Baltic Provinces and Russian Perceptions in Late Imperial Russia'. In *Russland an der Ostsee: imperiale Strategien der Macht und kulturelle Wahrnehmungsmuster (16. bis 20. Jahrhundert) = Russia on the Baltic: Imperial Strategies of Power and Cultural Patterns of Perception (16th–20th Centuries)*, edited by Karsten Brüggemann and Bradley D. Woodworth (Vienna: Böhlau, 2012), pp. 111–41.

Brüggemann, Karsten. 'Ein Fall von "Verschmelzung" mit Russland? Zur nationalen Frage in der Orthodoxen Kirche der Ostseeprovinzen im späten Zarenreich'. *Nordost-Archiv: Zeitschrift für Regionalgeschichte* 22 (2014), pp. 89–111.

Brüggemann, Karsten. 'Imperiale und lokale Loyalitäten im Konflikt: Der Einzug Russlands in die Ostseeprovinzen in den 1840er Jahren'. *Jahrbücher für Geschichte Osteuropas* 62, no. 3 (2014), pp. 321–44.

Brüggemann, Karsten. 'Representing Empire, Performing Nation? Russian Officials in the Baltic Provinces (Late Nineteenth/Early Twentieth Centuries)'. *Ab Imperio* no. 3 (2014), pp. 231–66.

Brüggemann, Karsten. *Licht und Luft des Imperiums: Legitimations- und Repräsentationsstrategien russischer Herrschaft in den Ostseeprovinzen im 19. und 20. Jahrhundert* (Wiesbaden: Harrassowitz, 2018).

Brüggemann, Karsten. *On Saints, Migrants and Communists: Transnational Explorations in Estonian History*, Special Issue, *Acta Historica Tallinnensia* 27, no. 1 (2021).

Brüggemann, Karsten and Bradley D. Woodworth, eds. *Russland an der Ostsee: Imperiale Strategie der Macht und kulturelle Wahrnehmungsmuster (16th bis 20. Jahrhudnert)/ Russia on the Baltic: Imperial Strategies of Power and Cultural Patterns of Perceptions (16th–20th Centuries)* (Vienna: Böhlau, 2012).

Brüggemann, Karsten and Katja Wezel. 'Nationally Indifferent or Ardent Nationalists? On the Options for Being German in Russia's Baltic Provinces, 1905–17'. *Kritika: Explorations in Russian and Eurasian History* 20, no. 1 (2019), pp. 39–62.

Bukšs, Miķelis. *Die Russifizierung in den baltischen Ländern* (Munich: Latgalischer Verlag, 1964).

Bulmer, Martin, Kevin Bales, and Kathryn Kish Sklar, eds. *The Social Survey in Historical Perspective, 1880–1940* (Cambridge: Cambridge University Press, 1991).

Burbank, Jane. 'Supervising the Supervisors: Bureaucracy, Personality and Rule of Law in Kazan Province at the Start of the 20th Century'. *Acta Slavonica Iaponica* 38 (2015), pp. 1–21.

Burbank, Jame and Mark von Hagen. *Russian Empire: Space, People, Power, 1700–1930* (Bloomington: Indiana University Press, 2007).

Burke, Peter. *Eyewitnessing: The Uses of Images as Historical Evidence* (London: Reaktion Books, 2001).

Butkus, Zenonas. 'Great Britain's Mediation in Establishing the Lithuanian-Latvian Frontier, 1920–1921'. *Journal of Baltic Studies* 24, no. 4 (1993), pp. 359–68.

Cadiot, Juliette. 'Searching for Nationality: Statistics and National Categories at the End of the Russian Empire (1897–1917)'. *Russian Review* 64 (2005), pp. 440–55.

Cadiot, Juliette. *Le laboratoire impérial: Russie-URSS, 1860–1940* (Paris: CNRS Éditions, 2007).

Cadiot, Juliette 'Un empire "un et indivisible"? La question de la représentation politique des non Russes à la Douma après la révolution de 1905'. *Cahiers Du Monde Russe* 48, nos. 2–3 (2007), pp. 221–2.

Campbell, Elena I. *The Muslim Question and Russian Imperial Governance* (Bloomington: Indiana University Press, 2015).

Case, Holly. *The Age of Questions. Or, A First Attempt at an Aggregate History of the Eastern, Social, Woman, American, Jewish, Polish, Bullion, Tuberculosis, and Many Other Questions over the Nineteenth Century, and Beyond* (Princeton: Princeton University Press, 2018).

Caune, Andris. *Latvijas zemju robežas 1000 gados* (Riga: Latvijas vēstures institūta apgāds, 1999).

Chambers, David Wade and Richard Gillespie. 'Locality in the History of Science'. *Osiris* 15 (2001), pp. 221–40.

Chuchvaha, Hanna. *Art Periodical Culture in Late Imperial Russia (1898–1917): Print Modernism in Transition* (Leiden; Boston: Brill, 2016).

Chumakova, Tat'iana. 'Pravoslavnye sviashchenniki-issledovateli religii narodov Rossii (XIX-nachalo XX v.)'. *Gosudarstvo, religiia, tserkov' v Rossii i za rubezhom* 36, no. 1 (2018), pp. 12–32.

Ciancia, Kathryn. *On Civilization's Edge: A Polish Borderland in the Interwar World* (Oxford: Oxford University Press, 2020).

Cimermanis, Saulvedis. 'The Livs of Svētciems Pagasts in the late 18th and 19th century'. *Pro Ethnologia* 15 (2003), pp. 11–27.

Coen, Deborah. 'Big Is a Thing of the Past: Climate Change and Methodology in the History of Ideas'. *Journal of the History of Ideas* 77, no. 2 (2016), pp. 305–21.

Coen, Deborah. *Climate in Motion: Science, Empire, and the Problem of Scale* (Chicago: University of Chicago Press, 2018).

Conrad, Sebastian. *Globalisation and the Nation in Imperial Germany*, trans. Sorcha O'Hagan (New York: Cambridge University Press, 2010).

Cook, Karen Severud. 'From False Starts to Firm Beginnings: Early Colour Printing of Geological Maps'. *Imago Mundi* 47 (1995), pp. 155–72.

Cornford, Ieva Mara Megnis. *Latvian Immigration to São Paulo, Brazil: 1890–1970. Immigration and Assimilation*. PhD dissertation, University of British Columbia, 1973.

Cosgrove, Denis. 'Maps, Mapping, Modernity: Art and Cartography in the Twentieth Century'. *Imago Mundi* 57, no. 1 (2005), pp. 35–54.

Crampton, Jeremy W. 'The Cartographic Calculation of Space: Race Mapping and the Balkans at the Paris Peace Conference of 1919'. *Social and Cultural Geography* 7, no. 5 (2006), pp. 731–52.

Crampton, Jeremy W. 'Maps as Social Constructions: Power, Communication and Visualization'. *Progress in Human Geography* 25, no. 2 (2001), pp. 235–52.

Crooks, Peter and Timothy H. Parsons, eds. *Empires and Bureaucracy in World History: From Late Antiquity to the Twentieth Century* (Cambridge: Cambridge University Press, 2016).

Cvetkovski, Roland and Alexis Hofmeister, eds. *An Empire of Others: Creating Ethnographic Knowledge in Imperial Russia and the USSR* (Budapest; New York: Central European University Press, 2013).

Daija, Pauls and Benedikts Kalnačs. '"Small" Germans and "Half"-Germans in the Baltic Provinces at the Turn of the Twentieth Century'. In *Identities In-Between in East-Central Europe*, edited by Jan Dr. Fellerer, Robert Pyrah, and Marius Turda (London: Routledge 2019), pp. 211–28.

Dando, Christina E. *Women and Cartography in the Progressive Era* (Abingdon: Routledge, 2017).

Darrow, David W. 'Census as a Technology of Empire'. *Ab Imperio* no. 4 (2002), pp. 145–76.

Daston, Lorraine and Peter Galison. *Objectivity* (New York: Zone Books, 2007).

Daston, Lorraine and Elizabeth Lunbeck, eds. *Histories of Scientific Observation* (Chicago: University of Chicago Press, 2011).

Delbourgo, James and Staffan Müller-Wille. 'Listmania: Introduction'. *Isis* 103 (2012), pp. 710–15.

DenBeste, Michelle. 'NOT Finding Women in the Archives: The Case of Evgeniia Serebrennikova, Pioneering Woman Physician in Late Nineteenth-Century Russia'. *Aspasia* 10, no. 1 (2016), pp. 1–17.

Dennis, Richard. *Cities in Modernity: Representations and Productions of Metropolitan Space, 1840–1930* (Cambridge: Cambridge University Press, 2008).

D'Ignazio, Catherine and Lauren F. Klein. *Data Feminism* (Cambridge: MIT Press, 2020).

Dolbilov, Mikhail. 'Russification and the Bureaucratic Mind in the Russian Empire's Northwestern Region in the 1860s'. *Kritika: Explorations in Russian and Eurasian History* 5, no. 2 (2004), pp. 245–71.

Dolbilov, Mikhail. 'The Stereotype of the Pole in Imperial Policy: The "Depolonization" of the Northwestern Region in the 1860s'. *Russian Studies in History* 44, no. 2 (2005), pp. 44–88.

Dolbilov, Mikhail. 'Vicissitudes of Cyrillization: The Ban on Latin Alphabet and Bureaucratic Russification of the Lithuanians in the Vilno General-Governorship, 1864–1882'. *Ab Imperio* no. 2 (2005), pp. 255–96.

Dolbilov, Mikhail. *Russkii krai, chuzhaia vera: Ethnokonfessional'naia politika imperii v Litve i Belorussii pri Aleksandre II* (Moscow: Novoe literaturnoe obozrenie, 2010).

Dolbilov, Mikhail and Darius Staliūnas. 'Vvedenie k forumu: "Alfavit, iazyk i natsional'naia identichnost" v Rossiiskoi Imperii'. *Ab Imperio* no. 2 (2005), pp. 123–34.

Donaldson, John W. 'Politics and Scale in Boundary-Making: The Work of Boundary Commissions'. *Journal of Historical Geography* 34, no. 3 (2008), pp. 393–6.

Duančić, Vedran. *Geography and Nationalist Visions of Interwar Yugoslavia* (New York: Palgrave Macmillan, 2020).

Dukhanov, M. M. *Ostzeitsy: Iav' i vymysel: o roli nemetskikh pomeshchikov i biurgerov v istoricheskikh sud'bakh latyshskogo i estonskogo narodov v seredine XIX veka* (Riga: Leisma, 1970).

Dukhanov, M. M. *Ostzeitsy: Polityka ostzeiskogo dvorianstva v 50–70x gg. XIX v. i kritika ee apologeticheskoi istoriogragrafii* (Riga: Leisma, 1978).

Dullin, Sabine. *La frontière épaisse: Aux origines des politiques soviétiques (1920–1940)* (Paris: Editions de l'EHESS, 2014).

Dunlop, Catherine T. 'Mapping a New Kind of European Boundary: The Language Border between Modern France and Germany'. *Imago Mundi* 65, no. 2 (2013), pp. 253–67.

Dunlop, Catherine Tatiana. *Cartophilia: Maps and the Search for Identity in the French-German Borderland* (Chicago: University of Chicago Press, 2015).

Dziarnovič, Aleh. 'Gudas as a Historical Name of Belarusians in the Lithuanian Language: "Goths" or "Barbarians"?' In *Belarus and Its Neighbors: Historical Perceptions and Political Constructs. International Conference Papers*, edited by Aleś Łahviniec and Taciana Čulickaja (Warsaw: Uczelnia Łazarskiego, 2013), pp. 56–68.

Dzlyuma, M. 'Vklad absolventov Tartuskogo Universiteta v razvitie latyshkoi geografii v xix veke (K. Valdemar, K. Baron, Iu. Alunan)'. In *Tartuskii Gosudarstvennyi Universitet: Istoriia razvitiia, podgotovka kadrov, nauchnye issledovaniia. Tochnye i estestvennye nauki*, vol. 2.1 (Tartu: Tartu gosudarstvennyi universitet, 1982), pp. 86–90.

Eberhardt, Piotr. 'Zasięg terytorialny Polski w koncepcjack badaczy polskich (1864–1921)'. *Przegląd Geograficzny* 71, nos. 1–2 (1999), pp. 3–23.

Eberhardt, Piotr. *Polska i jej granice: z historii polskiej geografii politycznej* (Lublin: Wydawnictwo Uniwersytetu Marii Curie-Sklodowskiej, 2004).

Edney, Matthew H. *Mapping an Empire: The Geographical Construction of British India, 1765–1843* (Chicago: University of Chicago Press, 1997).

Edney, Matthew H. 'The Irony of Imperial Mapping'. In *The Imperial Map: Cartography and the Mastery of Empire*, edited by James R. Akerman (Chicago: Chicago University Press, 2009), pp. 11–45.

Edney, Matthew H. *Cartography: The Ideal and Its History* (Chicago: University of Chicago Press, 2019).

Elden, Stuart. 'Secure the Volume: Vertical Geopolitics and the Depth of Power'. *Political Geography* 34 (2013), pp. 35–51.

Etkind, Alexander, *Internal Colonisation: Russia's Imperial Experience* (Cambridge: Polity Press, 2011).

Evtuhov, Catherine. *Portrait of a Russian Province: Economy, Society, and Civilization in Nineteenth-Century Nizhnii Novgorod* (Pittsburgh: University of Pittsburgh Press, 2011).

Evtuhov, Catherine and Stephen Kotkin, *The Cultural Gradient: The Transmission of Ideas in Europe, 1789–1991* (Lanham: Rowman and Littlefield, 2003).

Fedorowich, Kent and Andrew S. Thompson, eds. *Empire, Migration and Identity in the British World* (Manchester: Manchester University Press, 2015).

Fehrenbach, Lenka. *Bildfabriken: Industrie und Fotografie im Zarenreich (1860–1917)* (Paderborn: Ferdinand Schöningh, 2019).

Felten, Sebastian and Christine von Oertzen. 'Bureaucracy as Knowledge'. *Journal for the History of Knowledge* 1, no. 1 (2020), pp. 1–16.

Finlay, Mark R. 'International Science and Local Conditions on the Ground: The Agricultural Sciences and Baltic German Identity, 1845–1905'. *Journal of Baltic Studies* 44, no. 3 (2013), pp. 339–62.

Finnegan, Diarmid A. 'The Spatial Turn: Geographical Approaches in the History of Science'. *The Journal of the History of Biology* 41, no. 2 (2008), pp. 395–409.

Fletcher, William A. 'The British Navy in the Baltic, 1918–1920: Its Contribution to the Independence of the Baltic Nations'. *Journal of Baltic Studies* 7, no. 2 (1976), pp. 134–44.

Foucault, Michel. *The Archaeology of Knowledge* (London: Tavistock, 1972).

Franklin, Simon. 'Mapping the Graphosphere: Cultures of Writing in Early 19th-Century Russia (and Before)'. *Kritika: Explorations in Russian and Eurasian History* 12, no. 3 (2011), pp. 531–60.

Franklin, Simon. 'Printing Social Control in Russia 3: Blank Forms'. *Russian History* 42, no. 1 (2015), pp. 114–35.

Friendly, Michael. 'Milestones in the History of Data Visualization: A Case Study in Statistical Historiography'. In *Classification: The Ubiquitous Challenge*, edited by Claus Weihs and Wolfgang Gaul (New York: Springer, 2005), pp. 34–52.

Frierson, Cathy A. *Peasant Icons: Representations of Rural People in Late Nineteenth-Century Russia* (Oxford; New York: Oxford University Press, 1993).

Fujikane, Candance. *Mapping Abundance for a Planetary Future: Kanaka Maoli and Critical Settler Cartographies in Hawai'i* (Durham: Duke University Press, 2021).

Fyfe, Aileen and Bernhard Lightman, eds. *Science in the Marketplace: Nineteenth-Century Sites and Experiences* (Chicago: University of Chicago Press, 2007).

Gal, Susan. 'Polyglot Nationalism: Alternative Perspectives on Language in 19th Century Hungary'. *Langage et Société* 136 (2011), pp. 31–54.

Galiona, V. A. 'Etnicheskaia prigranichnogo regiona Pytalovo/Abrene: polevye issledovaniia 2006 goda i dannye E. A. Vol'tera 1918 goda'. In *Etnokonfessional'naia karta Leningradskoi oblasti i sopredel' nykh territorii. Vtorye Shergenovskie chteniia. Sbornik statei* (St. Petersburg: Evropeiskii Dom, 2008), pp. 199–33.

Gatrell, Peter. *A Whole Empire Walking: Refugees in Russia during World War I* (Bloomington: Indiana University Press, 1999).

Gavrilin, A. V. *Ocherki istorii Rizhskoi eparkhii. 19 vek.* (Riga: Filokaliia, 1999).

Geraci, Robert. 'On "Colonial" Forms and Functions'. *Slavic Review* 69, no. 1 (2010), pp. 180–4.

Gerasimov, I., S. Glebov, A. Kaplunovski, M. Mogilner, and A. Semyonov. 'In Search of a New Imperial History'. *Ab Imperio* no. 1 (2005), pp. 33–56.

Gerasimov, Ilya. *Plebeian Modernity: Social Practices, Illegality, and the Urban Poor in Russia, 1906–1916* (Rochester, NY: University of Rochester Press, 2018).

Gibson, Catherine. 'The Polish Livonian Legacy in Latgalia: The Confluence of Slavic Ethnolects in the Baltic-Slavic Borderland'. In *The Palgrave Handbook of Slavic Languages, Identities and Borders*, edited by Tomasz Kamusella, Motoki Nomachi, and Catherine Gibson (New York: Palgrave Macmillan, 2016), pp. 57–80.

Gibson, Catherine. 'Shading, Lines, Colors: Mapping Ethnographic Taxonomies of European Russia'. *Nationalities Papers* 46, no. 4 (2018), pp. 592–11.

Gibson, Catherine. 'Mapmaking in the Home and Printing House: Women and Cartography in Late Imperial Russia'. *Journal of Historical Geography* 67 (2020), pp. 71–80.

Gibson, Catherine. 'Experiencing Enumeration: Local Reactions and Resistance to Censuses in Imperial Russia, 1863–81'. *Journal of Social History* (Online First 8 June 2021).

Gibson, Catherine, 'Problema sozdaniia etnograficheskogo atlasa Evropeiskoi Rossii v perepiske Petra Keppena s Akademiei nauk'. *Quaestio Rossica* 9, no. 3 (2021), pp. 977–94.

Gibson, Catherine and Irina Paert. 'Apostasy in the Baltic Provinces: Religious and National Indifference in Imperial Russia'. *Past & Present* (Online First 15 December 2021).

Göderle, Wolfgang. *Zensus und Ethnizität: zur Herstellung von Wissen über soziale Wirklichkeiten im Habsburgerreich zwischen 1848 und 1910* (Göttingen: Wallstein, 2016).

Göderle, Wolfgang. 'Administration, Science, and the State: The 1869 Population Census in Austria-Hungary'. *Austrian History Yearbook* 47 (2016), pp. 61–88.

Godlewska, Anne Marie Claire. *Geography Unbound: French Geographic Science from Cassini to Humboldt* (Chicago: University of Chicago Press, 1999).

Goffart, Walter. *Historical Atlases: The First Three Hundred Years, 1570–1870* (Chicago: University of Chicago Press, 2003).

Golubev, Vadim. 'Stanovlenie Belorussko-Latviiskoi granitsy (1918–1926 gg.)'. In *Latviia–Belarus', 1918–2018*, edited by M. Korolev (Minsk: Chetyre chetverti, 2018), pp. 61–8.

Golubinskii, Aleksei. 'New Technology and the Mapping of Empire: The Adoption of the Atrolabe'. In *Information and Empire: Mechanisms of Communication in Russia 1600–1850*, edited by Simon Franklin and Katherine Bower (Cambridge: Open Book Publishers, 2017), pp. 59–74.

Gooday, Graeme. '"Re-writing the 'Book of Blots": Critical Reflections on the Histories of Technological "Failure"'. *History and Technology* 14 (1998), pp. 265–91.

Gordin, Michael D. 'The Importation of Being Earnest: The Early St. Petersburg Academy of Sciences'. *Isis* 91, no. 1 (2000), pp. 1–31.

Gordin, Michael. 'Measure of All the Russias: Metrology and Governance in the Russian Empire'. *Kritika: Explorations in Russian and Eurasian History* 4, no. 4 (2003), pp. 783–815.

Gorizontov, Leonid. 'The "Great Circle" of Interior Russia: Representations of the Imperial Center in the Nineteenth and Early Twentieth Centuries'. In *Russian Empire: Space, People, Power, 1700–1930*, edited by Jane Burbank, Mark von Hagen, and Anatolyi Remnev (Bloomington: Indiana University Press, 2007), pp. 67–93.

Gornischeff, Feliks. *Venemaa suhted Preisimaaga Napoleoni sõdade ajal baltisaksa päritolu diplomaatide tegevuse näitel (1806–1812)*. PhD dissertation, Tartu Ülikool, 2020.

Górny, Maciej. '"Futuristic Geography": The Role of Geographers in Shaping the Borders of East Central and South-Eastern Europe, 1914–1920'. *Studia z Dziejów Rosji i Europy Środkowo-Wschodniej* 48 (2013), pp. 79–102.

Górny, Maciej. *Kreślarze ojczyzn: Geografowie i granice międzywojennej Europy* (Warsaw: Instytut Historii PAN, 2017).

Górny, Maciej. *Science Embattled: Eastern European Intellectuals and the Great War* (Paderborn: Ferdinand Schöningh, 2019).

Gorshkov, Boris B. 'Serfs on the Move: Peasant Seasonal Migration in Pre-Reform Russia, 1800–61'. *Kritika: Explorations in Russian and Eurasian History* 1, no. 4 (2000), pp. 627–56.

Green, Sara. 'Lines, Traces, and Tidemarks: Further Reflections on Forms of Border'. In *The Political Materialities of Borders: New Theoretical Directions*, edited by Olga Demetriou and Rozita Dimova (Manchester: Manchester University Press, 2018), pp. 67–83.

Gueslin, Julien. *La France et les petits états Baltes: réalites baltes, perceptions françaises et ordre européen (1920–1932)*. PhD dissertation, Université Panthéon-Sorbonne, 2004.

Güttler, Nils Robert. 'Scaling the Period Eye: Oscar Drude and the Cartographical Practice of Plant Geography, 1870s–1910s'. *Science in Context* 24, no. 1 (2011), pp. 1–41.

Güttler, Nils Robert. 'Unsichtbare Hände: die Koloristinnen des Perthes Verlags und die Verwissenschaftlichung der Kartographie im 19. Jahrhundert'. *Archiv für Geschichte des Buchwesens* 68 (2013), pp. 133–53.

Hackmann, Jörg, ed. *Vereinskultur und Zivilgesellschaft in Nordosteuropa: Regionale Spezifik und europäische Zusammenhänge/Associational Culture and Civil Society in North Eastern Europe: Regional Features and the European Context* (Vienna: Böhlau Verlag, 2012).

Hackmann, Jörg. '"Wo liegt das Baltikum?" Enstehung Verwendung und Semantik des Begriffs seit dem 19. Jahrhundert'. In *Das Baltikum also Konstrukt (18.–19 Jahrhunder): Von einer Kolonialwahrnehmung zu einem nationalen Diskurs*, edited by Anne Sommerlat-Michas (Wurzburg: Konigshausen and Naumann, 2015), pp. 23–44.

Hackmann, Jörg. *Geselligkeit in Nordosteuropa: Studien zu Vereinskultur, Zivilgesellschaft und Nationalisierungsprozessen in einer polykulturellen Region (1770–1950)* (Wiesbaden: Harrassowitz, 2020).

Hachten, Elizabeth A. 'In Service to Science and Society: Scientists and the Public in Late-Nineteenth-Century Russia'. *Osiris* 17 (2002), pp. 171–209.

Von Hagen, Mark. 'The Entangled Eastern Front in the First World War'. In *The Empire and Nationalism at War*, edited by Eric Lohr, Vera Tolz, Alexander Semyonov, and Mark von Hagen (Bloomington: Slavica, 2014), pp. 9–48.

Hansen, Jason D. *Mapping the Germans: Statistical Science, Cartography and the Visualisation of the German Nation, 1848–1914* (Oxford: Oxford University Press, 2015).

Harley, J. B. 'Silences and Secrecy: The Hidden Agenda of Cartography in Early Modern Europe'. *Imago Mundi* 40 (1988), pp. 57–76.

Harley, J. B. *The New Nature of Maps: Essays in the History of Cartography* (Baltimore: Johns Hopkins University Press, 2002).

Haslinger, Peter. *Nation und Territorium im tschechischen politischen Diskurs 1880–1938* (Munich: R. Oldenbourg Verlag, 2010).

Haslinger, Peter. 'Dilemmas of Security: The State, Local Agency, and the Czechoslovak-Hungarian Boundary Commission, 1921–25'. *Austrian History Yearbook* 49 (2018), pp. 187–206.

Hatzel, Michael H. 'The Baltic Germans'. In *Russification in the Baltic Provinces and Finland 1855–1914*, edited by Edward C. Thaden (Princeton: Princeton University Press, 1981), pp. 111–204.

Hearne, Siobhán. 'To Denounce or Defend? Public Participation in the Policing of Prostitution in Late Imperial Russia.' *Kritika: Explorations in Russian and Eurasian History* 19, no. 4 (2018), pp. 717–744.

Hearne, Siobhán. *Policing Prostitution: Regulating the Lower Classes in Late Imperial Russia* (Oxford: Oxford University Press, 2021).

Heffernan, Michael. 'Geography, Cartography and Military Intelligence: The Royal Geographical Society and the First World War'. *Transactions of the Institute of British Geographers* 21, no. 3 (1996), pp. 504–33.

Henriksson, Anders. *The Tsar's Loyal Germans: The Riga German Community: Social Change and the Nationality Question, 1855–1905* (Boulder: Columbia University Press, 1983).

Henze, Charlotte E. *Disease, Healthcare, and Government in Late Imperial Russia: Life and Death on the Volga, 1823–1914* (London: Routledge, 2010).

Hiden, John. *The Baltic States and Weimar Ostpolitik* (Cambridge: Cambridge University Press, 1987).

Hilton, Marjorie L. *Selling to the Masses: Retailing in Russia, 1880–1930* (Pittsburgh: University of Pittsburgh Press, 2014).

Hirsch, Francine. 'Towards an Empire of Nations: Border-Making and the Formation of Soviet National Identities'. *Russian Review* 59, no. 2 (2000), pp. 201–26.

Hirsch, Francine. *Empire of Nations: Ethnographic Knowledge and the Making of the Soviet Union* (Ithaca: Cornell University Press, 2005).

Von Hirschhausen, Ulrike. *Die Grenzen der Gemeinsamkeit: Deutsche, Letten, Russen und Juden in Riga 1860–1914* (Göttingen: Vandenhoeck and Ruprecht, 2006).

Von Hirschhausen, Béatrice, Hannes Grandits, Claudia Kraft, Dietmar Müller, and Tomas Serrier, eds. *Phantomgrenzen: Räume und Akteure in der Zeit neu denken* (Göttingen: Wallstein, 2015).

Holquist, Peter. 'To Count, to Extract, and to Exterminate: Population Statistics and Population Politics in Late Imperial Russia and Soviet Russia'. In *A State of Nations: Empire and Nation-Making in the Age of Lenin and Stalin*, edited by Ronald Grigor Suny and Terry Martin (Oxford: Oxford University Press, 2001), pp. 111–44.

Holquist, Peter. *Making War, Forging Revolution: Russia's Continuum of Crisis, 1914–1921* (Cambridge, MA: Harvard University Press, 2002).

Van Houtum, Henk and Rodrigo Bueno Lacy. 'The Migration Map Trap: On the Invasion Arrows in the Cartography of Migration'. *Mobilities* 15, no. 2 (2020), pp. 196–219.

Iannàccaro, Gabriele and Vittorio Dell'Aquila. 'Mapping Languages from Inside: Notes on Perceptual Dialectology'. *Social and Cultural Geography* 2 (2001), pp. 265–80.

Ījabs, Ivars. 'Another Baltic Postcolonialism: Young Latvians, Baltic Germans, and the Emergence of Latvian National Movement'. *Nationalities Papers* 42, no. 1 (2014), pp. 88–107.

Ījabs, Ivars. 'Als "Latwija" zu einem politischen Begriff wurde: Die Revision von 1882/83 durch Senator Nikolaj Manassein und das politische Denken der lettischen Nationalbewegung'. *Zeitschrift für Ostmitteleuropa-Forschung* 65, no. 3 (2016), pp. 373–406.

Isakov, S. G. *Ostzeiskii vopros v russkoi pechati 1860-x godov* (Tartu: Uchenye zapiski tartuskogo gosudarstvennogo universiteta, 1961).

Isakov, S. G. and T. K. Shor. *Vlastiteli rossiiskoi imperii na estonskoi zemle* (Tallinn: TLÜ Kirjastus, 2009).

Iturrioz, Teresa and Monica Wachowicz. 'An Artistic Perspective for Affective Cartography'. In *Mapping Different Geographies*, edited by Karel Kriz, William Cartwright, and Lorenz Hurni (Berlin; Heidelberg: Springer, 2010), pp. 74–92.

Jaadla, Hannaliis and Allan Puur. 'The Impact of Water Supply and Sanitation on Infant Mortality: Individual-Level Evidence from Tartu, Estonia, 1897–1900'. *Population Studies* 70, no. 2 (2016), pp. 163–79.

Jääts, Indrek. 'Üks kuulus välitöö ja selle pikk vari: Oskar Kallas, Paulopriit Voolaine ja Lutsi maarahvas'. *Eesti Rahva Muuseumi aastaraamat*, no. 57 (2014), pp. 14–41.

Janicki, Arkadiusz. *Kurlandia w latach 1795–1915* (Gdańsk: Wydawnictwo Uniwersytetu Gdańskiego, 2011).

Jankowiak, Mirosław. 'Belarusian Dialects in Latvian Latgale—Transitional or Mixed?'. *Slavica Bergensia* 12 (2014), pp. 343–72.

Von Jannau, H. G. *Vene-õigeusu tulekust Lõuna-Eestis 1845.–46 a.* (Tartu: Loodus, 1927).

Jansen, Ea. '"Baltus", baltisakslased, eestlased I'. *Tuna* 2 (2005), pp. 35–44.

Jantson, Signe. 'Booksellers, Publishers and Press Workers in the Second Half of the 19th and Early 20th Century in Estonia'. *Knygotyra* 52 (2009), pp. 231–41.

Jaremko-Porter, Kristina. 'The Latvian Era of Folk Awakening: From Johann Gottfried Herder's Volkslieder to the Voice of an Emergent Nation'. In *The Voice of the People: Writing the European Folk Revival, 1760–1914*, edited by Matthew Campbell and Michael Perraudin (London: Anthem Press, 2013), pp. 141–56.

Jēkabsons, Ēriks. 'Stosunki polsko-łotewskie na przestrzeni dziej'. In *Polacy na Łotwie*, edited by Edward Walewander (Lublin: Wydawnictwo Katolickiego Uniwersytetu Lubelskiego, 1993).

Jēkabsons, Ēriks. 'Sešu pagastu un Grīvas pilsētas problēma Latvijas un Polijas attiecībās 20.–30. gados'. *Latvijas Vēstures Institūta Žurnāls*, no. 1 (1995), pp. 80–101.

Jēkabsons, Ēriks. 'Baltkrievi Latvijā 1918.–1940. gadā'. *Latvijas Vēstures Institūta Žurnāls* 4 (2001), pp. 104–33.

Jēkabsons, Ēriks. *Piesardzīgā draudzība: Latvijas un Polijas attiecības 1919. un 1920. gadā* (Riga: Latvijas Universitātes Akadēmiskais apgāds, 2007).

Jēkabsons, Ēriks. *Polijas ārējā politika 1918.–1937. gadā: vēstures avotu krājums* (Riga: Latvijas Universitātes Akadēmiskais apgāds, 2009).

Jēkabsons, Ēriks. 'Latviešu piedalīšanās Krievijas-Japānas karā 1904.–1905.gadā'. *Tēvijas Sargs* 2 (2014), pp. 26–7.

Jeske, Martin. 'The Map Specialcharte von Livland: Georg Friedrich Parrot's Agenda or a New Perspective on Livland'. *Acta Baltica Historiae et Philosophiae Scientiarum* 6, no. 2 (2018), pp. 127–37.

Jeske, Martin. *Ein Imperium wird vermessen: Kartographie, Kulturtransfer und Raumerschliessung im Zarenreich (1797–1919)*. PhD dissertation, Universität Basel, 2019.

Jeske, Martin. 'Topographische Karten im Zarenreich: Zur Verfügbarkeit und Geheimhaltung im langen 19. Jahrhundert'. In *Tagungsband zum 19. Kartographiehistorischen Colloquium Gotha 2016*, ed. Markus Heinz (Bonn: Kirschbaum Verlag, forthcoming).

Jõekalda, Kristina, *German Monuments in the Baltic Heimat? A Historiography of Heritage in the 'Long Nineteenth Century'*. PhD dissertation, Estonian Academy of Arts, 2020.

Johnson, Robert E. 'Liberal Professionals and Professional Liberals: The Zemstvo Statisticians and Their Work'. In *The Zemstvo in Russia: An Experiment in Local Self-Government*, edited by Terence Emmons and Wayne S. Vucinich (Cambridge: Cambridge University Press, 1982), pp. 343–63.

Judson, Pieter M. 'Constructing Nationalities in East Central Europe: Introduction'. In *Constructing Nationalities in East Central Europe*, edited by Pieter M. Judson and Marsha L. Rozenblit (New York: Berghahn Books, 2005).

Judson, Pieter M. 'When Is a Diaspora Not a Diaspora? Rethinking Nation-Centered Narratives about Germans in Habsburg East Central Europe'. In *The Heimat Abroad: The Boundaries of Germanness*, edited by Krista O'Donnell, Renate Bridenthal, and Nancy Reagin (Ann Arbour: University of Michigan Press, 2005), pp. 219–47.

Judson, Pieter M. 'Do Multiple Languages Mean a Multicultural Society? Nationalist "Frontiers" in Rural Austria, 1880–1918'. In *Understanding Multiculturalism: The Habsburg Central European Experience*, edited by Johannes Fechtinger and Gary B. Cohen (New York; Oxford: Berghan Books, 2014), pp. 61–82.

Judson, Pieter M. *The Habsburg Empire: A New History* (Cambridge, MA: Harvard University Press, 2016).

Jürgenson, Aivar. *Ladina rahva seas: Argentina ja sealsed eestlased* (Tallinn: Argo, 2011).

Jürgenson, Aivar. 'The Role of Religion in the Connection between Estonian Settlements in Abkhazia and the Estonian Homeland'. *Acta Historica Tallinnensia* 27, no. 1 (2021), pp. 133–64.

Jürjo, Indrek. 'Ludwig August Krahv Mellin kui talurahva sõber ja estofiil'. *Tuna* 4 (2003), pp. 50–66.

Kalemaj, Ilir. *Contested Borders: Territorialisation, National Identity and 'Imagined Geographies' in Albania* (Bern: Peter Lang, 2014).

Kalkun, Andreas. 'Russkaia vera ili vizantiiskoe nasledie? Reprezentatsii pravoslaviia narodnosti setu v XX v'. In *Pravoslavie v Pribaltike: religiia, politika, obrazovanie 1840-e–1930-e gg.*, edited by Irina Paert (Tartu: University of Tartu Press, 2018), pp. 341–73.

Kalling, Ken and Lea Heapost. 'Racial Identity and Physical Anthropology in Estonia, 1800–1945'. In *Baltic Eugenics: Bio-Politics, Race and Nation in Interwar Estonia, Latvia and Lithuania, 1918–1940*, edited by Björn M. Felder and Paul J. Weindling (Amsterdam: Rodopi, 2013), pp. 83–114.

Kamusella, Tomasz. *Politics of Language and Nationalism in Modern Central Europe* (Basingstoke: Palgrave Macmillan, 2009).

Kane, Eileen M. *Russian Hajj: Empire and the Pilgrimage to Mecca* (Ithaca: Cornell University Press, 2015).

Kaplan, Vera. *Historians and Historical Societies in the Public Life of Imperial Russia* (Bloomington and Indianapolis: Indiana University Press, 2017).

Karjahärm, Toomas. 'The Problem of Reorganization of Provincial Self-Government in Estonia at the Beginning of the 20th Century. In *Bevölkerungsverschiebungen und sozialer Wandel in den baltischen Provinzen Russlands 1850–1914*, edited by Gert v.

Pistohlkors, Andrejs Plakans, and Paul Kaegbein (Lüneburg: Institut Nordostdeutsches Kulturwerk, 1995), pp. 249–60.

Karjahärm, Toomas. 'Balti küsimus Vene poliitikas monarhia kokkuvarisemise eel'. *Looming* 10 (1996), pp. 1372–83.

Karjahärm, Toomas. 'The Awakening of the Estonian Nation: From Its Origins to Independence'. *Revue d'Histoire Nordique = Nordic Historical Review* 3 (2007), pp. 57–74.

Karjahärm, Toomas. 'Terminology Pertaining to Ethnic Relations as Used in Late Imperial Russia'. *Acta Historica Tallinnensia* 15 (2010), pp. 24–50.

Karjahärm, Toomas. 'Ostzeitsy i baltiitsy: ekskurs v terminologiiu'. In *Rossiia i pribaltiiskii region v XIX–XX vv.: Problemy vzaimootnoshenii v meniaiushchemsia mire* (Moscow: Lenand, 2013), pp. 7–26.

Karma, Tenu and Vello Pātsi. 'Kā tapis pirmais ģeogrāfiskais atlants latviešu valodā'. In *Grāmata Latviešu Sabiedrībā 1856–1870*, edited by Aleksejs Apīnis (Rīga: Avots, 1987), pp. 27–36.

Karnups, Ādolfs. 'Matīss Siliņš'. *Senatne un Māksla*, no. 1 (1936), pp. 151–3.

Karskii, E. F. *K voprosu ob etnograficheskoi karte belorusskago plemeni* (St. Petersburg: Tipografiia Imperatorskoi Akademii Nauk, 1902).

Karulis, Konstantīns. 'Matīss Siliņš (1861–1942) atmiņās'. *Latvijas Zinātņu Akadēmijas Vēstis* 50, no. 6 (1996), pp. 73–4.

Kasmach, Lizaveta. 'Forgotten Occupation: Germans and Belarusians in the Lands of Ober Ost (1915–17)'. *Canadian Slavonic Papers* 58 (2016), pp. 321–40.

Kaufmann, Thomas DaCosta. 'Baltic Reflections'. *Baltic Journal of Art History* 9 (2015), pp. 11–22.

Kazachenok, K. '"Belorusskii protsess" v Latvii: Obvinenie, opravdanie, nakazanie'. In *Latyshi i belorusy: vmeste skvoz' veka. Sbornik nauchnykh statei*, edited by M. G. Korolev (Minsk: RIVSh, 2012), pp. 29–43.

Keating, Jennifer. 'Amid the Horrors of Nature: "Dead" Environments at the Margins of the Russian Empire'. In *Empty Spaces: Perspectives on Emptiness in Modern History*, edited by Courtney J. Campbell, Allegra Giovine and Jennifer Keating (London: University of London Press, 2019), pp. 33–57.

Khalid, Adeeb. 'Printing, Publishing, and Reform in Tsarist Central Asia'. *International Journal of Middle East Studies* 26, no. 2 (1994), pp. 187–200.

Ķikuts, Toms. 'Diskusijas par migrācijas jautājumiem latviešu sabiedrībā 19. gs. 2. pusē-20. gs. sākumā: Viedokļi, argumenti, emocijas'. In *Daugavpils Universitātes 53.starptautiskās zinātniskās konferences materiāli/Proceedings of the 53rd International Scientific Conference of Daugavpils University*, edited by Dmitrijs Oļehnovičs and Inese Zuģicka (Daugavpils: Daugavpils Universitātes Akadēmiskais apgāds 'Saule', 2014), n.p.

Ķikuts, Toms. '1896. gada Latviešu etnogrāfiskā izstāde—laikmeta lieciniece / The Latvian Ethnographic Exhibition of 1896—Witness to an Era'. In *Latviešu etnogrāfiskā izstāde 1896/Latvian Ethnographic Exhibition 1896*, edited by Sanita Stinkule, Toms Ķikuts, and Jānis Ciglis (Riga: Neptuns, 2016), pp. 17–87.

Kil'dushevskaia, L. K. 'Izdatel'skaia deiatel'nost' "Kartograficheskogo zavedeniia A. Il'ina"', *Knizhnoe delo v Rossii vo vtoroi polovine XIX–nachale XX veka* 6 (1992), pp. 97–111.

Kivelson, Valerie A. *Cartographies of Tsardom: The Land and Its Meanings in Seventeenth-Century Russia* (Ithaca: Cornell University Press, 2006).

Kivelson, Valerie. 'Early Mapping: The Tsardom in Manuscript'. In *Information and Empire: Mechanisms of Communication in Russia 1600–1850*, edited by Simon Franklin and Katherine Bower (Cambridge: Open Book Publishers, 2017), pp. 23–57.

Kļaviņas, Sarma. 'Die Werke von August Bielenstein—eine bedeutende Quelle der Indoeuropäistik des 19. Jahrhunderts'. *Res Balticae* 8 (2002), pp. 151–68.

Klier, John Doyle. *Imperial Russia's Jewish Question, 1855–1881* (Cambridge University Press, 1995).

Knight, Nathaniel. 'Science, Empire, and Nationality: Ethnography in the Russian Geographical Society, 1845–1855'. In *Imperial Russia: New Histories for the Empire*, edited by Jane Burbank and David L. Ransel (Bloomington; Indianapolis: Indiana University Press, 1998), pp. 108–41.

Knight, Nathaniel. 'Seeking the Self in the Other: Ethnographic Studies of Non-Russians in the Russian-Geographical Society, 1845–1860'. In *Defining Self: Essays on Emergent Identities in Russia Seventeenth to Nineteenth Centuries*, edited by Michael Branch (Helsinki: Finnish Literature Society, 2009), pp. 117–38.

Knight, Nathaniel. 'Russian Ethnography and the Visual Arts in the 1840s and 1850s'. In *Visualizing Russia: Fedor Solntsev and Crafting a National Past*, edited by Cynthia Hyla Whittaker (Leiden; Boston: Brill, 2010), pp. 127–44.

Koblitz, Ann Hibner. *Science, Women and Revolution in Russia* (Amsterdam; Abingdon: Harwood Academic Publishers, 2000).

Köhler, Marcus. *Russische Ethnographie und imperiale Politik im 18. Jahrhundert* (Göttingen: V&R Unipress, 2012).

Kolosova, Alison Ruth. *Narodnost' and Obshchechelovechnost' in 19th Century Russian Missionary Work: N. I. Il'minskii and the Christianization of the Chuvash*. PhD dissertation, Durham University, 2016.

Kordiš, Ivan. 'Peter Kozler and His Map of the Slovenian Land and Its Provinces (1849–1871)'. *Imago Mundi* 68, no. 2 (2016), pp. 212–31.

Korovushukin, D. G. *Latyshi i estontsy v zapadnoi Sibiri: rasselenie i chislennost' v kontse XIX—nachale XXI vv.* (Novosibirsk: Institut arkheologii i etnografii SO RAN, 2008).

Korsaka, Inara. 'Martha Bielenstein: Ein bedeutendes Mitglied der Familie Bielenstein'. *Jahrbuch des baltischen Deutschtums* (2005), pp. 79–84.

Kosonen, Katariina. 'Making Maps and Mental Images: Finnish Press Cartography in Nation-Building, 1899–1942'. *National Identities* 10, no. 1 (2008), pp. 21–47.

Kotenko, Anton. 'Ukraine's First Ethnographic Map: Made in the Russian Empire'. *Nationalities Papers* 48, no. 5 (2020), pp. 931–41.

Krasnikova, Ol'ga. 'Istoriia etnicheskogo kartografirovaniia v Rossii XVII-nachala XX v. v dokumentakh (itogi proekta)'. In *Etnokonfessional'naia karta Leningradskoi oblasti i sopredel'nykh territorii. Vtorye Shergenovskie chteniia: Sbornik statei* (St. Petersburg: Evropeiskii Dom, 2008), pp. 18–27.

Krasnikova, Ol'ga Alekseevna. 'Sankt-Peterburg i guberniia na "Odnoverstakh" XIX v.: k istorii pereimenovaniia naselennykh punktov'. In *Istoriko-kulturnyi landshaft Severo-Zapada 3: Shestye Shegrenovskie chiteniia: Sbornik statei* (St. Petersburg: Evropeiskii Dom, 2014), pp. 102–12.

Krokar, James. 'Constant Desjardins, Peripatetic Cartographer'. *Imago Mundi* 67, no. 1 (2015), pp. 33–57.

Krūmiņa, Guna. 'Zinātniskas un populārzinātniskās grāmatas'. In *Grāmata Latviešu Sabiedrībā 1856–1870*, edited by Aleksejs Apīnis (Riga: Avots, 1987), pp. 9–25.

Kruus, Hans. *Linn ja küla Eestis* (Tartu: Noor-Eesti, 1920).

Kruus, Hans. *Talurahva käärimine Lõuna-Eestis XIX sajandi 40-ndail aastail* (Tartu: Eesti Kirjanduse Selts, 1930).

Kurschat, Friedrich. *Grammatik der littauischen Sprache* (Halle: Verlag der Buchhandlung des Waisenhauses, 1876).

Kuznetsov, G. I., ed. *Vialiki Histarychny Atlas Belarusi*, vol. 3 (Minsk: Belkatografiia, 2009).

Labbé, Morgane. 'Le projet d'une statistique des nationalités discutés dans les sessions du Congrès International de Statistique (1853–1876)'. In *Démographie et Politique*, edited by F. Ronsin, H. Le Bras, and E. Zucker (Dijon: Editions Universitaires de Dijon, 1997), pp. 127–42.

Labbé, Morgane. 'La carte ethnographique de l'Empire Austrichien: La multinationalité dans "L'Ordre des Choses"'. *Bulletin du Comité français de cartographie* 180 (2004), pp. 71–84.

Labbé, Morgane. 'Institutionalizing the Statistics of Nationality in Prussia in the 19th Century (from Local Bureaucracy to State-level Census of Population)'. *Centaurus* 49 (2007), pp. 289–306.

Labbé, Morgane. 'Eugene Romer's 1916 Atlas of Poland: Creating a New Nation State'. *Imago Mundi* 70, no. 1 (2018), pp. 94–113.

Labbé, Morgane. *La nationalité, une histoire de chiffres: politique et statistiques en Europe centrale (1848–1919)* (Paris: Presses de Sciences Po, 2019).

Lajosi, Kriztina and Andreas Stynen, *Choral Societies and Nationalism in Europe* (Leiden: Brill, 2015).

Laptev, Vladimir. *Russkaia Infografika* (St. Petersburg: Izdatel'stvo Politekhnicheskogo Universiteta, 2018).

Latour, Bruno. 'Drawing Things Together'. In *Representation in Scientific Practice*, edited by Michael Lynch and Steve Woolgar (Cambridge: MIT Press, 1990), pp. 19–68.

Lebow, Katherine, Małgorzata Mazurek, and Joanna Wawrzyniak, 'Making Modern Social Science: The Global Imagination in East Central and Southeastern Europe after Versailles'. *Contemporary European History* 28, no. 2 (2019), pp. 137–42.

Le Foll, Claire. *Vitebskaia khudozhestvannaia shkola (1897–1923)* (Minsk: Propilei, 2007).

Le Foll, Claire. *La Biélorussie dans l'histoire et l'imaginaire des Juifs de l'Empire russe, 1772–1905* (Paris: Honoré Champion, 2017).

Lembit, Andrezen. *Estonskie narodnye shkoly v XVII–XIX vekakh* (Tallinn: Valgus, 1980).

Leopold, Joan. 'Lithuanian Made "Visible" through German Linguists: August Friedrich Pott and August Schleicher'. In *Invisible Languages in the Nineteenth Century*, edited by Anna Havinga and Nils Langer (Bern: Peter Lang, 2015), pp. 211–38.

Leppik, Lea. 'Social Mobility and Career Patterns of Estonian Intellectuals in the Russian Empire'. *Historical Social Research* 33, no. 2 (2008), pp. 42–62.

Leskinen, Maria. *Poliaki i finny v Rossiiskoi nauke vtoroi poloviny XIX v.: 'Drugoi' skvoz' prizmu identichnosti* (Moscow: Indrik, 2010).

Lincoln, Bruce W. 'The Daily Life of St. Petersburg Officials in the Mid-Nineteenth Century'. *Oxford Slavonic Papers: New Series* 8 (1975), pp. 82–100.

Lincoln, Bruce W. *In the Vanguard of the Reform: Russia's Enlightened Bureaucrats, 1825–1861* (DeKalb: Northern Illinois University Press, 1982).

Lindsay, Debra. 'Intimate Inmates: Wives, Households, and Science in Nineteenth-Century America'. *Isis* 89 (1998), pp. 631–52.

Liulevicius, Vejas G. *The German Myth of the East: 1800 to the Present* (Oxford: Oxford Unversity Press, 2009).

Liulevicius, Vejas Gabriel. *War Land on the Eastern Front: Culture, National Identity, and German Occupation in World War I* (Cambridge: Cambridge University Press, 2000).

Livingstone, David. *The Geographical Tradition: Episodes in the History of a Contested Enterprise* (Malden: Blackwell, 1993).

Livingstone, David N. *Putting Science in Its Place: Geographies of Scientific Knowledge* (Chicago: University of Chicago Press, 2003).

Livingstone, David N. and Charles W. J. Withers, eds. *Geographies of Nineteenth-Century Science* (Chicago: University of Chicago Press, 2011).

Lohr, Eric. 'The Russian Army and the Jews: Mass Deportation, Hostages and Violence during World War I'. *Russian Review* 60, no. 3 (2001), pp. 404–19.

Lohr, Eric. *Nationalizing the Russian Empire: The Campaign against Enemy Aliens during World War I* (Cambridge, MA: Harvard University Press, 2003).

Loskutova, Marina. 'S chego nachinaetsia rodina? Prepodavanie geografii v dorevoliutsionnoi shkole v regional'noe samosoznanie (xix-nachalo xx v.)'. *Ab Imperio* no. 3 (2003), pp. 159–98.

Loskutova, Marina. 'Uezdnye uchenye: samoorganizatsiia nauchnoi obshchestvennosti v Rossiiskoi provintsii vo vtoroi polovine XX—pervye desiatiletiia XX v'. *Ab Imperio* no. 3 (2009), pp. 119–69.

Loskutova, Marina. '"Svedeniia o klimate, pochve, obraze khoziaistva i gospodstvuiushschikh rasteniiakh dolzhny byt' sobrany...": Prosveshchennaia biurokratiia, gumbol'dtovskaia nauka i mestnoe znanie v Rossiiskoi imperii vtoroi chetverti XIX v'. *Ab Imperio* no. 4 (2012), pp. 111–56.

Loskutova, Marina. 'Early Research on Insect Pests in the Russian Empire: Bureaucracy, Academic Community and Local Knowledge in the 1830s–1840s'. *Centaurus* 56, no. 4 (2014), pp. 229–53.

Loskutova, Marina V. and Anastasia A. Fedotova. 'The Rise of Applied Entomology in the Russian Empire: Governmental, Public, and Academic Responses to Insect Pest Outbreaks from 1840 to 1894'. In *New Perspectives on the History of Life Sciences and Agriculture*, edited by Denise Phillips and Sharon Kingsland (Cham: Springer, 2015), pp. 139–62.

Lotz, Christian. 'Gestrichelte Linien und schattierte Flächen. Darstellungen von Teilung und Einheit in ost- und westdeutschen Landkarten (1945–1972)'. In *Die geteilte Nation: Nationale Verluste und Identitäten im 20. Jahrhundert*, edited by Andreas Hilger and Oliver von Wrochem (Oldenbourg: De Gruyter, 2013), pp. 53–69.

Lust, Kersti. 'The Impact of the Baltic Emancipation Reforms on Peasant-Landlord Relations: A Historiographical Survey'. *Journal of Baltic Studies* 44, no. 1 (2013), pp. 1–18.

Lust, Kersti. 'Feeding the Landless in Post-Emancipation Livland during Times of Famine (1840s–1860s)'. *The Slavonic and East European Review* 92 (2014), pp. 81–106.

Maandi, Peeter. 'Land Reforms and Territorial Integration in Post-Tsarist Estonia, 1918–1940'. *Journal of Historical Geography* 36 (2010), pp. 441–52.

Mačiulis, Dangiras and Darius Staliūnas, eds. *Lithuanian Nationalism and the Vilnius Question* (Marburg: Herder Institut, 2015).

Maddrell, Avril. *Complex Locations: Women's Geographical Work in the UK 1850–1970* (New York: Wiley-Blackwell, 2009).

Maier, Charles S. *Once within Borders: Territories of Power, Wealth, and Belonging since 1500* (Cambridge, MA: Harvard University Press, 2016).

Mainardi, Patricia. *Another World: Nineteenth-Century Illustrated Print Culture* (New Haven: Yale University Press, 2017).

Majorova, Olga. 'Die Schlüsselrolle der "deutschen Frage" in der russischen patriotischen Presse der 1860er Jahre'. In *Deutsche und Deutschland aus russischer Sicht. 19./20. Jahrhundert: Von den Reformen Alexanders II. bis zum Ersten Weltkrieg*, edited by Dagmar Herrmann (Munich: Wilhelm Fink, 2006), pp. 81–101.

Manz, Stefan. *Constructing a German Diaspora: The 'Greater German Empire', 1871–1914* (London: Routledge, 2014).

Mardiste, Heino, Vello Paatsi, and Jüri Jagomägi. 'The Development of Estonian National Cartography'. *Estonia: Geographical Studies* 9 (2004), pp. 107–23.

Marker, Gary. *Publishing, Printing, and the Origins of Intellectual Life in Russia, 1700–1800* (Princeton: Princeton University Press, 1985).

Marten, Heiko F. '"Latgalian Is Not a Language": Linguistic Landscapes in Eastern Latvia and How They Reflect Centralist Attitudes'. In *Minority Languages in the Linguistic Landscape*, edited by Durk Gorter, Heiko F. Marten, and Luk Van Mensel (London: Palgrave Macmillan, 2011), pp. 19–35.

Martin, Terry. 'The German Question in Russia, 1848–1896'. *Russian History* 18, no. 4 (1991), pp. 373–434.

Mashkina, Olga. 'The Pulp and Paper Industry Evolution in Russia: A Road of Many Transitions'. In *The Evolution of Global Paper Industry 1800–2050: A Comparative Analysis*, edited by Juha-Antti Lamberg, Jari Ojala, Mirva Peltoniemi, and Timo Särkkä (Dordrecht; Heidelberg: Springer, 2012), pp. 285–306.

Mattisen, Edgar. *Eesti-Vene piir* (Tallinn: Ilo, 1993).

Maxwell, Alexander. 'Taxonomies of the Slavic World since the Enlightenment: Schematizing Perceptions of Slavic Ethnonyms in a Chart'. *Language & History* 58, no. 1 (2015), pp. 24–54.

Maxwell, Alexander. 'Effacing Panslavism: Linguistic Classification and Historiographic Misrepresentation', *Nationalities Papers* 46, no. 4 (2018), pp. 633–53.

Maxwell, Alexander and Sacha E. Davis. 'Germanness beyond Germany: Collective Identity in German Diaspora Communities'. *German Studies Review* 39 (2016), pp. 1–15.

Mayhew, Robert and Charles W. J. Withers, eds. *Geographies of Knowledge: Science, Scale, and Spatiality in the Nineteenth Century* (Baltimore: Johns Hopkins University Press, 2020).

McCain, Stewart. *The Language Question under Napoleon* (Cham: Springer International Publishing; Palgrave Macmillan, 2017).

Mazanik, Anna. *Sanitation, Urban Environment, and the Politics of Public Health in Late Imperial Moscow*. PhD dissertation, Central European University, 2015.

McReynolds, Louise. *The News under Russia's Old Regime: The Development of a Mass Circulation Press* (Princeton: Princeton University Press, 1991).

Mela, Marjo. *Läti eestlased: Ajalugu, keel ja kultuur* (Tallinn: Eesti Keele Sihtasutus, 2007).

Merritt, Harry C. 'The Colony of the Colonized: The Duchy of Courland's Tobago Colony and Contemporary Latvian National Identity'. *Nationalities Papers* 38, no. 4 (2010), pp. 491–508.

Mespoulet, Martine, *Statistique et révolution en Russie: un compromis impossible, 1880–1930* (Rennes: Presses universitaires de Rennes, 2001).

Mespoulet, Martine. 'Pratique de l'enquête et construction du savoir statistique en Russie à la fin du xixe siècle'. *Genèses* 52 (2003), pp. 96–118.

Meyer, James H. *Turks Across Empires: Marketing Muslim Identity in the Russian-Ottoman Borderlands, 1856–1914* (Oxford: Oxford University Press, 2019).

Mezhevich, Nikolai Maratovich. 'Rossiisko-Estonskaia granitsa: istoriia formirovaniia i sovremennoe znachenie dlia razvitiia severo-zapada Rossii'. *Pskovskii regionologicheskii zhurnal* 4 (2007), pp. 131–45.

Michaluk, Dorota. *Belaruskaia Narodnaia Respublika u 1918–1920 gg.: lia vytokau belaruskai dziarzhaunastsi* (Smalensk: Inbelkult, 2015).

Miller, Alexey. 'Natsiia, Narod, Narodnost' in Russia in the 19th Century: Some Introductory Remarks to the History of Concepts'. *Jahrbücher für Geschichte Osteuropas* 56, vol. 3 (2008), pp. 379–90.

Miller, Alexey. '"Official Nationality"? A Reassessment of Count Sergei Uvarov's Triad in the Context of the Politics of Nationalism, 1820–1850'. In *Defining the Self: Essays on Emergent Identities in Russia. Seventeenth to Nineteenth Centuries*, edited by Michael Branch (Helsinki: Studia Fennica Ethnologica, 2009), pp. 224–43.

Minins, Aldis. 'Latvia, 1918–1920: A Civil War?'. *The Journal of Baltic Studies* 36, no. 1 (2015), pp. 49–63.

Minnik, Taavi, 'The Cycle of Terror in Estonia, 1917–1919: On Its Preconditions and Major Stages'. *Journal of Baltic Studies* 46, no. 1 (2015), pp. 35–47.

Mogilner, Marina. *Homo Imperii: A History of Physical Anthropology in Russia* (Lincoln; London: University of Nebraska Press, 2013).

Mokyr, Joel. *The Gifts of Athena: Historical Origins of the Knowledge Economy* (Princeton University Press, 2002).

Monmonier, Mark. 'The Rise of the National Atlas'. *Cartographica* 31, iss. 1 (1994), pp. 1–15.

Monmonier, Mark. *How to Lie with Maps*, 3rd edn. (Chicago: University of Chicago Press, 2018).

Morrissey, Susan K. 'Mapping Civilization: The Cultural Geography of Suicide Statistics in Russia'. *Journal of Social History* 46, no. 3 (2013), pp. 651–67.

Morrison, Alexander. *The Russian Conquest of Central Asia: A Study in Imperial Expansion, 1814–1914* (Cambridge: Cambridge University Press, 2020).

Müller-Wille, Staffan and Isabelle Charmantier. 'Lists as Research Technologies'. *Isis* 103 (2012), pp. 743–52.

Murawska-Muthesius, Katarzyna. 'Mapping Eastern Europe: Cartography and Art History'. *Artl@s Bulletin* 2, no. 2 (2013), pp. 14–25.

Murawska-Muthesius, Katarzyna. *Imaging and Mapping Eastern Europe: Sarmatia Europea to Post-Communist Bloc* (New York: Routledge, 2021).

Musaev, Vadim Ibragimovich. 'Russko-estonskie pogranichnye problemy v 1917–1920 gg.'. In *Istoriko-kulturnyi landshaft Severo-Zapada 1: Chetvertye mezhdunarodnye Shegrenovskie chiteniia: Sbornik statei* (St. Petersburg: Evropeiskii Dom, 2011), pp. 121–7.

Naranch, Bradley D. 'Inventing the Auslandsdeutsche: Emigration, Colonial Fantasy, and German National Identity, 1848–71'. In *Germany's Colonial Pasts*, edited by Eric Ames, Lora Wildenthal, and Marcia Klotz (Lincoln: University of Nebraska Press, 2005), pp. 21–40.

Nelson, Robert L. 'The Baltics as Colonial Playground: Germany in the East, 1914–1918'. *Journal of Baltic Studies* 42, no. 1 (2011), pp. 9–19.

Nemes, Robert. 'Mapping the Hungarian Borderlands'. In *Shatterzones of Empires: Coexistence and Violence in the German, Habsburg, Russian, and Ottoman Borderlands*, edited by Omar Bartov and Eric D. Weitz (Bloomington: Indiana University Press, 2013), pp. 209–27.

Nenartovič, Tomaš. *Kaiserlich-russische, deutsche, polnische, litauische, belarussische und sowjetische kartographische Vorstellungen und territoriale Projekte zur Kontaktregion von Wilna 1795–1939* (Munich: Collegium Carolinum, 2016).

Nol'de, Boris Emmanuilovich. *Iurii Samarin i ego vremia* (Paris: Société anonyme imprimerie de Navarre, 1926).

Novikova, Irina. 'Imagining Africa and Blackness in the Russian Empire: From Extra-Textual *arapka* and Distant Cannibals to Dahomey Amazon Shows—Live in Moscow and Riga'. *Social Identities* 19, no. 5 (2013), pp. 571–91.

Oberländer, Erwin and Kristine Wohlfart, eds. *Riga: Porträt einer Vielvölkerstadt am Rande des Zarenreiches 1857–1914* (Paderborn: Ferdinand Schöningh, 2004).

Von Oertzen, Christine, Maria Rentetzi, and Elizabeth S. Watkins. 'Finding Science in Surprising Places: Gender and the Geography of Scientific Knowledge. Introduction to "Beyond the Academy: Histories of Gender and Knowledge"'. *Centaurus* 55, no. 2 (2013), pp. 73–80.

Von Oertzen, Christine. 'Prussian Census Box: Moving and Freezing Data'. In *Boxes: A Field Guide*, edited by Susanne Bauer, Martina Schlünder, and Maria Rentetzi (Manchester: Mattering Press, 2020), pp. 473–9.

O'Keeffe, Brigid. 'Gypsies as a Litmus Test for Rational, Tolerant Rule: Fin-de-siècle Russian Ethnographers Confront the Comparative History of Roma in Europe'. *International Journal of Comparative and Applied Criminal Justice* 38, no. 2 (2014), pp. 109–31.

O'Neill, Kelly. *Claiming Crimea: A History of Catherine the Great's Southern Empire* (New Haven: Yale University Press, 2017).

Ophir, Adi and Steven Shapin. 'The Place of Knowledge: A Methodological Survey'. *Science in Context* 4, no. 1 (1991), pp. 3–21.

Opitz, Donald L., Staffan Bergwick, and Brigitte van Tiggelen, eds. *Domesticity in the Making of Modern Science* (Basingstoke: Palgrave Macmillan, 2016).

Orlova, Galina, 'Karty dlia slepykh: politika i politizatsiia zreniia v stalinskuyu epokhu'. In *Visual'naiia antrpologiia: rezhimy vidimosti pri sotsializme*, edited by E. Iarskoi Smirnovoi and P. Romanova (Moscow: OOO "Variant", TsSPGI, 2009), pp. 57–105.

Osowska, Anna and Dariusz Przybytek. 'Thematic Maps in Eugeniusz Romer's Geographical and Statistical Atlas of Poland from 1916—The Historical and Methodological Perspective (on the 100-Year Anniversary of the Publication)'. *Polish Cartographical Review* 48, no. 2 (2016), pp. 77–86.

Pääbo, Heiko. 'Constructing Historical Space: Estonia's Transition from the Russian Civilization to the Baltic Sea Region'. *Journal of Baltic Studies* 45, no. 2 (2014), pp. 187–205.

Pae, Taavi. 'Eesti-Vene piir Vasknarvas—Euroopa Liit Narva jõe paremkaldal'. *Akadeemia* 10 (2014), pp. 1731–46.

Pae, Taavi, ed. *Eesti Rahvusatlas/Estonian National Atlas* (Tartu: Regio, 2019).

Paert, Irina. '"Two or Twenty Million?": The Languages of Official Statistics and Religious Dissent in Imperial Russia'. *Ab Imperio* no. 3 (2006), pp. 75–98.

Paert, Irina. 'The Russian Imperial Practices of Language and Religion in the Classroom in the Baltic Provinces: The Case of the Orthodox Schools in Livland (1880–1914)'. *Procedia—Social and Behavioral Sciences* 236 (2016), pp. 362–7.

Palsky, Gilles. 'The Debate on the Standardization of Statistical Maps and Diagrams (1857–1901): Elements for the History of Graphical Language'. *Cybergeo: European Journal of Geography* (1999), http://cybergeo.revues.org/148.

Palsky, Gilles. 'Emmanuel de Martonne and the Ethnographical Cartography of Central Europe (1917–1920)'. *Imago Mundi* 54 (2002), pp. 111–19.

Palsky, Gilles. 'Connections and Exchanges in European Thematic Cartography: The Case of 19th Century Choropleth Maps'. *Belgeo* nos. 3–4 (2008), pp. 413–26.

Patriarca, Silvana. *Numbers and Nationhood: Writing Statistics in Nineteenth-Century Italy* (Cambridge: Cambridge University Press, 1996).

Päts, Peeter. *Eesti Vabariigi ja Läti Vabariigi vaheline piir*. Master's dissertation, University of Tartu, 1927. University of Tartu Library, R Diss.Tart.272079.

Pearson, Karen S. 'The Nineteenth-Century Colour Revolution: Maps in Geographical Journals'. *Imago Mundi* 32 (1980), pp. 9–20.

Pedley, Mary Sponberg. *The Commerce of Cartography: Making and Marketing Maps in Eighteenth-Century France and England* (Chicago: University of Chicago Press, 2005).

Peluso, Nancy Lee. 'Whose Woods Are These? Counter-Mapping Forest Territories in Kalimantan, Indonesia'. *Antipode* 27, no. 4 (1995), pp. 383–406.

Petronis, Vytautas, *Constructing Lithuania: Ethnic Mapping in Tsarist Russia, ca. 1800–1914* (Stockholm: Södertörn Doctoral Dissertations, 2007).

Pi, C. T. 'Beyond the Isogloss: Isographs in Dialect Topography'. *The Canadian Journal of Linguistics/La revue canadienne de linguistique* 51 (2006), pp. 177–84.

Pickles, John. *A History of Spaces: Cartographic Reason, Mapping and the Geo-Coded World* (London: Routledge, 2004).

Piirimäe, Pärtel. 'The Baltic'. In *European Regions and Boundaries: A Conceptual History*, edited by Diana Mishkova and Balázs Trencsényi (New York: Berghahn Books, 2017), pp. 57–78.

Von Pistohlkors, Gert. '"Russifizierung" und die Grundlagen der deutsch-baltischen Russophobie'. *Zeitschrift für Ostforschung* 25 (1976), pp. 618–31.

Plakans, Andrejs and Charles Wetherel. 'The 1881 Census in the Russian Baltic Provinces: An Inventory and an Assessment'. *The History of the Family* 9, no. 1 (2004), pp. 47–61.

Plath, Ulrike. '"Euroopa viimased metslased": Eestlased saksa koloniaaldiskrusis 1770–1870'. In *Rahvuskultuur ja tema teised*, edited by Rein Undusk (Tallinn-Tartu: Underi ja Tuglase Kirjanduskeskus, 2006), pp. 37–65.

Plath, Ulrike. 'Rändrahnud ja majaisandad. Baltisakslaste suhe maaga ja kodutunde teke 19. sajandil'. *Vikerkaar* nos. 7–8 (2008), pp. 113–22.

Plath, Ulrike. '*Heimat*: Rethinking Baltic German Spaces of Belonging'. *Kunstiteaduslikke uurimusi* 23, nos. 3–4 (2014), pp. 55–78.

Põldvee, Aivar. 'Russwurmist ja "Eibofolkest", tõlke ilmumise puhul'. *Keel ja Kirjandus* 4 (2016), pp. 312–17.

Porter, Theodore M. *The Rise of Statistical Thinking 1820–1900* (Princeton: Princeton University Press, 1986).

Porter, Theodore M. *Trust in Numbers: The Pursuit of Objectivity in Science and Public Life* (Princeton: Princeton University Press, 1996).

Porter, Theodore. 'Revenge of the Humdrum: Bureaucracy as Profession and as a Site of Science'. *Journal for the History of Knowledge* 1, no. 1 (2020), pp. 1–5.

Porter-Szűcs, Brian. 'The Birth of the Polak-Katolik'. *Sprawy Narodowościowe* 49 (2017), pp. 1–12.

Postnikov, Alexei *Russia in Maps: A History of the Geographical Study and Cartography of the Country* (Moscow: Nash Dom—L'Age d'Homme, 1996).

Postnikov, Alexey V. 'Maps for Ordinary Consumers versus Maps for the Military: Double Standards of Map Accuracy in Soviet Cartography, 1917–1991'. *Cartography and Geographic Information Science* 29, no. 3 (2002), pp. 243–60.

Potepalov, S. G. 'Puteshestvie P. I. Keppena po slavianskim zemliam'. In *Iz istorii russko-slavianskikh literaturnikh sviazei XIX v.* (Moscow; Leningrad: Akademiia nauk SSSR, 1963), pp. 5–22.

Pourchier-Plasseraud, Suzanne. *Arts and a Nation: The Role of Visual Arts and Artists in the Making of the Latvian Identity, 1905–1940* (Leiden: Brill Rodopi, 2015).

Powell, Richard C. 'Geographies of Science: Histories, Localities, Practices, Future'. *Process in Human Geography* 31, no. 3 (2007), pp. 309–29.

Prokopovych, Markian, Carl Bethke, and Tamara Scheer, eds. *Language Diversity in the Late Habsburg Empire* (Leiden: Brill, 2019).

Prott, Volker. *The Politics of Self-Determination: Remaking Territories and National Identities in Europe, 1917–1923* (Oxford: Oxford University Press, 2016).

Psianchin, Aibulat Valievich. *Istoriia etnicheskoi kartografii v Rossii (do 30-x gg. XX v.)*. PhD dissertation, Institut istorii estestvoznaniia i tekhniki im. S. I. Vavilova, 2004.

Psianchin, Aibulat Valievich. *Komissiia po izucheniiu plemennogo sostava naseleniia: ot etnokartografii k perepisi naseleniia* (Ufa: Ufimskii nauchnyi tsentr RAN, 2010).

Pugliano, Valentina. 'Artisanal Writing or Natural Historical Paperwork?' *Isis* 103, no. 4 (2012), pp. 716–26.

Purs, Aldis. 'The Price of Free Lunches: Making the Frontier Latvian in the Interwar Years. *Global Review of Ethnopolitics* 1, no. 4 (2002), pp. 60–73.

Purs, Aldis. '"Unsatisfactory National Identity": School Inspectors, Education and National Identity in Interwar Latvia.' *Journal of Baltic Studies* 35, no. 2 (2004), pp. 97–125.

Purs, Aldis. 'Working towards "An Unforeseen Miracle" Redux: Latvian Refugees in Vladivostok, 1918–1920, and in Latvia, 1943–1944'. *Contemporary European History* 16, no. 4 (2007), pp. 479–94.

Purs, Aldis. 'Orphaned Testimonies: The Place of Displaced Children in Independent Latvia, 1918–26'. In *Displaced Children in Russia and Eastern Europe, 1915–1953*, edited by Nick Baron (Leiden: Brill, 2017), pp. 40–69.

Pushaw, Bart. 'The Visual Whitening of Estonians'. *Eurozine*, 20 November 2020, https://www.eurozine.com/the-visual-whitening-of-estonians/. Originally published in Estonian in *Vikerkaar*, Oct. 2020.

Pycior, Helena M. *Creative Couples in the Sciences* (New Brunswick, NJ: Rutgers University Press, 1996).

Pyle, Emily E. 'Peasant Strategies for Obtaining State Aid: A Study of Petitions during World War I'. *Russian History/Histoire Russe* 24, nos. 1–2 (1997), pp. 41–64.

Rabow-Edling, Susanna. *Slavophile Thought and the Politics of Cultural Nationalism* (Albany: State University of New York Press, 2006).

Rabow-Edling, Susanna. *Married to the Empire: Three Governors' Wives in Russian America 1829–1864* (Fairbanks: University of Alaska Press, 2015).

Rainbow, David, ed. *Ideologies of Race: Imperial Russia and the Soviet Union in Global Context* (Montreal: McGill-Queen's University Press, 2019).

Ramaswamy, Sumathi. *Terrestrial Lessons: The Conquest of the World as Globe* (Chicago: Chicago University Press, 2017).

Randeraad, Nico. *States and Statistics in the Nineteenth Century: Europe by Numbers* (Manchester: Manchester University Press, 2010).

Randeraad, Nico. 'The International Statistical Congress (1853–1876): Knowledge Transfers and Their Limits'. *European History Quarterly* 41, no. 1 (2011), pp. 50–65.

Raun, Toivo U. 'The Revolution of 1905 in the Baltic Provinces and Finland'. *Slavic Review* 43, no.3 (1984), pp. 453–67.

Raun, Toivo U. 'Estonian Emigration within the Russian Empire, 1860–1917'. *Journal of Baltic Studies* 17, no. 4 (1986), pp. 350–63.

Raun, Toivo U. 'The Petseri Region of the Republic of Estonia'. *Jahrbücher Für Geschichte Osteuropas* 39, no. 4 (1991), pp. 514–32.

Raun, Toivo. 'Violence and Activism in the Baltic Provinces during the Revolution of 1905'. *Acta Historica Tallinnensia* 10 (2006), pp. 48–9.

Raun, Toivo. 'The All-Estonian Congress in Tartu, November 1905: A Reassessment'. *Journal of Baltic Studies* 38, no. 4 (2007), pp. 383–400.

Raun, Toivo. 'Riga and Tallinn in the Late Tsarist Era: Multiethnicity and Social and Political Change'. *Hungarian Studies* 29, nos. 1–2 (2015), pp. 63–72.

Raun, Toivo U. 'Literacy in the Russian Empire in the Late 19th Century: The Striking Case of the Baltic Provinces'. *Act Historica Tallinnensia* 23 (2017), pp. 65–77.

Rebane, Leida. *Usuwahetuslik liikumine Põhja Eestis eriti kuberner Polivanovi ajal 1883–1885*, 2 vols. Master's dissertation, Tartu Ülikool, 1932.

Richter, Klaus. *Fragmentation in East Central Europe Poland and the Baltics, 1915–1929* (Oxford: Oxford University Press, 2020).

Richter, Klaus and Heidi Hein-Kircher. 'Too Small to Succeed? East Central Europe and the Historical Study of State Assessment'. *Zeitschrift für Ostforschung/Journal of East Central European Studies* 69, no. 4 (2020), pp. 519–32.

Riegg, Stephan Badalyan. *Russia's Entangled Embrace: The Tsarist Empire and the Armenians, 1801–1914* (Ithaca: Cornell University Press, 2020).

Roasto, Margo, 'The Political Debate about the Land Question in the Estonian Area of the Baltic Provinces, 1905–1914'. *Journal of Baltic Studies* 51, no. 4 (2020), pp. 611–30.

Rodogno, Davide, Bernhard Struck, and Jakob Vogel, eds. *Shaping the Transnational Sphere: Experts, Networks and Issues from the 1840s to the 1930s* (New York: Berghahn Books, 2015).

Romaniello, Matthew P. and Tricia Starks, eds. *Russian History through the Senses: From 1700 to the Present* (London: Bloomsbury: 2016).

Rose, Gillian. *Feminism and Geography: The Limits of Geographical Knowledge* (Cambridge: Cambridge University Press, 1993).

Rosenberg, Tiit. 'From National Territorial Autonomy to Independence of Estonia: The War and Revolution in the Baltic Region, 1914–1917'. In *Imperiology: From Empirical Knowledge to Discussing the Russian Empire*, edited by K. Matsuzato (Sapporo, Japan: Hokkaido University, Slavic Research Center, 2007), pp. 199–222.

Rowley, Alison. *Open Letters: Russian Popular Culture and the Picture Postcard 1880–1922* (Toronto: University of Toronto Press, 2013).

Rowney, Don K. 'Imperial Russian Officialdom during Modernization'. In *Russian Bureaucracy and the State: Officialdom from Alexander III to Vladimir Putin*, edited by Don K. Rowney and Eugene Huskey (Basingstoke: Palgrave Macmillan, 2009), pp. 26–45.

Rudling, Per A. *The Rise and Fall of Belarusian Nationalism, 1906–1931* (Pittsburgh: University of Pittsburgh Press, 2015).

Rudwick, Martin J. S. 'The Emergence of a Visual Language for Geological Science 1760–1840'. *History of Science* 14 (1976), pp. 149–95.

Ruud, Charles A. *Russian Entrepreneur: Publisher Ivan Sytin of Moscow 1851–1934* (Montreal: McGill Queen's University Press, 1990).

Ruud, Charles A. *Fighting Words: Imperial Censorship and the Russian Press, 1804–1906* (Toronto: University of Toronto Press, 2009).

Ryan, Daniel Cavender. *The Tsar's Faith: Conversion, Religious Politics, and Peasant Protest in Imperial Russia's Baltic Periphery, 1845–1870s*. PhD dissertation, UCLA, 2008.

Sach, Maike. 'Symbols, Conventions, and Practices: Visual Representation of Ethnographic Knowledge on Siberia in Early Modern Maps and Report'. In *An Empire of Others: Creating Ethnographic Knowledge in Imperial Russia and the USSR*, edited by Roland Cvetkovski and Alexis Hofmeister (Budapest: Central European University Press, 2013), pp. 171–210.

Safronovas, Vasilijus. 'Zum Wandel des räumlichen Begriffs "Litauen" im deutschsprachigen Diskurs während und nach dem Ersten Weltkrieg'. *Forschungen zur Baltischen Geschichte* 10 (2015), pp. 109–35.

Safronovas, Vasilijus. *The Creation of National Spaces in a Pluricultural Region: The Case of Prussian Lithuania* (Boston: Academic Studies Press, 2016).

Salmond, Wendy R. *Arts and Crafts in Late Imperial Russia: Reviving the Kustar Art Industries, 1870–1917* (Cambridge: Cambridge University Press, 1996).

Samborska-Kukuca, Dorota. '1863. gada Poļu sacelšanās notikumi Latgalē aculiecinieku liecībās un atmiņās'. *Latvijas Vēstures Institūta Žurnāls* 85 (2012), pp. 50–73.

Sanborn, Joshua. *Imperial Apocalypse: The Great War and the Destruction of the Russian Empire* (Oxford: Oxford University Press, 2014).

Sanborn, Joshua. 'War of Decolonization: The Russian Empire in the Great War'. In *The Empire and Nationalism at War*, edited by Eric Lohr, Vera Tolz, Alexander Semyonov, and Mark von Hagen (Bloomington: Slavica, 2014), pp. 49–72.

Ščerbinskis, Valters. 'Lauri Latvians: Between Latvia, Estonia, and Russia'. In *Narva und die Ostseeregion/Narva and the Baltic Sea Region*, edited by Karsten Brüggemann (Narva: University of Tartu Narva College, 2004), pp. 305–9.

Schenk, Frithjof Benjamin. *Russlands Fahrt in die Moderne: Mobilität und sozialer Raum im Eisenbahnzeitalter. Quellen und Studien zur Geschichte des östlichen Europa* (Stuttgart: Steiner, 2014).

Schulten, Susan. 'Emma Willard and the Graphic Foundations of American History'. *Journal of Historical Geography* 33 (2007), pp. 542–64.

Schulten, Susan. *Mapping the Nation: History and Cartography in Nineteenth-Century America* (Chicago: University of Chicago Press, 2012).

Schulten, Susan. 'Map Drawing, Graphic Literacy, and Pedagogy in the Early Republic'. *History of Education Quarterly* 57 (2017), pp. 185–220.

Schulz, Thomas. 'The Statistical Atlases of the Baltic States 1918–1940: The First National Atlases of the Three Newly Independent Countries'. *The Cartographic Journal* 55, no. 2 (2018), pp. 187–95.

Schunka, Alexander. 'Cartographic Commentaries and Map Descriptions of the 19th Century as Historical Sources'. In *History of Cartography. Lecture Notes in Geoinformation and Cartography*, edited by E. Liebenberg, P. Collier, and Z. Török (Berlin; Heidelberg: Springer, 2014), pp. 277–94.

Scott, James C. *Seeing Like a State: How Certain Schemes to Improve the Human Condition Have Failed* (New Haven: Yale University Press, 1998).

Scott, Joan W. 'Gender: A Useful Category of Historical Analysis'. *The American Historical Review* 91, no. 5 (1986), pp. 1053–75.

Sdvižkov, Denis. 'Russian and German Ideas of the West in the Long Nineteenth Century: Entanglements of Spatial Identities'. In *Germany and 'the West'. The History of a Modern Concept*, edited by Riccardo Bavaj and Martina Steber (New York: Berghahn, 2015), pp. 97–110.

Secord, Anne. 'Corresponding Interests: Artisans and Gentlemen in Nineteenth-Century Natural History'. *British Journal for the History of Science* 27 (1994), pp. 383–408.

Seegel, Steven. *Mapping Europe's Borderlands: Russian Cartography in the Age of Empire* (Chicago: University of Chicago Press, 2012).

Seegel, Steven. *Map Men: Transnational Lives and Deaths of Geographers in the Making of East Central Europe* (Chicago: University of Chicago Press, 2018).

Semyonov, Alexander. '"The Real and Live Ethnographic Map of Russia": The Russian Empire in the Mirror of the State Duma'. In *Empire Speaks Out: Languages of Rationalization and Self-Description in the Russian Empire*, edited by Ilya Gerasimov, Jan Kusber, and Alexander Semyonov (Leiden; Boston: Brill, 2009), pp. 191–228.

Semyonov, Alexander M. 'Imperial Parliament for a Hybrid Empire: Representative Experiments in the Early 20th-Century Russian Empire'. *Journal of Eurasian Studies* 11, no. 1 (2020), pp. 30–9.

Shor, Tat'iana. 'Pravoslavnye sholy v Pribaltiiskikh guberniakh v 1870-e gg.: obzor zakonodatel'stva i arkhivnykh istochnikov'. In *Pravoslavie v Pribaltike: religiia, politika, obrazovanie 1840-e–1930-e gg.*, edited by Irina Paert (Tartu: University of Tartu Press, 2018), pp. 97–110.

Siegelbaum, Lewis H. & Leslie Page Moch, *Broad is My Native Land: Repertoires and Regimes of Migration in Russia's Twentieth Century* (Ithaca: Cornell University Press, 2015).

Sinanoglou, Penny. *Partitioning Palestine: British Policymaking at the End of Empire* (Chicago: University of Chicago Press, 2019).

Slezkine, Yuri. *Arctic Mirrors: Russia and the Small Peoples of the North* (Ithaca: Cornell University Press, 1994).

Slezkine, Yuri. 'Naturalists Versus Nations: Eighteenth-Century Russian Scholars Confront Ethnic Diversity'. *Representations* 47 (1994), pp. 170–95.

Slocum, John W. 'Who, and When, Were the Inorodtsy? The Evolution of the Category of "Aliens" in Imperial Russia'. *The Russian Review* 57, no. 2 (1998), pp. 173–90.

Smith, Alison K. 'Documents, Migration, and Governance in Imperial Russian Towns'. In *Migration Policies and Materialities of Identification in European Cities: Papers and Gates,*

1500–1930s, edited by Hilde Greefs and Anne Winter (New York: Routledge, 2019), pp. 132–50.

Smith, Jeremy, ed. *Beyond the Limits: The Concept of Space in Russian History and Culture* (Helsinki: Suomen Historiallinen Seura, 1999).

Smith-Peter, Susan. 'Defining the Russian People: Konstantin Arsen'ev and Russian Statistics before 1861'. *History of Science* 45, no. 1 (2007), pp. 47–64.

Smith-Peter, Susan. 'Bringing the Provinces into Focus: Subnational Spaces in the Recent Historiography of Russia'. *Kritika: Explorations in Russian and Eurasian History* 12, no. 4 (2011), pp. 835–48.

Smith-Peter, Susan. 'Making Empty Provinces: Eighteenth-Century Enlightenment Regionalism in Russian Provincial Journals'. *REGION: Regional Studies of Russia, Eastern Europe, and Central Asia* 4, no. 1 (2015), pp. 7–29.

Smith-Peter, Susan. *Imagining Russian Regions: Subnational Identity and Civil Society in Nineteenth-Century Russia* (Leiden; Boston: Brill, 2017).

Snyder, Timothy. *The Reconstruction of Nations: Poland, Ukraine, Lithuania, Belarus, 1569–1999* (New Haven: Yale University Press, 2008).

Soboļeva, Liene, ed. *Sava ceļa gājējs: Matīss Siliņš* (Riga: Jumava, 2019).

Sommerlat, Anne. 'L'élément national dans les descriptions de Mitau autour de 1800'. *Revue Germanique Internationale* 11 (2010), pp. 103–24.

Sommerlat-Michas, Anne. *Das Baltikum also Konstrukt (18.–19 Jahrhunder): Von einer Kolonialwahrnehmung zu einem nationalen Diskurs* (Wurzburg: Konigshausen & Naumann, 2015).

Sooman, Imbi, Jesma McFarlane,Valdis Tēraudkalns, and Stefan Donecker. 'From the Port of Ventspils to Great Courland Bay: The Couronian Colony on Tobago in Past and Present'. *Journal of Baltic Studies* 44, no. 4 (2013), pp. 503–26.

Soroko, Svetlana. *Vitebskie gubernskie vedomosti (Ofitial'naia chast' i neofitsial'naia chast')* (Moscow; Novopolotsk: Nauchnyi tsentr slaviano-germanskikh issledovanii ISl RAN, 2004).

Stachel, Peter. 'Die Harmonisierung national-politischer Gegensätze und die Anfänge der Ethnographie in Österreich'. In *Geschichte der österreichischen Humanwissenschaften*, edited by Karl Achim (Vienna: Passagen, 2002), pp. 323–67.

Staliūnas, Darius. *Making Russians: Meaning and Practice of Russification in Lithuania and Belarus after 1863* (Amsterdam; New York: Rodopi, 2007).

Staliūnas, Darius. 'National Census in the Service of the Russian Empire: The Western Borderlands in the Mid-Nineteenth Century, 1830–1870'. In *Defining Self: Essays on Emergent Identities in Russia Seventeenth to Nineteenth Centuries*, edited by Michael Branch (Helsinki: Finnish Literature Society, 2009), pp. 435–48.

Staliūnas, Darius, ed. *Spatial Concepts of Lithuania in the Long Nineteenth Century* (Boston: Academic Studies Press, 2016).

Stanziani, Alessandro. 'Statisticens, zemstva et État dans la Russie des années 1880'. *Cahiers du monde russe et soviétique* 32, no. 4 (1991), pp. 445–67.

Steinberg, Mark, *Moral Communities: The Culture of Class Relations in the Russian Printing Industry, 1867–1907* (Berkeley: University of California Press, 1992).

Stergar, Rok and Tamara Scheer. 'Ethnic Boxes: The Unintended Consequences of Habsburg Bureaucratic Classification'. *Nationalities Papers* 46, no. 4 (2018), pp. 575–91.

Stinkule, Sanita. 'Skolotājam, literātam un tulkotājam, grāmatizdevējam, kartogrāfam un etnogrāfam Matīsam Siliņam—150'. *Latvijas Zinātņu Akadēmijas Vēstis* 68, nos. 5–6 (2014), pp. 34–52.

Stinkule, Sanita, ed. *Indivīds. Vēsture. Nācija. Latviešu etnogrāfiskajai izstādei–120: rakstu krājums* (Riga: Latvijas Nacionālais Vēstures Muzejs, 2017).

Stinkule, Sanita, Toms Ķikuts, and Jānis Ciglis. *Latviešu etnogrāfiskā izstāde 1896/Latvian Ethnographic Exhibition 1896* (Riga: Neptuns, 2016).

Stites, Richard. *The Women's Liberation Movement in Russia: Feminism, Nihilism, and Bolshevism, 1860–1930* (Princeton: Princeton University Press, 1978).

Stoler, Ann Laura and Carole McGranahan. 'Introduction: Refiguring Imperial Terrains'. In *Imperial Formations*, edited by Ann Laura Stoler, Carole McGranahan, and Peter C. Perdue (Santa Fe, NM: School for Advanced Research Press, 2007), pp. 3–42.

Štrāle, Aina. 'Dobeles apkārtnes mācītājmuižu bibliotēku un arhīvu liktenis 1905.gada revolūcijā: Augusta Bīlenšteina liecība Dobeles evaņģēliski luteriskās baznīcas grāmatā'. In *Starptautiska konference "Informācija, revolūcija, reakcija: 1905–2005": 2005. gada 23.-25. novembris, Rīga (Latvija): Materiālu krājums* (Riga: Latvijas Nacionālā bibliotēka: Latvijas Bibliotekāru biedrība, 2005), pp. 76–82.

Štrauhmanis, Jānis. *Matīss Siliņš: Pirmais latviešu kartogrāfs* (Riga: Latvijas Universitātē; Latvijas Kultūras fonds, 1994).

Štrauhmanis, Jānis. *Latvijas kartogrāfijas vēsture: bibliogrāfiskais rādītājs* (Riga: Mācību grāmata, 1994).

Štrauhmanis, Jānis. *Latvijas kartogrāfijas vēsture no XIII gadsimta līdz XX gadsimta 90. gadu sākumam*. PhD dissertation, Latvijas Universitāte, 1997.

Štrauhmanis, Jānis. 'Matīsa Siliņa ietekme uz Latvijas nacionālās kartogrāfijas veidošanos'. *Scientific Journal of Riga Technical University* 9 (2012), pp. 16–19.

Strods, Hinrisch [Heinrihs]. 'Rozruchy chłopskie we wschodniej Łotwie i Kurlandii w roku 1863 roku'. *Zapiski Historyczne* 55 (1990), pp. 105–20.

Sukhova, Nataliia G. 'Eshche raz o predystorii Russkogo geograficheskogo obschestva'. *Izvestiia RGO* 122 (1990), pp. 403–8.

Sukhova, Natalia G. 'Petr Ivanovich Keppen kak geograf: k 200-letiu so dnia rozhdeniia'. *Izvestiia RGO* 125, no. 5 (1993), pp. 1–11.

Sukhova, N. G. 'Russkoe geograficheskoe obshchestvo i Akademiia nauk v XIX v'. In *Akademiia nauk v istorii kultury Rossii XVIII-XX vekov*, edited by Zh. I. Alferov (St. Petersburg: Nauka, 2010), pp. 560–613.

Sunderland, Willard. 'The "Colonization Question": Visions of Colonization in Late Imperial Russia'. *Jahrbücher für Geschichte Osteuropas* 28, no. 2 (2000), pp. 210–32.

Sunderland, Willard. 'Imperial Space: Territorial Thought and Practice in the Eighteenth Century'. In *Russian Empire: Space, People, Power, 1700–1930*, edited by Jane Burbank, Mark von Hagen, and Anatolyi Remnev (Bloomington: Indiana University Press, 2007), pp. 37–55.

Sunderland, Willard. *The Baron's Cloak: A History of the Russian Empire in War and Revolution* (Ithaca: Cornell University Press, 2014).

Surgailis, Gintautas. 'Boi Litovskoi armii s Krasnoi armiei v Ilukstskom uezde v 1919 godu'. *Vēsture: avoti un cilvēki* 22 (2019), pp. 281–90.

Surman, Jan. 'Paris-Wien-St. Petersburg oder Alger-Brno-Charkiv? Wissenstransfer und die "composite states"'. *Quaestio Rossica* no. 3 (2015), pp. 98–118.

Talts, Mait. 'The Role of Popular Science Literature in Shaping Estonians' World Outlook'. *Istoriko-Biologicheskie Issledovaniia* 5, no. 2 (2013), pp. 59–83.

Tammiksaar, Erki. 'The Contribution of Karl Ernst von Baer to the Study of ethnic Minorities in the Russian Empire, 1819–1878. A Case Study'. In *Defining the Self: Essays on Emergent Identities in Russia Seventeenth to Nineteenth Centuries*, edited by Michael Branch (Helsinki: Finnish Literature Society, 2009), pp. 139–51.

Tammiksaar, Erki. 'Liivimaa Üldkasulik ja Ökonoomiline Sotsieteet ning teadmiste areng Eesti ala kõrgussuhete kohta 19. sajandil'. *Tartu Ülikooli ajaloo küsimusi* 41 (2013), pp. 130–45.

Tammiksaar, Erki. 'Das Verhältnis von Wissenschaft und Nationalität im Russländischen Imperium im 19. Jahrhundert am Beispiel Karl Ernst von Baers'. In *Deutsch als Wissenschaftssprache im Ostseeraum—Geschichte und Gegenwart*, edited by Michael Prinz and Jarmo Korhonen (Frankfurt am Main: Peter Lang, 2016), pp. 254–72.

Tammiksaar, Erki and Natalia G. Sukhova. 'Avgust Peterman v Rossii'. *Voprosy istorii estestvoznaniia i tekhniki* 1 (2015), pp. 46–75.

Tēraudkalns, Valdis. 'Latviešu izceļošana uz Brazīliju: cēloņi un ideoloģija'. *Religiski-filozofiski raksti* 9, no. 1 (2005), pp. 167–84.

Thaden, Edward C. 'N. A. Manaseins Senatorenrevision in Livland und Kurland während der Zeit von 1882 bis 1883'. *Jahrbücher Für Geschichte Osteuropas* 17, no. 1 (1969), pp. 45–58.

Thaden, Edward C. 'Samarin's "Okrainy Rossii" and Official Policy in the Baltic Provinces'. *The Russian Review* 33, no. 4 (1974), pp. 405–15.

Thaden, Edward C., ed. *Russification in the Baltic Provinces and Finland 1855–1914* (Princeton: Princeton University Press, 1981).

Thaden, Edward C. 'Iurii Fedorovich Samarin and Baltic History'. *Journal of Baltic Studies* 17 no. 4 (1986), pp. 321–8.

Tokarev, S. A. *Istoriia russkoi etnografii (Dooktiabr'skii period)* (Moscow: Akademiia Nauk SSSR; Institut ethnografii im. N. N. Mikluho-Maklaia, 1966).

Tøllefsen Trond O. and James M. White. 'Navigating an Orthodox Conversion: Community, Environment, and Religion on the Island of Ruhnu, 1866–7'. *Scandinavian Journal of History* 46, no.5 (2021): 642–664.

Tolz, Verz. 'Orientalism, Nationalism, and Ethnic Diversity in Late Imperial Russia'. *The Historical Journal* 48, no. 1 (2005), pp. 127–50.

Tolz, Verz. *Russia's Own Orient: The Politics of Identity and Oriental Studies in the Late Imperial and Early Soviet Periods* (Oxford: Oxford University Press, 2011).

Tolz, Vera. 'Constructing Race, Ethnicity and Nationhood in Imperial Russia: Issues and Misconceptions'. In *Ideologies of Race: Imperial Russia and the Soviet Union in Global Context*, edited by David Rainbow (Montreal: McGill-Queen's University Press, 2019), pp. 29–58.

Tooming, Kadri. 'Nälg ja ränne: 1868.–1869. aasta väljarändeliikumine Eestimaa kubermangus'. *Ajalooline Ajakiri* 156, no. 2 (2016), pp. 185–214.

Trumpa, Edmundas, '"Karte des lettischen Sprachgebiets" (1881)—pirmā latviešu valodas dialektu karte'. *Baltu filoloģija* 19, nos. 1–2 (2009): pp. 79–102.

Tyner, Judith A. *Stitching the World: Embroidered Maps and Women's Geographical Education* (Burlington: Ashgate, 2015).

Ureña Valerio, Lenny A. *Colonial Fantasies, Imperial Realities: Race Science and the Making of Polishness on the Fringes of the German Empire, 1840–1920* (Athens: Ohio University Press, 2019.

Valkova, Olga. 'The Conquest of Science: Women and Science in Russia, 1860–1940'. *Osiris* 23 (2008), pp. 136–65.

Vaughan, Laura. *Mapping Society: Spatial Dimensions of Social Cartography* (London: UCL Press, 2018).

Varep, Endel. *Jooni Eesti kartograafia ajaloost* (Tallinn: Eesti NSV Teaduste Akadeemia, 1960).

Vaskul, Anastasiia I. 'Etnograficheskaia programma Russkogo Geograficheskogo Obshchestva'. In *Russkii fol'klor XXXVI: Materialy i issledovaniia* (St. Petersburg: Nauka, 2012), pp. 460–71.

Veiss-Vendt, Anton. 'Komu prinadlezhit Narva? K voprosu o territorial'no-administrativnoi prinadlazhnosti goroda 1858–1917 gg.'. In *Baltiiskii Arkhiv: Russkaia Kul'tura v Pribaltike*, edited by Irina Belovrovtseva (Tallinn: Avenarius, 1997), pp. 22–47.

Venken, Machteld. *Peripheries at the Centre: Borderland Schooling in Interwar Europe* (New York: Berghahn, 2021).

Vermeulen, Hans F. *Before Boas: The Genesis of Ethnography and Ethnology in the German Enlightenment* (Lincoln: University of Nebraska Press, 2015).

Verner, Andrew. 'Discursive Strategies in the 1905 Revolution: Peasant Petitions from Vladimir Province'. *The Russian Review* 54, no. 1 (1995), pp. 65–90.

Vetter, Jeremy. 'Introduction: Lay Participation in the History of Scientific Observation'. *Science in Context* 24 (2011), pp. 127–41.

Vishlenkova, Elena. *Vizual'noe narodovedenie imperii, ili 'Uvidet' russkogo dano ne kazhdomu'* (Moscow: Novoe literaturnoe obozrenie, 2011).

Weber, Karen. 'Religion and Law in the Russian Empire: Lutheran Pastors on Trial, 1860–1917'. PhD dissertation, New York University, 2013.

Weeks, Theodore R. 'Defining Us and Them: Poles and Russians in the "Western Provinces," 1863–1914'. *Slavic Review* 53, no. 1 (1994), pp. 26–40.

Weeks, Theodore R. 'Religion, Nationality, or Politics: Catholicism in the Russian Empire, 1863–1905'. *Journal of Eurasian Studies* 2 (2011), pp. 52–9.

Weeks, Theodore R. *Vilnius between Nations, 1795–2000* (DeKalb: North Illinois University Press, 2015).

Weske, M. 'Über die Witebskischen Esten'. *Verhandlungen der Gelehrten Estnischen Gesellschaft zu Dorpat* 8, no. 4 (1877), pp. 29–33.

Wezel, Katja. 'Introduction: German Community—German Nationality? Baltic German Perceptions of Belonging in the Nineteenth and Twentieth Century'. *Journal of Baltic Studies* 48, no. 1 (2017), pp. 1–11.

Whelan, H. W. 'The Debate on Women's Education in the Baltic Provinces, 1850 to 1905'. In *Bevölkerungsverschiebungen und sozialer Wandel in den baltischen Provinzen Russlands 1850–1914/Population Shifts and Social Change in Russia's Baltic Provinces 1850–1914*, edited by Gert von Pistohlkors, Andrejs Plakans, and Paul Kaegbein (Lüneburg: Institut Nordostdeutsches Kulturwerk, 1995), pp. 163–80.

Whelan, H. W. *Adapting to Modernity: Family, Caste and Capitalism among the Baltic German Nobility* (Cologne: Böhlau, 1999).

Wilhelmi, Anja. 'Hauswirtschaft fur Frauen? Erfahrungswelten von Frau der deutschen Minderheit in den Ostseeprovinzen des Russlandischen Reiches um 1900'. *L'homme* 22, no. 2 (2011), pp. 121–9.

Wilhelmi, Anja. 'Mädchenbildung: Bildungspraktiken und diskurse in der deutschbaltischen Bevölkerungsgruppe (1850–1900)'. In *Bildungskonzepte und Bildungsinitiativen in Nordosteuropa (19. Jahrhundert)*, edited by Anja Wilhelmi (Wiesbaden: Harrassowitz, 2011), pp. 309–22.

Wilkinson, Henry Robert, *Maps and Politics: A Review of the Ethnographic Cartography of Macedonia* (Liverpool: Liverpool University Press, 1951).

Winichakul, Thongchai. *Siam Mapped: A History of the Geo-body of a Nation* (Honolulu: University of Hawaii Press, 2009).

Withers, C. W. J. 'Place and the "Spatial Turn" in Geography and in History'. *Journal of the History of Ideas* 70, no. 4 (2009), pp. 637–58.

Withers, Charles W. J. 'On Trial—Social Relations of Map Production in Mid-Nineteenth-Century Britain'. *Imago Mundi* 71, no. 2 (2019), pp. 173–95.

Withers, Charles W. J. 'Map Making, Defamation and Credibility: The Case of the *Athenaeum*, Charles Tilstone Beke, and W. & A. K. Johnston's *Edinburgh Educational Atlas* (1874)'. *Imago Mundi* 73, no. 1 (2021), pp. 46–63.

Wohlfart, Kristine. *Der Rigaer Letten Verein und die lettische Nationalbewegung von 1868 bis 1905* (Marburg: Herder-Institut, 2006).

Wolf, Michaela. *The Habsburg Monarchy's Many-Languaged Soul: Translating and Interpreting, 1848–1918* (Amsterdam; Philadelphia: John Benjamins Publishing Company, 2015).

Wolff, Larry. *Woodrow Wilson and the Reimagining of Eastern Europe* (Stanford: Stanford University Press, 2020).

Wood, Denis. *Rethinking the Power of Maps* (New York: The Guildford Press, 2010).

Woodworth, Bradley D. *Civil Society and Nationality in the Multiethnic Russian Empire: Tallinn/Reval, 1860–1914*. PhD dissertation, Indiana University, 2003.

Woodworth, Bradley D. 'Patterns of Civil Society in the Moderning Multiethnic City: A German Town in the Russian Empire Becomes Estonian'. *Ab Imperio* no. 2 (2006), pp. 135–62.

Woodworth, Bradley D. 'Multi-Ethnicity and Estonian Tsarist State Officials in Estland Province, 1881–1914'. In *Russian Bureaucracy and the State*, edited by Eugene Huskey and Don Karl Rowney (Basingstoke: Palgrave Macmillan, 2009), pp. 72–88.

Wortman, Richard. 'The "Integrity" (Tselost') of the State in Imperial Russia'. *Ab Imperio* no. 2 (2011), pp. 20–47.

Wortman, Richard. *Visual Texts, Ceremonial Texts, Texts of Exploration: Collected Articles on the Representation of the Russian Monarchy* (Boston: Academic Studies Press, 2014).

Zabele, Angelīna. 'Latvijas ģeoloģiskas kartēšanas aizsākumi'. *Zinātņu Vēsture un Muzejniecība* 780 (2012), pp. 278–307.

Zahra, Tara. 'Imagined Noncommunities: National Indifference as a Category of Analysis'. *Slavic Review* 69, no. 1 (2010), pp. 93–119.

Zahra, Tara. *The Great Departure: Mass Migration from Eastern Europe and the Making of the Free World* (New York; London: W. W. Norton & Company, 2016).

Zajas, Krzysztof. *Absent Culture: The Case of Polish Livonia*, trans. Agnieszka Marczyk (Frankfurt am Main: Peter Lang, 2013).

Zajc, Marko. 'Contemporary Borders as "Phantom Borders": An Introduction'. *Südosteuropa* 67.3 (2019), pp. 297–303.

Zariņš, Klāvs. 'Das Land Ober Ost: Saksa okupatsioonipoliitika kuramaal i maailmasõja ajal (1915–1917)'. *Eesti sõjaajaloo aastaraamat* 5 (2015), pp. 96–118.

Zavarina, A. A. *Russkoe naselenie vostochnoi Latvii vo vtoroi polovine xix-nachale xx veka: istoriko-etnograficheskii ocherk* (Riga: Zinatne, 1986).

Zelče, Vita. *Latviesu avizniecіba: Laikraksti sava laikmeta un sabiedriba (1822–1865)* (Riga: Zinātne, 2009).

Zemlia obetovannaia, ili terra incognita. Iz istorii zemleustroitel'nykh ekspeditsii 30-x gg. XX v. Katalog dokumental'noi vystavki (Ekaterinburg: Izdatel'skii dom Basko, 2020).

Index

Note: Locators that refer to maps and other cartographic material are presented in **boldface type**. Key place names with different language versions are listed under each variant. Individuals are listed under their most commonly used name.